Remote Space in the Architecture of Cosmology

Daniela Bertol
Sky Spirals Institute, USA

IGI Global
Scientific Publishing
Publishing Tomorrow's Research Today

Published in the United States of America by
IGI Global Scientific Publishing
701 East Chocolate Avenue
Hershey, PA, 17033, USA
Tel: 717-533-8845 | Fax: 717-533-7115
Website: https://www.igi-global.com E-mail: cust@igi-global.com

Library of Congress Cataloging-in-Publication Data

Names: Bertol, Daniela, DATE- author
Title: Remote space in the architecture of cosmology / Daniela Bertol.
Description: Hershey, PA : IGI Global Scientific Publishing, [2026] | Series: Earth sky geometries ; 1 | Includes bibliographical references and index. | Summary: "This book explores human engagement with celestial and terrestrial spaces, interpreting ancient and modern constructions through autoethnography, geography, and deep mapping. It examines how the "architecture of cosmology" reflects consciousness, creativity, and the quest to understand time and space, bridging monuments, art, and everyday human experience"-- Provided by publisher.
Identifiers: LCCN 2025054907 (print) | LCCN 2025054908 (ebook) | ISBN 9798337333991 hardcover | ISBN 9798337334004 paperback | ISBN 9798337334011 ebook
Subjects: LCSH: Architecture and cosmology
Classification: LCC NA2543.C67 B47 2026 (print) | LCC NA2543.C67 (ebook)
LC record available at https://lccn.loc.gov/2025054907
LC ebook record available at https://lccn.loc.gov/2025054908

British Cataloguing in Publication Data
A Cataloguing in Publication record for this book is available from the British Library.

Table of Contents

Preface

I make places that create awareness of space and time

This trilogy was inspired by my life of the past four years: in June 2020 I moved to S.U.N.F.A.R.M, the earth art and architecture project I have been working on for the past two decades. Observing the sky and recording celestial events has become my daily routine, almost an existential need, integrating research with creative work, durational performance and contemplative practices, which became interwoven with in-situ methodologies. While I practice daily mindfulness dynamic meditations, my body moves in the liminal space between earth and sky, serving as a compass while aligning to the sun and casting shadows like a sundial. The simple built interventions in the landscape I have designed and built for over twenty years, guide and orient my movements and postures. These daily practices are video recorded in a living inquiry and durational performance is staged in an ever-changing natural landscape—transformed by the presence of light, and never looking the same at different times of the day and days of the year cycle.

The books of the trilogy are a narrative about discovering and creating places in the natural landscape, with meaning beyond their geographical location and which make us aware of where we are in space and time. Such places are often characterized by artistic expression and the search for meaning connects the perception of local and remote space. Experiencing, recording and writing about a landscape from a phenomenological perspective goes beyond the local space of the first-person experience. When the visual perception reaches from the earth to the remote space of the sky a connection is established between the mind-body of the subject and the cosmos, object of perception.

The principle that the self and the universe are complementary aspects of reality is not only a spiritual quest of mindfulness practices, but can also be situated in scientific inquiries, such as system theory. In my research, it represents a living inquiry of Husserl's lifeworld (Lebenswelt) and Heidegger's being-in-the-world / Dasein; the interconnectedness between the self, consciousness of the perceiving subject, and the outside object of perception, reaches from the local space of the landscape to the remote space of the sky. Such interconnectedness is also confirmed by the knowledge produced by positional (spherical) astronomy. In my practice-based research I document the apparent path of the sun, moon stars and constellations with photography and video, where spatial embodiment is an essential component of the local space of observation, in a place which creates awareness of space and time.

While my living inquiry—developed in a methodology weaving autoethnography with egogeography and deep maps in creative practices—is specific to S.U.N.F.A.R.M., the quest for understanding remote space can be situated in places—or just built constructions in the landscape—humankind has created for thousands of years and which I define "architecture of cosmology." These structures, studied by archaeoastronomy, are here interpreted through first-person perceptions. In the embodied space of observations of the sky, human consciousness is expressed in the built environment by the architecture of cosmology. The object of the first book of the trilogy is in understanding how these humankind expressions can go beyond the iconic mysterious mythical fascination of monuments to become part of de Certeau's "practice of everyday life"—from artist's creations to the individual's experience, as it has been shown for millennia by the multitude of examples of the collective quest for understanding space and time beyond immediate reach.

db
Roma, December 8, 2024

Introduction

Human constructions have long extended their meaning beyond the immediate physical landscape in which they are built. From prehistoric monuments such as Stonehenge to contemporary astronomical installations and radio telescopes, the built environment has often functioned as a medium through which humans orient themselves toward the remote space of the sky. By aligning structures to celestial bodies or events—such as the rising and setting of the Sun, Moon, stars, and planets—many cultures have created architectural and landscape interventions that interpret astronomical phenomena while also expressing cosmological beliefs.

Across cultures and historical periods, the observation of the sky has generated both practical and symbolic responses. Early constructions used geometrical alignments to track seasonal cycles or celestial events, while simultaneously embedding these observations within mythological, spiritual, and ritual frameworks. In this way, the sky above—perceived from specific locations on Earth—became a tangible image of the universe. Built structures designed in relation to celestial phenomena therefore established links between the local environment and the remote cosmic realm.

In this book, such constructions are examined through the concept of the Architecture of Cosmology. This term refers to human-made interventions in the landscape or urban environment whose geometric configuration reflects the geometry of celestial objects or phenomena. These constructions include not only buildings but also monuments, earthworks, sculptural markers, and other spatial devices that frame the observation of the sky. By embodying geometric relationships between Earth and sky, they connect local space—the position of the observer—with the remote space of celestial events.

The analysis of these works is approached through a hermeneutic phenomenological perspective. Phenomenology offers a framework for understanding how humans experience space and orientation in the world. In particular, Martin Heidegger's notion of the *fourfold*—the gathering of earth, sky, mortals, and divinities—provides a conceptual foundation for examining the relationship between built structures and cosmic order. While this study is not a philosophical treatise, key phenomenological concepts help illuminate how architectural constructions mediate the relationship between the observing subject on Earth and the celestial phenomena that structure human understandings of the universe.

Within this framework, the geometry of astronomy becomes central. Observations of celestial bodies are inherently geometric: the apparent paths of the Sun, Moon, and stars define spatial orientations and temporal cycles that can be translated into architectural form. Alignments, axes, and geometric configurations embedded in built environments often reflect these astronomical relationships. Through such design strategies, architectural constructions transform abstract celestial motions into tangible spatial experiences.

The connections between geometry, astronomy, and cosmology have deep historical roots. In many ancient cultures, the study of geometry was inseparable from the observation of the sky. The traditions of Babylonian, Vedic, and Greek mathematics integrated terrestrial measurement with celestial observation. Many of the most influential mathematicians of antiquity—including Pythagoras, Euclid, and Ptolemy—were also astronomers. Philosophical reflections on the structure of the cosmos often relied on geometric models, as exemplified by Plato's association between the elements of nature and the regular polyhedra, later known as the Platonic solids. Even in early modern science, geometry continued to structure cosmological speculation, as seen in Johannes Kepler's attempt to relate planetary orbits to geometric solids.

These historical relationships between geometry and cosmology also appear in architectural form. The armillary sphere, for example, served as a model of celestial motion and appears in architectural and scientific contexts across cultures. Similarly, monuments, temples, and urban spaces aligned with cardinal directions or solar paths express geometric relationships between Earth and sky. Such constructions often function as cosmograms—spatial diagrams that symbolize the relationship between the human observer and the cosmos.

Remote Space in the Architecture of Cosmology investigates these relationships across a range of historical and contemporary examples. The book explores how architectural and landscape constructions mediate the perception of remote cosmic space from specific earthly locations. Case studies range from prehistoric astronomical monuments and archaeological sites to urban spaces, cosmographic representations, and contemporary artistic interventions in the landscape.

Through diagrams, maps, drawings, and photographs, the book examines how geometric principles link built structures on Earth with celestial phenomena. Particular attention is given to constructions designed for naked-eye observation of the sky, where alignments and spatial orientation play a central role in shaping architectural form. These examples reveal how geometry serves not only as a scientific tool but also as a cultural and symbolic language through which humans articulate their relationship with the universe.

By examining the architectural manifestations of Earth–Sky relationships, the book seeks to illuminate a recurring human impulse: the desire to situate ourselves within the cosmos. Through geometry, observation, and construction, architecture becomes a means of expressing the fundamental condition that Heidegger described as *being-in-the-world*—a lived relationship between human existence, the Earth we inhabit, and the sky that frames our understanding of the universe.

Part I
Sky, Perception, and Geometry

Chapter 1
Introduction:
Geometry and Astronomy as Being-in-the-World

ABSTRACT

This introduction states the theme of the book and defines the Architecture of Cosmology within its historical and geographical context. It introduces the book's transdisciplinary approach, whose content—filtered through autoethnography—spans phenomenology, astronomy, and architecture. While outlining the chapters and organization of the book, this introduction highlights the challenges of engaging fields often considered unrelated and shows how scholarly research, when pursued through transdisciplinary methodologies, can extend into creative and design-based practices.

DOI: 10.4018/979-8-3373-3399-1.ch001

There is ... only being.
— Albert Einstein

BACKGROUND

The built environment extends beyond its immediate physical presence to encompass remote spaces and events. From early human constructions such as Stonehenge to contemporary radio and space telescopes, humankind has consistently expressed a desire to communicate beyond the local landscape, reaching toward the remote space of the sky above. Ancient cultures shaped interventions in the landscape to interpret celestial events by observing the rising and setting of stars and planets; structures were aligned to celestial objects and phenomena whose observation often transcended perceptual experience to acquire metaphysical, spiritual, or religious meaning. Associated with divinities, skywatching became a source of myths and rituals. In ancient cosmologies, the remote space of the sky, perceived within the local landscape, became an image of the universe.

The built environment has long functioned as a mediator between immediate perception and distant reality. Architecture does not merely shelter; it frames, directs, and orients attention. When aligned to celestial events, it becomes a cognitive instrument. In such moments, architecture exceeds functionality and becomes cosmographic — a spatial inscription of humanity's attempt to interpret its position within a larger order.

Both utilitarian works and artistic expressions—human-made structures aligned with celestial events—often evolved into dwellings or ritual places, establishing a tangible link between sky and earth. By framing what is otherwise unreachable, they became material signs of a connection with the universe and fostered awareness of a realm beyond direct physical access. Such constructions constitute what is defined throughout the *Earth–Sky Geometries* trilogy as the architecture of cosmology: human-made constructions in natural or urban environments whose geometric configuration reflects the geometry of celestial objects and phenomena. Spanning from prehistoric observatories to contemporary radio telescopes and land art, these works connect remote and local space by embodying the geometry of the sky in design.

The architecture of cosmology is not restricted to monumental constructions. It includes minimal interventions, subtle alignments, and geometric inscriptions that might be overlooked without attentive observation. What unites these examples is not scale but intention: the deliberate embedding of astronomical geometry into the built environment. In this embedding, architecture becomes both epistemic device and existential marker.

The architecture of cosmology includes not only buildings but, more prominently, monuments, sculptures, earthworks, and minimal landscape interventions that serve as markers or aids for observing and interpreting celestial phenomena. The examples presented in this book are examined through a hermeneutic phenomenological lens. A phenomenological approach to the remote space of the sky draws on Martin Heidegger's concept of the *fourfold* (*das Geviert*), articulated in his 1951 essay *Building Dwelling Thinking* as the gathering of earth, sky, mortals, and divinities (Heidegger, 1971). In *Earth–Sky Geometries*, mortals are reframed as humankind and divinities as spirituality, and the interrelation of the fourfold is explored from the perspective of geometry.

Although the books in the *Earth–Sky Geometries* series are not philosophy texts, key concepts from hermeneutic phenomenology provide an interpretive framework for the central inquiry: the definition of the architecture of cosmology as the geometric mediation between buildings, earth, and sky. Architectural constructions provide physical interpretations of celestial observations, establishing a link between the local space of the observer and the remote space of the observed. Heidegger's notions of *being-in-the-world* (*In-der-Welt-sein*) and *Dasein*, articulated through "building" and "dwelling," are expressed here as human-made interventions that connect local places on earth with the remote space of the sky. The fourfold is further developed into consciousness-body-Earth-Sky[1], expanding from the microcosm of the self to the macrocosm of the Universe.

This departure from the Heideggerian fourfold begins with the introduction of astronomical geometry as situated within the architecture of cosmology. The experience of remote space in local places is contextualized within the geometry of the apparent paths of celestial objects as perceived from Earth. The architecture of cosmology presented in *Earth–Sky Geometries* focuses on naked-eye astronomical observatories and places in which geometric configurations guide architectural design and construction.

The reframing of mortals as humankind and divinities as spirituality does not secularize the fourfold; rather, it translates it into a contemporary vocabulary. The architecture of cosmology is not bound to theological doctrine. It instead acknowledges that spatial orientation toward the sky has historically carried symbolic and metaphysical dimensions. Geometry becomes the shared language between measurable astronomy and lived meaning.

Note on Terminology: "Wholism"

In this book I use the term "wholistic," rather than the more common "holistic," to emphasize the philosophical notion of wholeness as articulated by David Bohm in *Wholeness and the Implicate Order* (Bohm, 1980). Bohm's concept of wholeness does not simply refer to the integration of parts into a functional whole, but to an underlying, undivided totality in which distinctions such as subject and object, mind and matter, or microcosm and macrocosm, are understood as abstracted aspects of a deeper, implicate order. In this sense, "wholistic" is used here to foreground an ontological and epistemological position that aligns with the phenomenological and transdisciplinary approach adopted throughout this work.

The wholistic approach of *cosmography*, in its literal meaning of description of the cosmos (from the Greek κόσμος (kosmos, "universe/world/order") and γραφή (graphe, "writing/description")), filters the numerous built examples of links between geometry, astronomy and cosmology, which has been present in scientific inquiries since early mathematics (Heath 1913).

Architectural Treatises and Visual Epistemologies

My approach to writing and visual thinking has been shaped by a long-standing engagement with historical architectural treatises. Having spent the first twenty-six years of my life in Rome, formative spatial experiences—such as standing beneath the oculus of the Pantheon, tracing the shifting alignments of Bernini's colonnade in Piazza San Pietro, or inhabiting the concave and convex geometries of Santa Maria della Pace—instilled an early understanding of architecture as the manifestation of tangible geometry. These spaces conveyed how proportion, number, and geometric order could be experienced directly through movement, perception, and bodily orientation.

This experiential foundation later evolved into a more explicit intellectual inquiry through the study of Renaissance treatises by Leon Battista Alberti, Luca Pacioli, Filarete, and Sebastiano Serlio. In these works, architecture, geometry, perspective, typography, and graphic layout form an integrated epistemic system in which visual representation is not illustrative but constitutive of knowledge. The present book draws methodological inspiration from this tradition, reinterpreting the treatise not as a historical genre but as a mode of inquiry. Although the number of images is deliberately reduced, visual content remains essential as a methodological tool, enabling the articulation of spatial and cosmological relationships that exceed purely textual description. In this sense, the book aligns contemporary computational and transdisciplinary practices with historical foundations, reaffirming the treatise as a viable framework for integrating geometry, architecture, and cosmological thought.

WHY GEOMETRY?

As shown in the etymology of *geometry* (from the Greek γεωμετρία (geometría), 'land measurement'), the disciplines and practice concerned with Earth and sky were integrated into Babylonian, Vedic and Greek mathematics. Geometry precedes instrumentation. Before telescopes and satellites, orientation was achieved through the naked eye, memory, and repetition. The celestial sphere is an abstraction derived from perception—a conceptual model grounded in embodied observation. Geometry thus emerges not only from mathematical deduction but from phenomenological encounter.

The most notable mathematicians were also astronomers—the long list includes Pythagoras, Plato, Euclid, Eratosthenes, Archimedes, Apollonius of Perga, Pappus of Alexandria, and Ptolemy. According to Plato, mathematics, which was subdivided into arithmetic, geometry, stereometry and astronomy, was also necessary in the formation of philosophers and the ruling class, as stated in the inscription at the entrance of his Academy: "let no one destitute of geometry enter my doors". Plato, in his symbolic view of the Universe discussed in the dialogue *Timaeus* (Cornford 2014), established several relationships between geometry and cosmology, including the famous association of the primary elements of nature (fire, air, earth, water) to the five regular convex polyhedra, which were named after him *Platonic solids*: the tetrahedron was linked to fire, the octahedron to air, the hexahedron to earth, the icosahedron to water and the dodecahedron to ether—which was, according to Plato, the upper atmosphere where deities live.

Such strong connections between geometry and cosmological views are found again in cosmology and even in scientific views, most remarkably in Kepler's cosmological theories linking the planetary orbits to the five Platonic solids. While sometimes, like in Kepler's theories of the planets, geometrical associations are more of metaphorical nature, geometry is embedded in astronomy. As it will be discussed in this book, geometry is at core of the interpretation of celestial objects and phenomena, as synthesized by the famous sentence of Galileo about the Universe written in the language of mathematics[2] (Galilei, 1623). Galileo's assertion that the universe is written in the language of mathematics does not negate perception; rather, it presupposes it. Mathematical language formalizes patterns first observed. The architecture of cosmology translates these patterns into spatial form. It materializes celestial order without dissolving its experiential dimension.

The geometry of cosmology in architectural examples is literally represented by the armillary sphere, used as a model of the revolutions of the celestial bodies and present in astronomical buildings at different scales. It can also be found in the alignment and shape of built structures. Monuments, temples and dwellings following the cardinal directions and solar alignments often exhibit strong geometrical configurations and define *cosmograms*, symbols of the relationship between the observing subject on Earth and the object of observation in the sky. The geometry of the sky reflected in human-made built environments is found also in urban spaces, as presented in chapter 7 with the case study of St. Peter's Square in Rome.

ABOUT THE EARTH-SKY GEOMETRY TRILOGY

The *Earth Sky Geometries* series begins with a trilogy of books whose inquiries revolve on the phenomenology of astronomy. *Remote Space in the Architecture of Cosmology* is the first book of the series introducing the foundation and concepts behind spherical (positional) astronomy expressed in the human-made built environment. The trilogy explores being-in-the-world as the phenomenology of spherical astronomy, in the embodied cognition of the sky observed from Earth: a geometrical interpretation of the perceived celestial objects and phenomena leads to the perception of remote space. Geometry not only provides the ordering principles of astronomical space and time but is also a metaphor of the mathematics of the cosmos, embodied in the landscape. Geometry is also the organizing principle of the architectural constructions. The inquiries are not only based on text, but also visually illustrated with sketches, diagrams, drawings and, most of all, photographs, representing a documentation of the experience of places. It also includes case studies selected from renowned archaeoastronomical examples and explored from a unique perspective, because of the phenomenological approach and inclusion of visual inquiries. Architectural drawings, sketches, satellite imagery, maps and photographs illustrate the geometrical underpinning of how the architectural features weave local and remote space. The selected examples include human-made constructions shaped according to geometrical configurations reflecting the earth-sky geometries.

The second book is titled *S.U.N.F.A.R.M.*: *Stories of Cosmology, Land, and Community in the Hudson Valley* and explores how people understand place, time, and belonging through the relationship between land and sky in the Hudson Valley. Centered in Columbia County, it examines how cosmology, landscape, and embodied experience shape cultural meaning in the region. It focuses on *S.U.N.F.A.R.M.*, a contemporary case study of the architecture of cosmology. *S.U.N.F.A.R.M.*, is a place and an Earth art project based on the mapping of a mathematical theory of the universe in the landscape—as embodiment of a phenomenological approach to the study of astronomy and the cosmos. Through reflective writing and graphic inquiry, the project documents how solar alignments, cardinal orientation, walking paths, and earth-based artworks organize the landscape as a lived, experiential system rather than a static site. These observations are expanded through broader cultural and historical lenses, including Indigenous Mohican relationships to land

and sky, agrarian histories of the region. drawing on archival research, historical sources, and existing Indigenous scholarship to situate S.U.N.F.A.R.M. within a longer continuum of sky-based knowledge, land use, and cosmological orientation in the Hudson Valley. The book narrates how the embodied cosmologies of the Indigenous peoples, in which the sky functioned as a lived system of knowledge, orientation, and relational belonging, were disrupted by European colonization and forced displacement—and how in the postcolonial period, the sky largely entered regional culture through representational art rather than lived experience. Already introduced in the present book, *S.U.N.F.A.R.M.* is explored through essays, interviews and graphic inquiries about the several Earth art interventions shaping the landscape; the connections of these constructions in the site plan develops into a series of digital deep maps. The concepts and forms of circularity and astronomical cycles are the common thread of different space-time explorations of the geometric shaping of the landscape as connection to the sky.

A Place: Mind-Body /Self-Universe is the third book of the trilogy, exploring the relationship of the human body and senses with the landscape from a phenomenological standpoint which evolves into the awareness that the self and the universe are not separated entities but are part of a system. The microcosm of the human body encounters the macrocosm of the universe. Such awareness is achieved not only by contemplative practices but by the transformation of daily practices in art production. Explorations are conducted through rigorous research: the inquiries use location-based media, and the practice of art is based on interconnectedness—experienced in the art-in-nature interventions based on spherical astronomy, developing on the phenomenological first-person experience that relates the observing subject to the object of observation. resolving the dualism between self and universe through observations based on positional astronomy.

Together, the three volumes progressively shift scale: from historical analysis (Book I), to site-specific embodied inquiry (Book II), to phenomenological interiority and systemic awareness (Book III). The trajectory moves from observation, to construction, to integration — from perceiving geometry to living it.

The research of the trilogy is based on both qualitative and quantitative methodologies: text and numerical/computing processes often accompany the visual materials which are part of the inquiry. With the exception of this first book, in the other publications of the series there is a particular emphasis on art and other creative practices as a first-person way of knowing and not only presenting knowledge as often happens in scholarly works. The presented case studies will be selected according to the model of S.U.N.F.A.R.M., where a living inquiry leads to knowledge production, by first-person observation and recordings. Throughout the trilogy, concepts are expressed not only in writing, but also with diagrams, maps and two-dimensional representations and three-dimensional digital models. Visual content includes custom maps, realized with the geographic information program Google Earth and the web-based mapping platform Google My Maps, developed as enhanced digital "deep maps", based on GPS (Global Position System) and GIS (Global Information System) methodologies and technologies.

ABOUT THIS BOOK

Remote Space in the Architecture of Cosmology features different architectural expressions of "remote" space in the perception and interpretation of the local environment as characterized by geometrical properties. The book visually illustrates selected case studies which range from ancient astronomical observatories and archeological ruins to urban spaces, maps, cosmograms and unbuilt designs. It situates these architectural constructions on Earth within a framework linking them to a symbolic, often imaginary, meaning of the sky beyond the astronomical understanding. The common denominator of the different built constructions presented in *Remote Space in the Architecture of Cosmology* is in their functions of providing means and guidance in the observation of remote space. The intent of this book is to emphasize how these constructions, beyond their serving as scientific observation tools, can establish themselves as built expressions of humankind's desire to grasp infinity and establish a presence in the Universe, often associating a metaphysical meaning to the built architecture. The featured examples of architecture of cosmology are selected for their geometrical characteristics, which are associated with the perception of the remote space of the sky from the local environment. Several types of maps are presented for the selected case studies as an inquiry method, and span from historical cartography to digital geographic maps and "deep maps". In the context of this book, digital deep maps are personalized maps created from Google My Maps or Google Earth satellite imagery with superimposed diagrams and other metadata. This type of mapping can provide a visual presentation of the link between the geometry of the constructions on Earth and the geometry of the celestial objects in the sky the buildings relate to.

The term "remote space" does not refer merely to astronomical distance. It refers to conceptual distance — to that which lies beyond immediate grasp yet shapes human orientation. Architecture becomes the mediator between these scales. Through geometry, it renders the remote perceptible without collapsing its transcendence.

The architecture of cosmology finds in archaeastronomy the best-known examples of built observational astronomy, but these cases, even if they are the most identifiable, are not unique: visionary explorations from treatises or contemporary constructions from the land art movement also exhibit astonishing connections between their architectural features and observations of the sky. The archaeoastronomical examples presented here are framed by the lenses of phenomenology and autoethnography. The introductory historical research on a site is followed by a qualitative approach derived from "being-there." The selection has been narrowed not only by my IRL (In Real Life) experience of the place, but also by the geographical location. Although I have visited several archaeoastronomical areas in North and Central America, in this book I present only sites connected with the Mediterranean basin.

The examples of architecture of cosmology presented in this book include only structures designed for naked-eye (unaided) observations of the sky. Therefore, astronomical observatories incorporating telescopes are not included, since they are lacking the alignments which characterize instead the constructions designed for unaided astronomical observations. Built alignments to celestial events promotes awareness of the surrounding environment creating the experience of being part of a system in a continuity between the observer on the local space of Earth and the objects of observation, in the remote space of the sky.

The restriction to naked-eye observatories is methodological. It foregrounds the continuity between perception and construction. Telescopic observatories represent a shift toward instrumental mediation; naked-eye constructions preserve the architectural articulation of sky–earth relationships in spatial form.

The book is conceptually structured in three interrelated parts, each addressing a different mode through which humans have understood and spatialized their relationship with the sky. Part I, *Sky, Perception, and Geometry*, establishes the theoretical and phenomenological foundations of skywatching, examining how cosmological ideas emerge from naked-eye observation, bodily orientation, and geometric abstraction. The first five chapters define the key concepts for the book explorations, drawing references from broader transdisciplinary investigations, tracing the translation of celestial phenomena into symbolic systems, diagrams, and cosmograms, foregrounding perception as the primary interface between human experience and the cosmos. This part includes a discussion on the sky as a physical phenomenon contextualized in geometric definitions, which is the object of perception as well as carrier of symbolic values. Basic concepts from the philosophical movement of phenomenology are outlined, as they are relevant to naked-eye astronomy. Diagrams and mathematical definition of the celestial sphere as well as a few introductory concepts of observational astronomy are presented in this section, illustrating the progression from the sky as an object of phenomenological perception and "lived experience" (Merleau-Ponty, 1962) to its identification with the cosmos. Geometry in this context is not considered only as an instrument for observational astronomy but also as an approach to phenomenological explorations, where being-in-the-world is a presence defined by a geospatial location. Different kinds of cartography are introduced—from the celestial maps of astronomical treatises of several hundred years ago to contemporary "deep" and digital maps. These different kinds of maps relate to phenomenological orientation and they are both representational and process-based investigations of Earth, sky and their connection. In Part I, the no-longer extant observatory Uraniborg, which relied on naked-eye astronomical observations, introduces the architecture of cosmology, since it exhibits a strong geometric configuration derived by the study of the sky,

Part II, *Sky, Earth, and Culture*, investigates how these cosmological frameworks are embodied in architecture, landscape, and urban form across historical and cultural contexts. Through archaeoastronomy and traditional design systems, this section explores how local place is structured in relation to remote space through orientation, alignment, and ritualized spatial practices. This section discusses examples from archaeastronomy, Renaissance and Baroque architecture, presenting case studies of buildings, from past cultures and periods, devoted to sky observations. The Egyptian pyramids and the Neolithic temples of Malta are here discussed as some of the most remarkable constructions that deal with "remote" space. The focus of this section is on the aesthetics of the architecture built for astronomical purposes: satellite imagery, drawings, diagrams as well as photographs of the ruins are used to interpret each building in reference to its relationship with the celestial phenomena. Sundials, and other early astronomical instruments built at monumental scale, are presented with emphasis on their integration within the architectural and urban context. The case studies are selected according to three different criteria: the first is given by a phenomenological encounter, the "lived experience": my having been there and experienced, after a few centuries, or even thousands of years, the sky perceptions that motivated such constructions. The second criterion is the presence of clearly defined geometric configurations characterizing the built construction. Finally, the geographic proximity links these examples, which are instead chronologically so far apart, separated by several thousand years. There are other examples which have been object of my inquiries, but do not comply with these criteria and will be presented in another publication.

Part III, *Remote Space in the Contemporary World*, addresses the transition to modern and contemporary conditions, in which the relationship between architecture and the sky is increasingly mediated by scientific instruments and digital technologies. This part examines how architecture is no longer embodying the connection between local and remote space through the naked-eye observations, when the focus has shifted instead on the instruments currently used to connect with the sky: throughout the centuries, telescopes have developed into "black boxes" of electronics, while the architectural intent of the ancient naked-eye observatories has disappeared and the sky has lost the mystery and spiritual meaning which was present in traditional cultures. Part III introduces the renewed engagement with cosmic scale in land art and phenomenological practices, culminating in the introduction of S.U.N.F.A.R.M. as an ongoing exploration of embodied awareness of space and time, located in the Hudson Valley (New York, US). which will be the main subject of the following books in the *Earth Sky Geometries* series. It highlights how artists—more than contemporary architects—have investigated the metaphysics of tectonics, building works of architectural scale relating to the cosmos, often finding inspiration in archaeoastronomy.

Together, the three parts articulate a continuous trajectory from perception to construction, and from cosmological thought to lived spatial experience. The exploration of the geometry of space, which may seem untimely several decades after the "spatial turn", stems from a systemic approach that can be also relevant not only as a contribution to knowledge but also leading to practices devoted to sustainable living in harmony with the natural world.

A few words about the conceptual organization of this book may help clarify how each chapter is structured. The overall approach has been inspired by the 1921 seminal treatise *Tractatus Logico-Philosophicus* by Ludwig Wittgenstein (Wittgenstein, 1922), whose rigor and logical sequencing of propositions have long influenced my own way of developing and presenting ideas. In the *Tractatus*, concepts are articulated through a hierarchical and relational structure, in which statements unfold from more general to more specific levels; this logical architecture is even more evident in digital editions that allow navigation among propositions according to their conceptual relationships (Ludwig Wittgenstein Project, n.d.). The present book is not a philosophical treatise; however, it adopts a similarly systematic approach in which themes are introduced, revisited, and expanded across chapters at different levels of depth. This layered organization is intended to support the discussion of complex and interrelated topics by progressively building connections among architectural form, astronomical observation, geometry, and cultural meaning.

This book does not attempt encyclopedic coverage. Instead, it proposes a framework. It argues that geometry is the connective tissue linking perception, astronomy, architecture, and meaning. The selected examples function as case studies within a broader inquiry into how human beings spatialize cosmological awareness.

There are many disclaimers throughout this book: the breadth of the topics and disciplines is such that cannot be developed in a publication by a single author but would require a collaborative effort from researchers with expertise in the different field or studies approached, including philosophy, astronomy, architectural theory and archaeastronomy. This text does not claim to be an introduction to archaeastronomy or spherical astronomy or cosmology. The uniqueness is in bringing together different approaches to knowledge toward the understanding of peculiar humankind inquiries: skywatching from a phenomenological perspective highlighting geometric features of architectural constructions. Writing has presented many challenges: mainly in highlighting the thesis of how the geometry of astronomical phenomena can be translated into architecture. The methodology has always included visual tools and diverse means of expression, from sketching to making and computing, and even using my body as "living experience" to define a phenomenological understanding of the sky. In developing the initial outline into the final book several additional questions have been raised, which will be discussed in the other texts of the envisioned series.

Transdisciplinary work inevitably risks incompleteness. Yet incompleteness is not a flaw but a condition of inquiry. The architecture of cosmology, like the sky itself, exceeds singular disciplinary containment. Its study requires dialogue between mathematics and myth, measurement and memory, site and symbol.

Establishing continuity between archaeoastronomical constructions and contemporary GPS systems, between ancient temples and Earth art, requires bridging vast temporal and cultural distances. These works differ in technique and worldview, yet they converge in a shared impulse: to situate human existence within the cosmos. Remote space is significant not because it is distant, but because it reveals proportion — the scale of our presence within a larger order. When architecture embeds celestial geometry into the built environment, it becomes more than construction. It becomes orientation. It materializes our existential condition — our need, as Heidegger articulated, to dwell as beings who are always already under the sky.

REFERENCES

Bohm, D. (1980). *Wholeness and the implicate order*. Routledge & Kegan Paul.

Drake, S. (1957). *Discoveries and opinions of Galileo* (S. Drake, Trans.). Doubleday & Company, Inc. (Original work published 1610-1623).

Galilei, G. (1623). *Il saggiatore*. Appresso Giacomo Mascardi. (Original work published 1623)

Heath, T. L. (1913). *Aristarchus of Samos, the ancient Copernicus: a history of Greek astronomy to Aristarchus, together with Aristarchus's Treatise on the sizes and distances of the sun and moon: a new Greek text with translation and notes*. Clarendon Press.

Heidegger, M. (1971). Building, Dwelling, Thinking. In Poetry, Language, Thought (A. Hofstadter, Trans.). Harper and Row.

Ludwig Wittgenstein Project. (n.d.). *Tractatus Logico-Philosophicus* (English). Retrieved October 28, 2025, from https://www.wittgensteinproject.org/w/index.php/Tractatus_Logico-Philosophicus_(English)

Merleau-Ponty, M. (1962). Phenomenology of perception (C. Smith, Trans.). Routledge & Kegan Paul. (Original work published 1945)

Wittgenstein, L. (1922). Tractatus logico-philosophicus (C. K. Ogden, Trans.). Kegan Paul, Trench, Trubner & Co.

ADDITIONAL READING

de Certeau, M. (1984). *Practice of Everyday Life*. University of California Press.

Eliade, M. (1951). Images and Symbols: Studies in Religious Symbolism (P. Mairet, Trans.). Harvill Press.

Eliade, M. (1961). *The sacred and the profane: the nature of religion*. Harper & Brothers.

Embree, L. (1997). *Encyclopedia of Phenomenology*. Kluwer Academic Publishers.

Gibson, J. J. (1950). *The perception of the visible world*. The Riverside Press.

Husserl, E. (1907). *Things and Space: Lectures of 1907*. (R. Rojcewicz, Trans.). Kluwer Academic Publishers, 1997.

Husserl, E. (1970). The Crisis of European Science and Transcendental Phenomenology: an Introduction to Phenomenological Philosophy (D. Carr, Trans.). Northwestern University Press.

Husserl, E. (1901-2001). *Logical investigations*. (J. N. Findlay Trans.) Routledge.

Ingold, T. (2011). *Being Alive: Essays on Movement, Knowledge and Description*. Routledge, (Original work published 2001).

Lefebvre, H. (1991). *The production of space*. Blackwell.

Lefebvre, H. (2004). Rhythmanalysis: Space, Time and Everyday Life (S. Elden & G. Moore, Trans.). Continuum.

Liddell, H. G., Scott, R., Jones, H. S., & McKenzie, R. (1940). *A Greek-English lexicon*. Clarendon Press.

Malpas, J. (1999). *Place and Experience: A Philosophical Topography*. Cambridge University Press.

Malpas, J. 2008. Heidegger's Topology: Being, Place, World. MIT press.

Merleau-Ponty, M., Lefort, C., & Lingis, A. (1968). *The visible and the invisible: followed by working notes*. Northwestern University Press.

Norberg-Schulz, C. (1971). *Existence, Space & Architecture*. Praeger.

Pollio, V. (1914). Vitruvius, the ten books on architecture (M. H. Morgan, Trans.). Harvard University Press.

Tuan, Y. F. (1974). *Topophilia: a study of environmental perception, attitudes, and values*. Prentice-Hall.

Tuan, Y. F. (1977). *Space and place: the perspective of experience*. University of Minnesota Press.

KEY TERMS AND DEFINITIONS

Architecture of Cosmology: Human-made constructions, interventions, or spatial markers in the landscape or built environment whose geometric configuration reflects or is aligned with celestial objects or phenomena, thereby establishing a perceptual and symbolic connection between local earthly space and remote cosmic space. In this book, the term includes monuments, earthworks, observatories for naked-eye astronomy, urban alignments, and land art interventions.

ENDNOTES

1 Throughout this book, the terms *Earth* and *Sky* are capitalized when they refer to phenomenological and ontological dimensions—derived from Heidegger's notion of the fourfold—and are not merely descriptive of physical terrain or atmospheric conditions; when used in a strictly observational or empirical sense, the terms appear in lowercase.

2 The original complete sentence. in Italian: "La filosofia è scritta in questo grandissimo libro che continuamente ci sta aperto innanzi agli occhi (io dico l'universo), ma non si può intendere se prima non s'impara a intendere la lingua, e a conoscere i caratteri nei quali è scritto. Egli è scritto in lingua matematica, e i caratteri son triangoli, cerchi e altre figure geometriche, senza i quali mezi è impossibile a intenderne umanamente parola; senza questi è un aggirarsi vanamente per un oscuro laberinto" (Galilei, 1623: 25).

The English translation is: "Philosophy is written in this grand book, the universe, which stands continually open to our gaze. But the book cannot be understood unless one first learns to comprehend the language and read the letters in which it is composed. It is written in the language of mathematics, and its characters are triangles, circles, and other geometric figures without which it is humanly impossible to understand a single word of it; without these, one wanders about in a dark labyrinth" (Drake, 1957: 237-238).

Chapter 2
The Observer's Space, Perception, Geometry—and Place

ABSTRACT

Space, place, body, and perception are introduced from a phenomenological perspective. The Heideggerian terms "Dasein," "being-in-the-world," and "fourfold" are revisited through the lens of the phenomenology of architecture. Merleau-Ponty's seminal concepts of the "lived body" and spatial embodiment are contextualized within the built environment. This chapter grounds the geometry of spatial perception in phenomenology and Gestalt theory and introduces architecture of cosmology as constructions situated in local natural landscapes or urban contexts that establish perceptual and geometrical connections to remote space. It also contextualizes the attribute "remote" in relation to physical space, as distinct from its contemporary use in virtual reality and other digital environments.

DOI: 10.4018/979-8-3373-3399-1.ch002

Figure 1. An encounter between self, Earth, and sky.

INTRODUCTION: EXISTENCE AND SKY PERCEPTIONS (FROM EARTH)

The reference to phenomenology in this study derives from my own approach to explorations of the cosmos as perceived from Earth. As an architect and artist, I use my own perceptions in the creative process of establishing a presence of remote space—the sky—within interventions in the built environment that I design. As a scholar, I study constructions and constructs grounded in unaided human perceptions of the remote space of the sky, in which the interconnection between local and remote space is based on the embodied observations of the perceiver. This phenomenological approach differs from that of astronomers, meteorologists, or astrophysicists, whose perceptions of the sky are often mediated by observational instruments and methods and are directed toward different objectives. Furthermore, the reference to phenomenology derives from the fact that all examples of *architecture of cosmology* discussed in this book have been experienced in the first person.

Accordingly, the phenomenological approach adopted in this study is situated within the broader human inquiry into existence in the world—what Heidegger defined as being-in-the-world (Heidegger, 1927). More specifically, positional (spherical) astronomy is explored in the first person as embodied cognition of the sky, observed from Earth and framed by human interventions in the built environment. A geometrical interpretation of perceived celestial objects and phenomena thus provides a framework for understanding the perception of remote space. This geometrical framework does not abstract perception away from experience; rather, it emerges from embodied observation. Geometry is not imposed upon the sky but recognized within it through repeated, situated acts of looking. In this sense, celestial geometry becomes a cognitive articulation of lived experience.

The *Stanford Encyclopedia of Philosophy* defines phenomenology as "the study of structures of consciousness as experienced from the first-person point of view. The central structure of an experience is its intentionality, its being directed toward something, as it is an experience of or about some object" (Smith, 2018). The definition of phenomenology to which I primarily refer—both in this book and throughout my work—is articulated by Merleau-Ponty in the preface to his seminal text *Phenomenology of Perception*:

> Phenomenology is the study of essences; and according to it, all problems amount to finding definitions of essences: the essence of perception, or the essence of consciousness, for example. But phenomenology is also a philosophy which puts essences back into existence, and does not expect to arrive at an understanding of man and the world from any starting point other than that of their 'facticity'. It is a transcendental philosophy which places in abeyance the assertions arising

> out of the natural attitude, the better to understand them; but it is also a philosophy for which the world is always 'already there' before reflection begins—as 'an inalienable presence'; and all its efforts are concentrated upon re-achieving a direct and primitive contact with the world, and endowing that contact with a philosophical status. It is the search for a philosophy which shall be a 'rigorous science', but it also offers an account of space, time and the world as we 'live' them. (Merleau-Ponty, 1962, p. vi)

The geometry defined by celestial phenomena and objects populating the sky—at the core of my explorations—is not the first visual perception from a phenomenological standpoint, according to Merleau-Ponty: "As I contemplate the blue of the sky … I am the sky itself as it is drawn together and unified, and as it begins to exist for itself; my consciousness is saturated with this limitless blue …" (Merleau-Ponty, 1962, p. 249). The immersion in the limitless blue described by the French philosopher is a common daily experience, available to any observer open to mindfulness and intentionality in perception. Figure 2 illustrates my interpretation and enactment of Merleau-Ponty's perception of the limitless blue. The photograph, taken using a tripod and timer, records myself as the observer looking at the blue sky; both the observing subject and the object of observation are united within my perception of the sky. This unity between subject and sky precedes analytical interpretation; it is a pre-reflective condition of perception. Only after this immersion does the sky disclose measurable structure—its arcs, cycles, and orientations—through sustained attention.

Figure 2. Looking at the "limitless blue" of the sky.

Nevertheless, beyond the initial perception of the sky as an unbounded field of color, sustained observation across daily and seasonal cycles reveals the underlying geometry of celestial phenomena. The apparent emptiness of the night sky is structured by time-based geometric configurations generated by the motions of celestial bodies, including the Moon, planets, and stars. These apparent motions define circles and arcs whose orientation and curvature depend on the observer's position on Earth and the direction of observation.

During daylight hours, the apparent path of the Sun is perceived through time-based observation as an arc rising from the horizon at sunrise, reaching its maximum altitude at solar noon, and descending toward sunset. Figure 3 presents a composite sequence of photographs documenting the Sun's apparent motion on August 21, 2008, framed by the Sunrise Trellis structure at the S.U.N.F.A.R.M. site in New York, USA.

Figure 3. Apparent path of the Sun at S.U.N.F.A.R.M. (New York, USA), recorded on August 21, 2008.

ABOUT PHENOMENOLOGY

This chapter is not intended as a comprehensive introduction to phenomenology as a philosophical movement, but rather as a definition of a phenomenological methodological approach specific to the perception of remote space. Selected foundational approaches from phenomenology are introduced insofar as they are relevant to the main theme of this book—the geometric definition of the architecture of cosmology—outlining the principal phenomenological concepts employed in this context, including *being-in-the-world*, *fourfold*, and *lived body*. The chapter also incorporates theories and concepts from other disciplines relevant to establishing a framework for the architecture of cosmology, such as *axis mundi* and *genius loci*.

The etymology of the word *phenomenology* has roots in the Greek *φαινόμενον* (*phainómenon*), "that which appears," and *λόγος* (*logos*), "study." The term denotes an interpretation of reality grounded in the relationship between a perceiving subject and the world as the object of perception. Phenomenology as a philosophical movement emerged in the early twentieth century with the seminal works of the German philosopher Edmund Husserl (1859–1938), primarily the two volumes of *Logical Investigations* (*Logische Untersuchungen*), first published in 1901–1902 (Husserl, 1901/2001). Husserl's focus can be summarized in his oft-repeated phrase "the things themselves" (*Sachen selbst*), expressing the experience of phenomena as they appear to consciousness, without judgment or preconception. He states in the introduction to the second volume of *Logical Investigations*:

> Here phenomenological analysis must begin. Logical concepts, as valid thought-unities, must have their origin in intuition: they must arise out of an ideational intuition founded on certain experiences, and must admit of indefinite reconfirmation, and of recognition of their self-identity, on the reperformance of such abstraction. Otherwise put: we can absolutely not rest content with 'mere words', i.e. with a merely symbolic understanding of words', such as we first have when we reflect on the sense of the laws for 'concepts', 'judgements', 'truths' etc. (together with their manifold specifications) which are set up in pure logic. Meanings inspired only by remote, confused, inauthentic intuitions - if by any intuitions at all - are not enough: we must go back to the 'things themselves'. (Husserl, 1901/2001, p. 168)

The philosophical tradition initiated by Husserl is known as transcendental phenomenology, with its focus on consciousness and the key methodological concepts of *epoché* (bracketing or reduction). *Lifeworld* (*Lebenswelt*) is another seminal Husserlian concept, introduced in *The Crisis of European Sciences and Transcendental Phenomenology* (Husserl, 1936), expressing the experience of existing in the world and the contextual grounding of human consciousness. Naked-eye observation of the sky may also be situated within transcendental phenomenology, insofar as initial perceptions occur without preconceived interpretation.

Phenomenology has since developed into several branches, at times merging with other schools of thought and disciplines. Hermeneutic (interpretative) phenomenology was developed by the German philosopher Martin Heidegger (1889–1976), a student of Husserl, with a primary focus on *being-in-the-world*, that is, on the fact that phenomena and their perception always occur within a contextual horizon. Existential phenomenology integrates attention to individual freedom and other existential concerns with lived experience; this school includes thinkers such as Hannah Arendt (1906–1975), Jean-Paul Sartre (1905–1980), and Simone de Beauvoir (1908–1986) (Embree, 1997, p. 10). Maurice Merleau-Ponty (1908–1961) is often associated with existential phenomenology; however, his emphasis on the primacy of perception, spatiality, consciousness, and—most importantly—the body situates his work within embodied phenomenology. *The Encyclopedia of Phenomenology* states:

> The word "phenomenology" not only stands for a substantive philosophy (as in the writings of Husserl, in some of Heidegger's works, and in Sartre and Merleau-Ponty), but for a distinctive approach that, applied to a specific domain X, gives rise to what may be called the "phenomenology of X." Thus, we have phenomenology of religion, phenomenology of natural science, phenomenology of human science, phenomenology of literature, phenomenology of perception, and so forth. (Embree, 1997, p. 10)

In recent decades, phenomenology has played an important role in interdisciplinary studies integrating psychology and cognitive sciences with religion and contemplative practices, most prominently within the framework of embodied cognition (Varela et al., 1991). The interconnection between first-person experience and the world is established through an interpretation of reality that is also enacted, translating knowledge into lived experience. Although not discussed in detail in this book, phenomenology-based embodied cognition is referenced because it constitutes an important dimension of my work on the architecture of cosmology, presented in future volumes of the *Earth–Sky Geometries* series.

Of particular relevance to this study is hermeneutic phenomenology, which provides a framework for exploring how geometry shapes the phenomena of the "now" (time) and the "here" (space) of the subject's perception in local space, in relation to the object of observation in remote space. *Being-in-the-world* recurs as a common denominator across the examples of architecture of cosmology presented in this book. References to embodiment similarly underpin the human link established between Earth and sky, drawing on phenomenology of perception (Merleau-Ponty, 1962).

The following table summarizes the concepts relevant to the present inquiry. The chronological order is tentative, as significant overlap exists among concepts and schools of thought.

Table 1. Phenomenological schools and chronology.

School/Concepts	Context/Chronology
Transcendental Phenomenology	Edmund Husserl — 1901
Noema	Edmund Husserl — 1913
Figure-ground	Edgar Rubin — 1915 Gestalt psychology
Lifeworld (Lebenswelt)	Edmund Husserl — 1936
Hermeneutical Phenomenology	Martin Heidegger — 1962
Existential Phenomenology	
Intentionality	Edmund Husserl Maurice Merleau-Ponty
Epoché / Bracketing	Husserl — 1907 Transcendental Phenomenology
Dasein	Heidegger — *Being and Time*, 1962
Being-in-the-world (in-der-Welt-sein)	Martin Heidegger — *Being and Time*, 1962
Fourfold	Martin Heidegger — Building, Dwelling, Thinking
Axis Mundi	Mircea Eliade — 1951
Lived Experience	Maurice Merleau-Ponty

This study adopts a first-person phenomenological methodology grounded in embodied perception and direct observation. The approach foregrounds experiential engagement with celestial phenomena as they are perceived without instrumental mediation, acknowledging both the strengths and limits of such inquiry. While phenomenology necessarily reflects a situated and culturally inflected perspective, its value here lies in making explicit the conditions under which perception, geometry, and spatial meaning arise. The exclusion of digital, computational, and data-driven observational methods is intentional and serves to delimit the scope of the investigation to architectures conceived through unaided human perception. Rather than offering a comprehensive critique of phenomenology, this chapter employs it as an operative framework for interpreting spatial constructions that mediate between Earth and sky. Phenomenology here functions not as abstract philosophy but as a methodological discipline of attention. It structures how observation is conducted, how geometry is recognized, and how spatial meaning emerges from lived engagement with place. In this way, perception becomes the ground from which cosmological orientation is articulated architecturally.

Epoché

Epoché is one of the key concepts introduced by Husserl within transcendental phenomenology. Its roots can be traced to ancient Greek philosophy, specifically the skeptical school of Pyrrhonism in the fourth century BCE (Long, 1986). Derived from the Greek *ἐποχή*, meaning "suspension of judgment," *epoché* in phenomenology refers to the methodological suspension of belief in order to focus on perception as experienced by consciousness (Husserl, 1907). It is often described as "bracketing," that is, becoming free from dogmas, judgments, biases, and belief systems that may influence experience. Through bracketing, one transitions from the "natural attitude"—another Husserlian term—to the phenomenological attitude.

In the perception of the remote space of the sky, *epoché* enables freedom from judgment, assumptions, or learned interpretations, such as associations between color and weather or time of day. The sky thus encounters consciousness as a direct experience, akin to Merleau-Ponty's description of the "limitless blue," potentially leading to insight and discovery. Such suspension is not ignorance but disciplined openness; it allows celestial phenomena to disclose structure before they are subsumed under scientific or symbolic interpretation.

Dasein

Dasein is a German term literally meaning "there-being" (*da-sein*), commonly translated as "being-there," and is used to denote human existence. Introduced by Martin Heidegger in *Sein und Zeit* (*Being and Time*, 1962), it is a foundational concept of phenomenology. Heidegger defines *Dasein* in relation to being itself: "Being lies in the fact that something is, and in its Being as it is; in reality; in presence-at-hand; in subsistence; in validity; in Dasein; in the 'there is'" (Heidegger, 1962, p. 26). He further situates *Dasein* within ontology, stating, "Therefore fundamental ontology, from which alone all other ontologies can take their rise, must be sought in the existential analytic of Dasein" (Heidegger, 1962, p. 34).

Dasein is central to Heidegger's phenomenology, as the concept to which all others relate. The etymology itself underpins the analytical framework of *Being and Time*: the "there" in *Dasein* refers to existence in the world and denotes spatiality as an integral component of being. Spatiality in Heidegger is not geometric extension but existential situatedness; nevertheless, it provides a philosophical foundation for later geometric articulation within architectural space.

Being-in-the-World

Another seminal concept in Heidegger's phenomenology is *being-in-the-world*, closely related to *Dasein* and defined as its fundamental structure. Heidegger writes that the "compound expression 'Being-in-the-world' indicates in the very way we have coined it that it stands for a unitary phenomenon. This primary datum must be seen as a whole" (Heidegger, 1962, p. 68). *Dasein* and *being-in-the-world* are complementary, as *da-sein*—"there-being"—cannot exist without a *there*.

Being-in-the-world is also central to Merleau-Ponty's phenomenology of perception, in which the world is understood as an integral component of embodied experience arising from consciousness. Subject and object, self and world, mind and body are understood as a unity, in contrast to the dualism characteristic of Cartesian rationalism. *Being-in-the-world* is thus inherently spatial: "We have said that space is existential; we might just as well have said that existence is spatial" (Merleau-Ponty, 1962, p. 342). The spatial character of existence thus precedes constructed space; architecture does not create spatiality but gives it form.

The importance of *being-in-the-world* for understanding the built environment is emphasized by architectural theorist Christian Norberg-Schulz, a central figure in architectural phenomenology. For Norberg-Schulz, architectural space forms the basis of existential space and represents a "concretization of environmental schemata or images, which form a necessary part of man's general orientation or 'being in the world'" (Norberg-Schulz, 1971, p. 7). Architecture thus becomes the material expression of existential space.

Fourfold

The *fourfold* (*das Geviert*) is a concept introduced by Heidegger to denote the relationship between earth, sky, mortals, and divinities in his essay "Building Dwelling Thinking" (Heidegger, 1971). In this study, the fourfold serves as a fundamental framework for interpreting the architecture of cosmology. Three elements—earth, sky, and mortals—embody the principal themes, with humankind constructing links between earth and sky. Such links define what is understood here as the architecture of cosmology. The fourth element, divinities, also plays a role, sometimes literally, as in archaeoastronomical sites associated with ritual, and often metaphorically, insofar as remote space remains beyond human reach. Heidegger defines the fourfold as follows:

> But "on the earth" already means "under the sky." Both of these also mean "remaining before the divinities" and include a "belonging to men's being with

> one another." By a primal oneness the four-earth and sky, divinities and mortals-belong together in one.
>
> Earth is the serving bearer, blossoming and fruiting, spreading out in rock and water, rising up into plants and animals. When we say earth, we are already thinking of the other three along with it, but we give no thought to the simple oneness of the four.
>
> The sky is the vaulting path of the sun, the course of the changing, moon, the wandering glitter of the stars, the year's seasons and their changes, the light and dusk of day, the gloom and glow of night, the clemency and inclemency of the weather, the drifting clouds and blue depth of the ether. When we say sky, we are already thinking of the other three along with it, but we give no thought to the simple oneness of the four.
>
> The divinities are the beckoning messengers of the godhead. Out of the holy sway of the godhead, the god appears in his presence or withdraws into his concealment. When we speak of the divinities, we are already thinking of the other three along with them, but we give no thought to the simple oneness of the four.
>
> The mortals are the human beings. They are called mortals because they can die. To die means to be capable of death as death. Only man dies, and indeed continually, as long as he remains on earth, under the sky, before the divinities. When we speak of mortals, we are already thinking of the other three along with them, but we give no thought to the simple oneness of the four. (Heidegger, 1971, pp. 149-150)

The presence of the fourfold in the architecture of cosmology becomes evident in the case studies presented in subsequent chapters. These examples span diverse cultures, belief systems, historical periods, geographical regions, scales, and urban or architectural contexts. Despite this diversity, the oneness of the fourfold constitutes the unifying element motivating intentionality in physical construction.

Within this framework, the architecture of cosmology can be understood as a material condensation of the fourfold, making visible the relational unity of earth, sky, mortals, and the symbolic dimension attributed to the divine.

Axis Mundi

Axis mundi, Latin for "axis of the world," is a concept drawn from cultural anthropology and is fundamental to this study because it establishes a geometric interpretation of the architecture of cosmology. Although not explicitly addressed within phenomenology, the metaphorical axis connecting self, Earth, and sky may be related to the phenomenological body and *being-in-the-world*, expressing the first-person subject positioned at the center of an axis. Geometrically, it may be depicted as a vertical vector linking three points: the center of the Earth, the embodied self, and the sky. This vertical vector is not merely symbolic; it establishes an orientational schema that situates the body within a cosmological coordinate system.

Figure 4. A diagram of the Axis mundi.

The *axis mundi* was examined extensively by Mircea Eliade, who also referred to it as the cosmic axis. Closely associated with the concept of center, it transcends geometric definition to become sacred space—a point where sky is transformed into heaven and Earth into the underworld. Ancient cultures frequently translated scientific observation into mythological symbolism:

> Let us dwell for a moment upon this mythological image of the zenith which is at the same time the Summit of the World and the "Centre" par excellence, the infinitesimal point through which passes the Cosmic Axis (Axis Mundi). … A "Centre" represents an ideal point which belongs not to profane geometrical space,

> but to sacred space; a point in which communication with Heaven or Hell may be realized. (Eliade, 1951, p. 75)

The center is also associated with the *omphalos*, a stone of religious and symbolic significance in ancient Greece. Derived from the Greek *ὀμφαλός*, meaning "navel," it is often translated into Latin as *umbilicus mundi*. Eliade notes that settlements frequently included a built element representing cosmogony, the origin of the universe according to a given belief system:

> ...we can still better understand why every human establishment repeats the creation of the world from a central point (the navel). Just as the universe unfolds from a center and stretches out toward the four cardinal points, the village comes into existence around an intersection. In Bali, as in some parts of Asia, when a new village is to be built the people look for a natural intersection, where two roads cross at right angles. A square constructed from a central point is an imago mundi. (Eliade 1961, p. 45).

The *axis mundi* is commonly represented by vertical built elements connecting three cosmic levels—Earth, heaven, and the underworld—and is also referred to as the cosmic pillar or *universalis columna*. It appears in numerous forms, including pillars, poles, ladders, and trees, as well as natural features such as mountains. It is found across cultures worldwide, including India, ancient Rome, Saxon cultures, Australian Aboriginal traditions, and Indigenous cultures of the Americas and Indonesia (Eliade, 1961).

Norberg-Schulz emphasizes the verticality of the *axis mundi* as an expression of the desire to transcend earthly existence. He also associates the concept with architectural and urban elements that may be oriented horizontally, as in the ancient Roman city organized around the *cardo* (north–south axis) and *decumanus* (east–west axis) (Norberg-Schulz, 1971, p. 22).

PHENOMENOLOGY AND PERCEPTION: SUBJECT & OBJECT, BODY & WORLD

Perception is the process that enables awareness of the environment outside ourselves and is a key process in the phenomenological understanding of reality. The term derives from the Latin *perceptio*, meaning "receiving" or "gathering." The multidisciplinary study of perception spans numerous sciences and disciplines, from biology, physiology, and neuroscience to philosophy, psychology, and psychophysiology. From a basic physiological standpoint, human perception occurs through sensory stimulation processed by the nervous system.

The phenomenological approach to perception is grounded in the first-person encounter with the perceived world. Perception is always contextual: it involves the subject of perception and the object of perception, the self and the world, the encounter of the mind–body with the external environment. As philosopher of science Patrick Heelan explains:

> A perceptual object, for example, is given not atomically as an isolated experience unconnected with anything else, but as fulfilling certain enabling conditions, such as being located in time and space among the things and situations that comprise a World. Our World is the general background reality context that is experienced as given to our perception together with the individual objects that we perceive. (Heelan, 1983, p. 10)

Merleau-Ponty addresses perception extensively in *Phenomenology of Perception*. Perception and the perceiving body are inseparable, as he states in the title of one of the chapters, "The theory of the body is already a theory of perception" (Merleau-Ponty, 1962, p. 235). Perception is central to the subject's encounter with the world as object of perception: it is not merely sensory stimulation but an expression of the embodied mind. Perception participates in the disclosure of reality through a relational process in which the object of perception is apprehended within the structures of embodied experience. Body, mind, and perception are integrated within the holistic system of *being-in-the-world*. The dualisms of subject and object, self and world, mind and body, *res cogitans* and *res extensa* articulated in Cartesian thought are overcome within this unifying framework.

The interconnection between perception, embodied consciousness, and the perceived world—expressed in different forms—appears not only in Merleau-Ponty's work but throughout phenomenology and in other disciplines, including contemporary physics and contemplative sciences. A well-known example is the "observer effect" in quantum physics (Heisenberg, 1958), which articulates the interdependence between observer and object of observation, each affecting the other (Dirac, 1967). Although the epistemological implications differ across disciplines, both perspectives challenge the notion of a completely detached observer standing outside the world observed. In Buddhism, the principle of *pratītyasamutpāda*, meaning "dependent origination," similarly affirms that all phenomena arise through mutual interdependence (Bowker, 1997).

The architecture of cosmology expresses *being-in-the-world* at a further level by integrating "building" into this holistic system—here understood not only in its literal sense but also as a Heideggerian concept (Heidegger, 1971). Building expresses human engagement with the world through the transformation and articulation of space. The perception of remote space participates in this transformation, as constructions are guided by geometries defined through the Earth–sky relationship as observed by the unaided human eye.

If perception is embodied and relational, then clarification of the sensory modalities through which this relation is enacted becomes necessary. The architecture of cosmology depends not on abstract cognition alone but on the concrete operations of the senses that situate the body within local space while orienting it toward remote phenomena. A brief examination of the sensory systems therefore prepares the ground for understanding how celestial perception becomes architecturally articulated.

About the Senses

The following discussion of the senses involved in the perception of the world clarifies the physiological and embodied foundations of this inquiry, particularly how the phenomenological perception of remote space relates to the meaning and construction of the architecture of cosmology.

We perceive the external environment primarily through five senses, corresponding to complex anatomical, physiological, and neurological systems: sight (visual system), hearing (auditory system), taste (gustatory system), smell (olfactory system), and touch (tactile system). These five senses, first articulated by Aristotle in *De Anima* (c. 350 BCE), constitute the external sensory system. Additional sensory modalities also exist. Interoception provides awareness of internal bodily sensations, such as pain or heartbeat. Proprioception—originally termed the "muscular sense," and including kinesthesia—refers to awareness of body position, orientation, and movement in space, and is often associated with the vestibular system responsible for balance (Brandt et al., 2024).

Architectural phenomenology over recent decades has often challenged the primacy of vision as the dominant mode of sensory engagement, introducing *atmosphere* as a multisensory experience of built space. Atmosphere refers to moods and emotions generated by subtle, often invisible elements that make us "feel" a space. It extends beyond visual perception; as Italian philosopher Tonino Griffero observes, "atmospheres are not accessible to a representative-ocular-distal perception but to an ambulatory and synesthetic one" (Griffero, 2014, p. 30), and they can even be smelled (Griffero, 2021). In *The Eyes of the Skin: Architecture and the Senses*, Finnish architect Juhani Pallasmaa argues that architecture should engage all the senses rather than privileging vision alone (Pallasmaa, 2005). As the title suggests, Pallasmaa emphasizes tactility: "All the senses, including vision, are extensions of the tactile sense; the senses are specializations of skin tissue, and all sensory experiences are modes of touching" (Pallasmaa, 2005, pp. 10–11).

Nevertheless, in the present inquiry, visual perception constitutes the primary sensory mode for understanding remote space. Vision is augmented by proprioception, which provides awareness of bodily orientation and position on Earth relative to the object of observation—the sky. When the perception of the remote sky is actualized through constructions of architecture of cosmology, kinesthetic movement within built space becomes integrated into the embodied local spatiality of perceived remote space. Sensory engagement alone, however, does not yet explain how spatial relations are structured. Perception always unfolds within an organized field. To understand how local bodily orientation and remote celestial phenomena become geometrically related, the inquiry must turn from the senses themselves to the structure of space as it is lived and conceptualized.

Space: Phenomenology, Geometry, and Perception

The inquiry into the perception of space is fundamental to this study. The focus lies on the relationship between the local space of the perceiving subject and the remote space as object of perception. The space of perception and the perception of space are not treated as a dichotomy but as complementary experiences. Geometry provides a framework for both, unifying the perceiving subject in local space with objects of perception in the remote space of the sky. Geometry does not replace lived space but formalizes relations already operative in embodied perception. A phenomenological approach is particularly relevant, not only because phenomenology has developed extensive theories of space, but also because of its foundational emphasis on the primacy of the human body in perception and in the encounter with the world.

At the origins of transcendental phenomenology, Husserl distinguished four contexts of space: "(1) the space of everyday living, i.e., the intuitive space that is both before and beyond science; (2) the space of pure geometry or mathematics; (3) the space of applied geometry, i.e., the space of the natural sciences; and (4) the space of metaphysics" (Embree, 1997, p. 670). Husserl recognized the relevance of geometry in lived experience and addressed geometric space extensively in his essay "The Origin of Geometry," published as an appendix to *The Crisis of European Sciences and Transcendental Phenomenology* (Husserl, 1970). Jacques Derrida later analyzed this text in *Edmund Husserl's Origin of Geometry: An Introduction* (1978). In contrast to Kant's a priori conception of space, Husserl understood geometric space as grounded in perception, even though geometric constructions may become idealized abstractions. He critiqued modern science for losing contact with the immediacy of lived perception.

Husserl's separation of geometric space from everyday lived space differs from later phenomenological developments, particularly those of Merleau-Ponty, who regarded geometry and mathematics as embodied experiences simultaneously derived from and generated by perception. Merleau-Ponty defines space as a lived relationship rather than a purely epistemological construct:

> I catch space at its source... and pass from spatialized to spatializing space. In the first case, I am concerned with physical space... in the second with geometrical space having interchangeable dimensions, homogeneous and isotropic. (Merleau-Ponty, 1962, pp. 283–284)

Movement is essential both to geometry and to the perception of space. Merleau-Ponty illustrates this through the example of constructing a geometric figure, emphasizing the primacy of the subject: "The subject of geometry is a motor subject" (Merleau-Ponty, 1962, p. 450).

In this study, the concept of space is guided initially by Merleau-Ponty's embodied geometry. The relationship between remote and local space enacts the geometries of *being-in-the-world*, defined by a subject positioned within a local landscape and perceiving phenomena—light, solar cycles, stellar motion—occurring in the remote sky. The introduction of built interventions in local space marks a further step, as architecture creates an additional spatial framework that enters into and reshapes the system defined by perception.

Being-in-the-World: Orientation, Coordinate Systems, Cardinal Directions, Anatomical Planes

Being-in-the-world can be understood as a system defined by the interconnection of mind–body, perception, space, and geometry. This system can be expressed through six orientations: left/right, front/back, and up/down. These directions presuppose an embodied subject and are relative to bodily position and orientation in three-dimensional space; only left and right exhibit bilateral symmetry (Figure 5a). Geographer Yi-Fu Tuan emphasizes this bodily grounding of spatial orientation:

> Vertical–horizontal, top–bottom, front–back and right–left are positions and coordinates of the body that are extrapolated onto space... Awake and upright, man regains his world, and space is articulated in accordance with his corporeal schema. (Tuan, 1977, pp. 35–36)

Figure 5. Human body and reference systems.

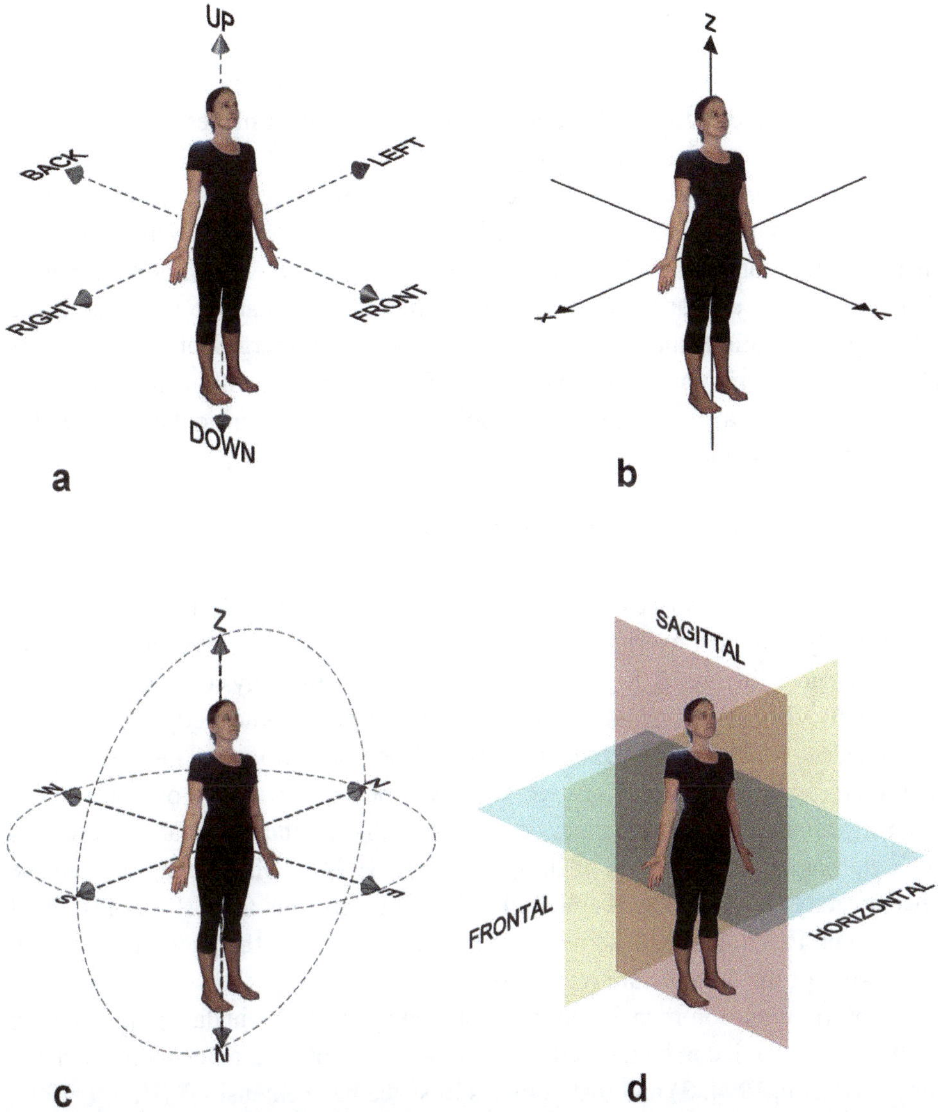

Geometrically, these six directions may be represented as a three-dimensional coordinate system. For a subject positioned at the origin of a right-handed coordinate system, left/right corresponds to the x-axis, front/back to the y-axis, and up/down to the z-axis (Figure 5b).

When the spatial world of being encompasses Earth and sky, the six directions align with the cardinal points (north, south, east, west), zenith, and nadir (Figure

5c). These directions can be represented through the horizontal coordinate system, a spherical framework centered on the observer and analogous to the geographic coordinate system of latitude, longitude, and altitude. As discussed in subsequent chapters, the cardinal points are defined by solar motion at the equinoxes, and astronomical reference systems are examined in detail in Chapter 3.

These spatial directions also correspond to anatomical planes used in human anatomy (Figure 5d): sagittal (left/right), frontal or coronal (front/back), and horizontal or transverse (upper/lower). The spatial framework of bodily orientation thus operates across multiple contexts that define reference systems for lived experience. In architectural phenomenology, orientation is fundamental. Norberg-Schulz notes that space is "an aspect of any orientation," requiring humans to understand and unify spatial relations into a coherent space concept (Norberg-Schulz, 1971, p. 9). Directions and orientations become constructed elements within the built environment and play a significant role in social and ritual life, particularly within the architecture of cosmology.

Space and Form in Perception: Dualities

The perception of space originating from lived experience develops into an object of study in almost every field of science, from formal and physical sciences to social sciences, including logic, mathematics, astronomy, physics, neurosciences, philosophy, chemistry, and architecture. All these fields of knowledge, disciplines, and practices deal with space, but each offers different approaches, methodologies, and interpretations. In a previous work, I developed my own view of space drawn from my approach as an architect dealing with digital and physical space not as separate but as related fields of design (Bertol, 1994). Thirty years later, in the present study on a different theme and field of inquiry, this view still holds, and I am presenting it again, with some revisions congruent with the phenomenological framework of the architecture of cosmology.

In my view, the common denominator shared by the different disciplines dealing with space is that it can be defined as "a field—a sort of 'container'—in which we move" (Bertol, 1994, 3) and that "objects in space have extension" (Bertol, 1994, 4). This view of space does not refer to Newtonian absolute space but to space as a relative field defined by our field of vision or by other sensorial inputs—for example, the space of a room, which is contained within the space of a cityscape or a landscape, which are themselves contained within the sky. Extension is the common property of both the containing field and the contained things, a perceptual experience that leads to the concept of form.

> Plato defined form (εἶδος (eidos), synonymous with ἰδέα (idea)) as the permanent ideal reality of things, as opposed to their temporary attributes. According to Aristotle, form (εἶδος) is what makes things emerge from an undifferentiated primordial matter. In our discussion, form means physical form or shape as a perceivable entity, characterized by spatial extension and dimension, that creates distinctions in an otherwise amorphous field. According to our perception (which provides us with our first and most immediate knowledge of the world outside ourselves), the world is a three-dimensional space ordered and defined by constant or changing forms. Forms are really what make space perceptible; they provide a series of contrasts, like solid and void, inside and outside, figure and ground, positive and negative, and contrast between different colors. We can perceive all these oppositions because of boundaries, which create discontinuities in an otherwise uniform field. Therefore, space cannot be defined without the boundaries that give shape to forms. An interesting definition of form (which finds application in Chapter III) can be found in a book of mathematical logic by G. Spencer Brown: 'A distinction is drawn by arranging a boundary with separate sides so that a point on one side cannot reach the other side without crossing the boundary ... Call the space cloven by any distinction, together with the entire content of the space, the form of the distinction. Call the form of the first distinction the form.' Even in this context the primary characteristics of form are given by boundaries, which determine distinctions. (Bertol, 1994, pp. 4-5)

Even in this context, the primary characteristics of form are given by boundaries, which determine distinctions. The perception of space therefore develops into forms and the boundaries that make forms perceivable. In three-dimensional space, boundaries are defined by the surfaces bounding solids, separating them from the surrounding void. "Anything beneath the most superficial layers is perceptually unknown. Previous knowledge or experience might suggest what a physical object is like in its totality, but the perceptual experience is only of its external surfaces" (Bertol, 1994, 11). From this approach, the perception of space can also be understood as a series of interrelating dualities:

- figure–ground
- solid–void
- inside–outside

I use the term *duality* instead of *dualism*. While dualism denotes a division into irreconcilable opposition, duality refers to different yet interrelating aspects of a system, once again pointing to a whole greater than the sum of its parts.

Figure and Ground

Among the dualities arising from the perception of form, figure–ground is one of the most fundamental for learning about the surrounding world through the visual system. Boundaries define an object (Figure 5) with respect to its surrounding background, or ground. Figure and background are perceptual opposites, but they complement each other as a whole, creating a system in which one form is shaped by the other (Wever, 1927). In a figure–ground configuration, the visual system recognizes a form in the foreground, defined by its boundaries, as emerging from a surrounding field. Boundaries themselves are often defined by color contrasts—most commonly black and white—and by shape.

The boundary itself can be perceived as a figure. When a closed boundary is filled, ambiguity may arise in discerning figure from ground, although black is usually perceived as figure and white as background. In Figure 6a, the black circular contour is the boundary; when filled with black (Figure 6b), it is perceived as a figure on a white background. The white circle in Figure 6c is typically perceived as a figure on a black background, while the surrounding black shape may be perceived either as background or as a cutout figure. The text shown in Figure 7 is perceived as a figure even when color is inverted with the background.

Figure 6. A closed boundary defines a figure.

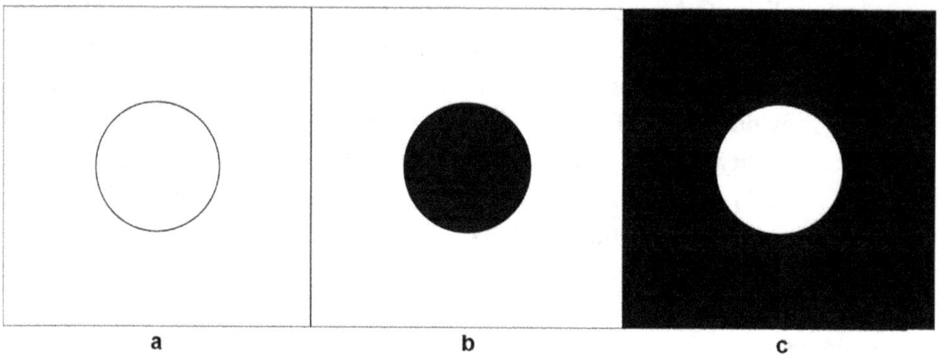

Figure 7. The text appears as ground independently from the color.

The figure–ground relationship plays an essential role in visual perception and was studied extensively by Gestalt theory, developed by a German school of psychology in the early twentieth century. Gestalt—translated as "form," "configuration," or "pattern"—is founded on a phenomenological approach that understands perceptual reality as a whole or system, often expressed by the statement that "the whole is greater than the sum of its parts.".

Figures 6 and 7 illustrate these perceptual principles.

Merleau-Ponty frequently refers to Gestalt theory. Figure–ground is fundamental to perception as a system—a whole that overcomes the subject–object dualism and expresses being-in-the-world. Figure–ground provides a "definition of the phenomenon of perception, that without which a phenomenon cannot be said to be perception at all" (Merleau-Ponty, 1962, p. 4). The figure–ground relation is not passive sensorial data but an active, intentional, embodied process. It lies at the core of Merleau-Ponty's view of perception as a spatial, embodied engagement with the world:

> If my body can be a 'form' and if there can be, in front of it, important figures against indifferent backgrounds, this occurs in virtue of its being polarized by its tasks, of its existence toward them, of its collecting together of itself in pursuit of its aims; the body image is finally a way of stating that my body is in-the-world. (Merleau-Ponty, 1962, p. 115)

Figure–ground relationships in the two-dimensional plane can be ambiguous. The face–vase illusion (Figure 8), discussed by the Danish psychologist Edgar Rubin (Rubin, 1915), is a well-known example: the same bounded region can be perceived either as a vase or as two facing human profiles. The illusion typically relies on strong color contrast and tends to diminish when detail or three-dimensionality is introduced.

Figure 8. The Rubin Vase: the illusion diminishes when the vase is three-dimensional.

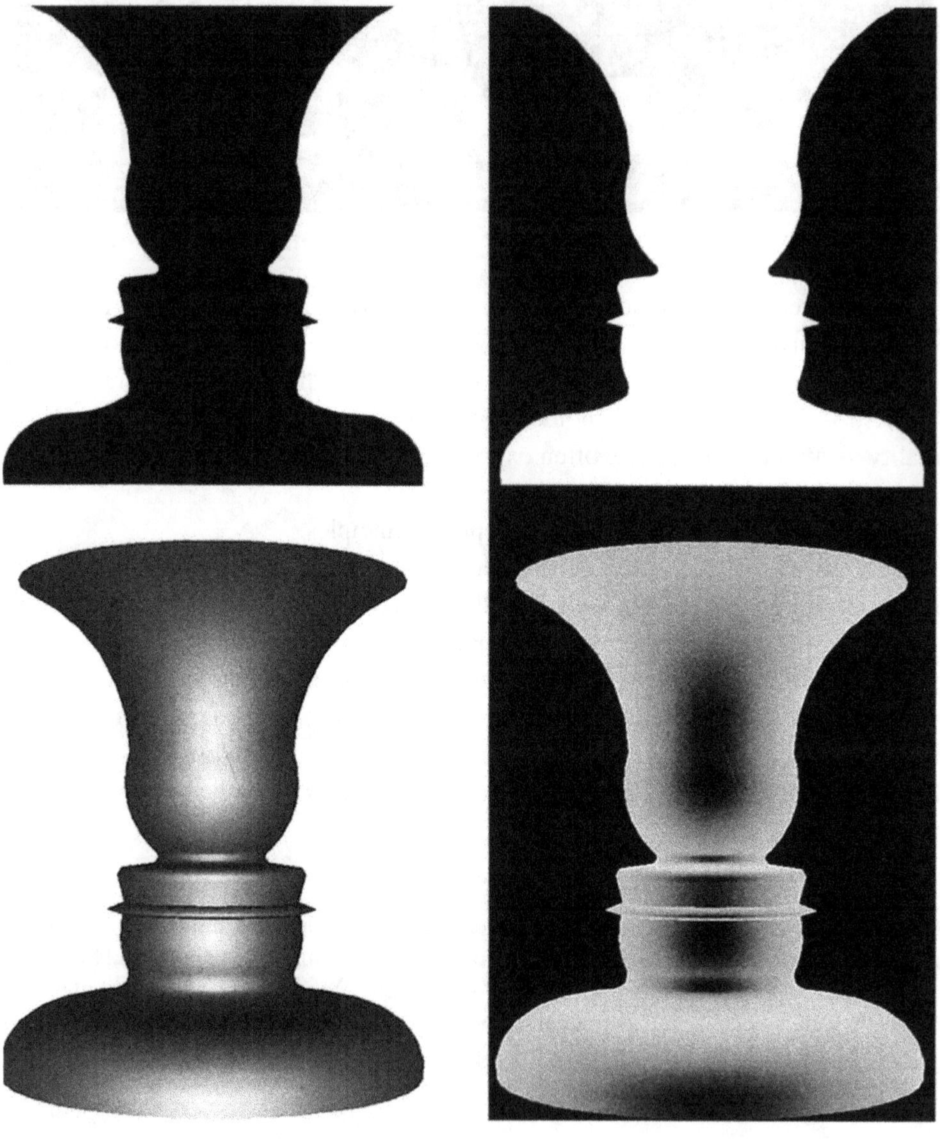

The shapes shown in Figure 9 illustrate further figure–ground ambiguities. In Figure 9a, the same square boundary produces different figure–ground perceptions depending on whether it is filled. In Figure 9b, smaller shapes are perceived as figures, whereas larger shapes tend to be perceived as ground (Arnheim, 1974). The meander patterns in Figure 9c demonstrate figure–ground reversal through complex boundary relationships.

Figure 9. Figure-ground ambiguities (Bertol, 1994).

Although figure–ground is often discussed in relation to two-dimensional images—both analog and digital—it is always present in the perception of three-dimensional space, in both natural and built environments. In this context, figure–ground is closely associated with the dualities of solid–void and inside–outside. Solid as mass and void as empty space provide the most fundamental information about three-dimensional space and can be perceived not only visually but also through other senses, particularly proprioception. As previously argued:

> If a solid occupies a position in space, no other solid can have the same position at the same time. In the same way that space can be viewed in terms of both container and extension, the perceptual field of space is articulated through a series of solids separated by voids. Often in architecture solid is defined as positive space and void as negative space; they are complementary to each other and one (Bertol, 1994, 7)

Architectural phenomenology, particularly through Norberg-Schulz's theories, extends figure–ground analysis to the other dualities (Norberg-Schulz, 1979). In three-dimensional space, a system is defined by the interrelation of solid–void, inside–outside, and figure–ground. Inside a building, solid walls typically form the background of the enclosure, while furniture and other elements appear in the foreground. Outside a building, the background may consist of streets in an urban context or landscape in a natural setting. Trees, vegetation, and water features may appear as figures within the landscape background, while the sky ultimately functions as the final background for both natural and built environments.

The examples that follow illustrate these relationships. A house in a natural setting (Figure 10) shows hierarchical layers of perception: the house and trees appear as solids; windows frame the landscape and may turn the exterior into a figure when their scale emphasizes the view. Outdoor sculpture (Figure 11) provides another example, in which human-made forms coexist with natural elements, both acting as figures against the landscape and sky. The landscape is a background and turns into figure into the sky-background.

Figure 10. Figure-ground in a house on a natural setting (S.U.N.F.A.R.M. 2024).

Figure 11. Figure-ground in outdoor sculpture (S.U.N.F.A.R.M. 2024).

Tempietto del Bramante, a Renaissance building in Rome, exemplifies a highly articulated series of figure–ground, solid–void, and inside–outside relationships. Designed by Donato Bramante (1444–1514) and built between 1502 and 1509, the Tempietto is situated at the center of the courtyard of San Pietro in Montorio. It features a circular plan approximately four meters in diameter, crowned by a dome and surrounded by sixteen columns—a number considered ideal by Vitruvius (Vitruvius Pollio, 1914).

The photographic sequences of Figures 12–15 show the Tempietto as focal point and figure, framed by the arches at the entrance of the courtyard—solid elements articulating the view toward the center, in an interplay between solid–void and figure–ground, forming a paradoxically “open enclosure.” These relationships are best perceived by a moving body, walking along the Tempietto’s bilateral symmetry axis from the entrance of the courtyard to its interior.

Figure 12. Tempietto di Bramante framed by an arch, at the entrance of the courtyard, 2025.04.29.

Figure 13. Tempietto del Bramante, with the bell tower of San Pietro in Montorio in the background, Rome, 2025.03.30.

Figure 14. Entrance of the Tempietto del Bramante, 2025.03.30.

Figure 15. Inside Tempietto del Bramante, 2025.03.30.

These spatial dualities are fundamental to the architecture of cosmology. Megalithic structures such as Stonehenge are perceived as solids against the background of landscape and sky, while at Newgrange the solar alignment is experienced within an interior void enclosed by solid stone. Obelisks—often functioning as sundials—appear as solids against the background of the cityscape and sky, as in the Obelisco di Montecitorio (Figure 16), an Egyptian obelisk (594–589 BCE) functioning as the gnomon of one of the largest sundials ever built. The Sun at solar noon likewise appears as a figure against the sky background.

The examples presented in this section demonstrate how spatial dualities operate hierarchically, shifting with context and scale. A background at one scale may become a figure at another. The sky is the exception: it functions as an absolute background—outside and void—populated by figures at multiple distances and scales, from the Sun (Figure 16) and Moon (Figure 17) to planets, stars, and galaxies.

Figure 16. Obelisco di Montecitorio at solar noon, Rome, 2025.06.23.

Figure 17. Full Moon over Rome, 2025.06.01.

Architectural plans and other two-dimensional representations also exhibit such hierarchical shifts. In a floor plan, walls appear as solids defining a void enclosure; in a site plan, the building itself becomes a figure against the background of the site.

LOCAL AND REMOTE SPACE

We live in an age of ubiquitous computing and digital realities in which, at the time of writing, local and remote space are most often contextualized within Information and Communication Technologies (ICT). The Internet has profoundly transformed our sense of space and social interaction, giving rise to notions of digital space and virtual environments. I explored digital space from an early design perspective in the 1990s, when I authored one of the first books on Virtual Reality (VR) in architecture (Bertol & Foell, 1997). At that time, *cyberspace*—a term coined in a work of science fiction (Gibson, 1984)—was entering common usage, and VR was still in its infancy, well before its integration into ubiquitous mobile devices. My interest then lay in developing a scholarly understanding of computing and spatial perception, closely intertwined with an art and design practice deeply engaged with digital media.

This early engagement with digital environments foregrounded a central question that remains operative in the present study: how does mediation transform spatial experience? In virtual environments, spatiality is constructed algorithmically and perceived through interfaces; it is programmable, scalable, and detachable from physical emplacement. The shift toward cosmological architecture did not abandon this question but redirected it toward a different order of mediation—one grounded not in code, but in celestial mechanics and embodied orientation.

In the mid- to late 1990s, my practice shifted toward what I describe as the *architecture of cosmology*: the design and construction of landscape interventions that guide the perception of remote celestial phenomena. This work responds to an existential need to situate human presence within a broader cosmic order. The present inquiry develops a scholarly investigation of how local and remote space are articulated through architectural constructions. Although grounded in phenomenology, it requires semantic clarification in order to distinguish this approach from the contemporary, everyday understanding of remote space shaped by digital technologies.

The distinction is not merely terminological. In everyday discourse shaped by ICT, "remote" implies technological mediation and screen-based representation. In the architecture of cosmology, by contrast, remote space refers to physically existing celestial phenomena that are spatially distant yet perceptually accessible without technological amplification. The difference is ontological: one remote space is representational, the other astronomical.

Within ICT, *local space* refers to the physical environment of the user and, more narrowly, to the immediate space of the screen and interface of computers or mobile devices. *Remote space* denotes the images—static or dynamic—displayed on the screen, representing places and events physically distant from the observer. Remote space is mediated through local digital devices and remains disconnected from the surrounding environment in which perception takes place. During videoconferences, digital displays function as windows onto remote space. Figure 18 documents a videoconferencing session in which local space is reduced to the laptop keyboard, screen, speakers, microphone, and camera, enabling interaction with a person located in Brooklyn, New York, over one hundred miles away. Similarly, webcams and livestreams offer views of distant locations far beyond the reach of unaided vision, as illustrated in the screen capture shown in Figure 19.

Figure 18. Videoconferencing session, local digital workspace, 2025.08.01.

Figure 19. Livestream webcam view, Cape Town, 2025.08.02 Source: https://www.earthcam.com/.

In digital experience, space becomes fundamentally disembodied when perception becomes spatially dislocated from bodily co-presence. Mediated perception privileges sight and hearing, while the body remains physically situated elsewhere. Disembodiment involves not only cognition but perception itself, as sensory engagement is detached from the spatial conditions of bodily presence. The experience of time and space in the digital realm differs radically from that of the analog, physical world. Being *here and now* assumes a meaning distinct from Heidegger's being-in-the-world, as "I am now here" is transformed into "I am now somewhere else." Digital "elsewhere" is not an ontological elsewhere but a representational elsewhere. Telepresence represents an extreme form of this condition: local agency produces effects in remote environments and social interactions. Agency may extend across distance, yet the architectural structure that enables mediation remains indifferent to the spatial content transmitted. The screen does not orient the body within the remote environment; it merely displays it. In terms of the built environment, there is no architectural relationship between the remote space and the physical structure mediating perception. For this reason, an architecture of cosmology cannot emerge within such a context.

Not all ICT media, however, contribute to disconnection and disembodiment. Location-based media (LBM), such as Global Positioning Systems (GPS), associate geolocation with the moving body carrying a mobile device. GPS relies on orbiting satellites transmitting signals to receivers on Earth, establishing a spatial relationship between the remote space of the satellite and the local space of the observer.

In this case, local and remote space are explicitly linked rather than separated. Here mediation does not replace orientation but supplements it. The moving body becomes the operative center of a coordinate system extending beyond immediate perception, integrating satellite geometry with terrestrial position. The relationship anticipates, in technological form, the cosmological alignment between Earth-bound observer and celestial bodies.

Returning to the present inquiry, local space can be defined in terms of distance and perceptual scale. The body and its sensory systems, including proprioception, provide measures of locality. Expressions from common language—such as "out of sight," "out of touch," and "out of reach"—may be understood not only metaphorically but literally. Local space is always within sight and may or may not be within reach, whereas remote space is always out of touch and may also be out of sight, though still accessible through unaided perception or mediatic means. In its simplest definition, local space emerges where the perceptual field of vision intersects with the principles of projective geometry: the three-dimensional world is projected onto the two-dimensional surface of the retina. Linear perspective offers an approximation of this perceptual field, with vanishing points and the horizon line providing fundamental visual cues.

The horizon—the meeting of sky and Earth—marks the boundary between local and remote space as perceived by unaided vision. The horizon is not merely a visual limit but an experiential threshold. It defines the maximum extension of embodied sight while simultaneously suggesting what exceeds it. As such, it is both boundary and invitation: a perceptual closure that opens toward astronomical vastness. In the architecture of cosmology, the horizon becomes a primary geometric and existential datum. Geometrically, it is a circle. The term *horizon* derives from the Greek verb ὁρίζω (to divide, to separate), itself derived from the noun ὅρος, meaning boundary or landmark (Liddell et al., 1940). The distance between an observer *O* and a point *G* on the horizon represents the maximum distance visible without refraction and is given by the approximation:

$$OG = \sqrt{2Rh + h^2},$$

where *R* is the radius of the Earth (6,378 km) and *h* is the height of the observer (Young, n.d.). From this relation, both mathematically and intuitively, the higher the observation point, the farther the horizon. At an observation height of approximately three meters, the distance to the horizon is about 6,318 meters. This value represents an ideal condition, as actual visibility is affected by atmospheric refraction and intervening elements.

The mathematical formulation makes explicit what perception intuits: visibility is conditioned by bodily elevation and planetary curvature. The Earth itself becomes a geometric partner in defining the limit between local and remote space. Thus, even the most immediate perceptual boundary is already cosmological in scale.

The most distant horizon views occur when the observer looks out over the ocean, where sightlines are unobstructed (Figure 20). In urban environments or forests, buildings, terrain, and vegetation interfere with visibility. Clear horizon lines may also be perceived from highways traversing open landscapes, as shown in Figure 21. Flat desert environments similarly offer extended, uninterrupted views of the horizon.

Figure 20. Ocean horizon view, Rockaway Beach, New York, 2019.06.09.

Figure 21. Horizon line from a highway, Garden State Parkway, New Jersey, 2017.12.25.

PLACE AND EMPLACEMENT

The centrality of embodiment to being-in-the-world has been addressed in the preceding phenomenological discussion. *Emplacement* extends embodiment by emphasizing being-in-place. Both space and place belong to the world of lived experience, extending beyond the self and forming an integral component of the perceptual system as a whole. While *space* can be defined through physical, geographic, geometric, and other measurable quantitative properties, *place* is characterized by qualitative attributes associated with individual emotion as well as cultural and social meaning. If local and remote space describe relational distances, place introduces qualitative density. Emplacement situates the perceiving subject not only geometrically but existentially. It is through emplacement that spatial orientation acquires meaning beyond measurement.

Among the many definitions of place, two are particularly relevant to this inquiry. The first, originating in a geometric framework but applicable more broadly, defines place as a relationship between two geometries, "relating a containing geometry to a contained geometry" (Schema.org, n.d.). The second definition identifies three key components of place: location, locale, and sense of place. Location refers to the position of a point on the Earth's surface; locale denotes the physical setting for social relations, such as a region or landscape; and sense of place describes the emotional attachments formed through experience. Place, in this sense, can operate at multiple scales and does not need to be fixed in either time or space (National Geographic Society, n.d.).

A place has identity and character, as emphasized by architectural phenomenology. Place is described as "a space which has a distinct character" (Norberg-Schulz, 1979, p. 5), and "whereas 'space' denotes the three-dimensional organization of the elements which make up a place, 'character' denotes the general 'atmosphere' which is the most comprehensive property of any place" (Norberg-Schulz, 1979, p. 11). While space can be defined according to the physical properties of the most immediate perceptions, place also brings emotions and recalls memories, being rooted in personal and collective experiences.

Tuan defines place as an "enclosed and humanized space" (Tuan, 1977, p. 54); differently from space, which can be defined by properties not involving human presence, place is constructed through people's experiences. Humankind always associates place with significance when space "is transformed into place as it acquires definition and meaning" (Tuan, 1977, p. 136). People have emotional attachment to places, and Tuan uses the term *topophilia*—meaning "love of place," from the Greek *τόπος* ("place") and *φιλία* ("love")—to denote the "affective bond between people and place or setting" (Tuan, 1974, p. 4).

Being-in-the-world turns into emplacement when the focus shifts to interaction with places as the environment of our experiential being—an interaction that involves not only sensory perceptions but also cultural identity and emotional attachments. Places are beyond the relationship between the self and the world, as they are constructed from social interaction or the lack thereof. Culture shapes space into place beyond its geometric and quantitative properties. Places start from the most immediate physical space surrounding us, associated with biological functions and daily activities, and expand from dwelling to streets, neighborhoods, and cities.

The word "place" in language has both a literal and a metaphorical meaning. *Genius loci* is one of the metaphorical elements associated with place. The term means, in Latin, "spirit of place" and is a concept present since ancient Rome. The *genius* was a spirit assigned to persons of power and to places; it was represented iconographically as a figure holding a cornucopia, a bowl, or a snake. The *genius* was not only a guardian, like the *Lares*, but embodied the place itself.

The expression has become common in architectural discourse, popularized by Norberg-Schulz, who defined *genius loci* as the character or identity of a place and stated that it "in many cases has even proved strong enough to dominate any political, social, and cultural changes. This, for instance, holds true for cities like Rome, Istanbul, Paris, Prague, and Moscow" (Norberg-Schulz, 1971, p. 27). The *genius loci* can be situated in natural as well as human-made places. Architecture can bring a *genius loci* to a place: "Through building, man-made places are created which possess their individual genius loci" (Norberg-Schulz, 1979, p. 58).

An additional contribution to the understanding of place can be found in *The Practice of Everyday Life* (de Certeau, 1984). Michel de Certeau makes a distinction between space (*espace*) and place (*lieu*) in terms of the "order" and stability of place versus the dynamism and time-dependent transformation of space, generated by "vectors of direction, velocities, and time variables," leading to the definition of space as "a practiced place" (de Certeau, 1984, p. 117). This distinction is particularly relevant for cosmological architecture. Celestial observation transforms an abstract site into a practiced place through repeated, intentional acts of orientation. Time-based return—solstice after solstice, equinox after equinox—converts geometric alignment into lived emplacement.

A thorough discussion of place is not within the scope of this inquiry; however, these introductory considerations are relevant to contextualizing the architecture of cosmology—the translation of perceptions of space beyond immediate reach into built expressions aiming to create places of meaning. As presented in the following chapters, these constructions were "monuments" to the sky, but they also served functions ranging from timekeeping to ritual. They exemplify how humankind, from time immemorial, has turned to architecture not only for shelter but to give character and identity to the space of existence—that is, to create places.

I AM NOW HERE: TOWARD A PHENOMENOLOGY OF ARCHITECTURE OF COSMOLOGY

This chapter has outlined the conceptual foundations for a phenomenological framework of the architecture of cosmology. The conscious experience of time-based observations of the remote space of the sky is interpreted through patterns—cyclical manifestations of celestial phenomena. Intentionality in observation is expressed through the making of physical constructions that modify local space in order to frame, guide, and stabilize perception. The architecture of cosmology thus refers not only to buildings or spatial artifacts, but to a phenomenological method in which perception, construction, and being-in-the-world are inseparably intertwined.

Throughout history, and as early as prehistory, humankind has produced physical constructions that act as markers within the local landscape, guiding perception toward the remote space of the sky. These constructions operate as gestalts—perceptual wholes—emerging as figures against the background of landscape and sky. The architecture of cosmology constitutes a spatial intervention that incorporates time into its geometry: time is present not only as historical duration embedded in place, but as an active dimension necessary for perceiving cyclical astronomical phenomena.

Through built form, astronomical time becomes integrated with existential time, uniting the object of perception and the perceiving subject within the wholeness of being-in-the-world.

The photographic sequence in Figure 22 illustrates and synthesizes the key concepts developed in this chapter. It documents bodily movements aligned with the Sun at Sunrise Trellis (S.U.N.F.A.R.M., New York), a contemporary example of what I define as architecture of cosmology. Sunrise Trellis integrates multiple planes of vision and operates as a phenomenological device mediating the relationship between local and remote space. Architectural elements—the trellis structure, the surrounding field, and the tree line—form the foreground, establishing a local spatial framework against which the Sun's apparent motion unfolds in remote space. The trellis serves both as a constructed reference for the Sun's path and as the site of a daily contemplative practice oriented toward the east.

Figure 22. "Finding the Axis Mundi," Sunrise Trellis (S.U.N.F.A.R.M., New York), 2025.08.09.

The existential, lived dimension of space is articulated through a sequence of bodily movements derived from yoga postures, which define a spiritual and experiential field—the genius loci of Sunrise Trellis as lived practice. The images are screen captures from a time-lapse video recorded on August 9, 2025, beginning at 6:15 a.m., shortly after sunrise at 5:55 a.m. The time-lapse interval was 120 seconds, with the exception of the movement sequence, recorded at 10-second intervals. The composite image consists of nine sequential frames (a–i), forming an experiential composition of local, remote, and lived space, integrated with time and place. The east–west orientation of the trellis (local geometry) interacts with the apparent solar trajectory (remote geometry) as time unfolds, a relationship further expressed through the evolving patterns of shadow. I refer to this movement sequence as *Finding the Axis Mundi*, denoting the alignment between my moving body, the architectural structure, the local landscape, and the sky.

In this experience, I inhabit the interconnectedness between myself—as a mindful, embodied subject—and the remote celestial phenomena framed by a constructed reference within the local environment. The architecture of cosmology is thus experienced not as representation, but as an enacted condition of being-in-the-world: I am, now and here.

The phenomenological framework developed in this chapter establishes the conceptual conditions for understanding the architecture of cosmology as embodied orientation. The following chapter shifts register from philosophical articulation to geometric and astronomical analysis. If this chapter has clarified the experiential and ontological dimensions of local and remote space, the next examines the mathematical structures that make Earth–sky alignment measurable, calculable, and constructible. The movement from phenomenology to geometry does not abandon embodiment; rather, it formalizes its spatial logic. The astronomical equations that follow do not introduce an abstract system external to lived experience; they formalize the geometric relations already operative in perception, orientation, and emplacement.

Methodological Note

The phenomenological framework developed in this chapter operates through a first-person, embodied mode of inquiry that integrates perception, movement, spatial configuration, and temporal observation. This approach does not pursue objectivity in a positivist sense, nor does it aim at exhaustive empirical generalization. Instead, it adopts situated experience as a legitimate epistemic ground, acknowledging both its strengths and its limitations. The method is intentionally selective: it privileges lived perception over computational modeling, qualitative spatial awareness over quantitative simulation, and experiential alignment over data-driven abstraction.

The architecture of cosmology is therefore investigated as a relational system involving body, environment, and sky, in which meaning emerges through repeated, time-based encounters with celestial phenomena. Gestalt perception is central to this approach: architectural constructions are understood as perceptual wholes—figures that organize orientation and attention against the background of landscape and sky. While this mode of inquiry necessarily reflects the positionality of the observing subject, it also allows for intersubjective resonance grounded in shared bodily capacities, environmental conditions, and astronomical cycles. The method does not claim universality, but proposes a transferable framework rooted in embodiment, spatial practice, and phenomenological description, applicable to both historical interpretation and contemporary architectural investigation.

Recent research in embodied cognition supports the premise that spatial perception and understanding are grounded in bodily experience rather than in abstract mental representation alone. Scholars such as Varela, Thompson, and Rosch, along with more recent work in cognitive science and neurophenomenology, argue that cognition emerges from the dynamic coupling of brain, body, and environment. In the context of spatial and astronomical perception, orientation, scale, and movement are enacted through posture, locomotion, and sensory feedback. The observer's bodily situatedness thus becomes integral to the interpretation of celestial phenomena, reinforcing the argument that architectural configurations can guide not only visual alignment but also embodied awareness of Earth–sky relationships.

Contemporary architectural phenomenology has extended earlier existential approaches by incorporating insights from sensory studies, environmental psychology, and landscape theory. Recent discourse emphasizes atmosphere, material presence, and multisensory engagement as critical dimensions of spatial experience. Within this expanded framework, landscape is understood not as a static backdrop but as a field of relational processes involving perception, memory, movement, and cultural meaning. This perspective aligns with the architecture of cosmology, where spatial form, environmental orientation, and temporal cycles interact to produce experiential knowledge of the sky. Architectural meaning thus emerges not solely through symbolic representation, but through repeated embodied encounters with celestial rhythms.

REFERENCES

Arnheim, R. (1974). *Art and visual perception: A psychology of the creative eye*. University of California Press.

Bertol, D. (1994). *Visualizing with CAD: An AutoCAD exploration of geometric and architectural forms*. TELOS.

Bertol, D., & Foell, D. (1997). *Designing digital space: An architect's guide to virtual reality*. John Wiley & Sons.

Brandt, Y., Dieterich, M., & Huppert, D. (2024). Human senses and sensors from Aristotle to the present. *Frontiers in Neurology*, *15*, 1404720. PMID: 39022724

de Certeau, M. (1984). *The practice of everyday life*. University of California Press.

Derrida, J. (1978). *Edmund Husserl's* Origin of geometry*: An introduction* (J. P. Leavey, Trans.). University of Nebraska Press.

Dirac, P. A. M. (1967). *The principles of quantum mechanics*. Oxford University Press.

Eliade, M. (1951). Images and symbols (P. Mairet, Trans.). Harvill Press.

Eliade, M. (1961). *The sacred and the profane*. Harper & Brothers.

Embree, L. (Ed.). (1997). *Encyclopedia of phenomenology*. Kluwer Academic Publishers.

Gibson, W. (1984). *Neuromancer*. Ace Books.

Griffero, T. (2014). Atmospheres and lived space. *Studia Phaenomenologica*, *14*, 29–51.

Griffero, T. (2021). Sniffing atmospheres: Observations on olfactory being-in-the-world. In *Olfaction: An interdisciplinary perspective from philosophy to life sciences* (pp. 75–90). Springer.

Heidegger, M. (1962). Being and time (J. Macquarrie & E. Robinson, Trans.). Harper & Row. (Original work published 1927)

Heidegger, M. (1971). Building, dwelling, thinking. In Poetry, language, thought (A. Hofstadter, Trans.). Harper & Row.

Heisenberg, W. (1958). *Physics and philosophy: The revolution in modern science*. Harper & Brothers.

Husserl, E. (1970). The crisis of European sciences and transcendental phenomenology (D. Carr, Trans.). Northwestern University Press.

Husserl, E. (2001). *Logical investigations* (J. N. Findlay, Trans.). Routledge. (Original work published 1901–1902)

Liddell, H. G., Scott, R., Jones, H. S., & McKenzie, R. (1940). *A Greek–English lexicon*. Clarendon Press.

Long, A. A. (1986). *Hellenistic philosophy: Stoics, Epicureans, sceptics*. University of California Press.

Merleau-Ponty, M. (1962). Phenomenology of perception (C. Smith, Trans.). Routledge & Kegan Paul. (Original work published 1945)

Merleau-Ponty, M., & Lingis, A. (1968). *The visible and the invisible*. Northwestern University Press.

Norberg-Schulz, C. (1971). *Existence, space and architecture*. Praeger.

Norberg-Schulz, C. (1979). *Genius loci: Towards a phenomenology of architecture*. Rizzoli.

Pallasmaa, J. (2005). *The eyes of the skin: Architecture and the senses*. John Wiley & Sons.

Smith, D. W. (2018). Phenomenology. In E. N. Zalta (Ed.), *The Stanford encyclopedia of philosophy* (Summer 2018 ed.). Retrieved June 26, 2025, from https://plato.stanford.edu/entries/phenomenology/

Tuan, Y.-F. (1974). *Topophilia*. Prentice-Hall.

Tuan, Y.-F. (1977). *Space and place*. University of Minnesota Press.

Varela, F. J., Thompson, E., & Rosch, E. (1991). *The embodied mind: Cognitive science and human experience*. MIT Press.

Vitruvius Pollio. (1914). The ten books on architecture (M. H. Morgan, Trans.). Harvard University Press.

Wever, E. G. (1927). Figure and ground in the visual perception of form. *The American Journal of Psychology, 38*(2), 194–226.

ADDITIONAL READING

Bachelard, G. (1994). The poetics of space (M. Jolas, Trans.). Beacon Press. (Original work published 1958)

Bohm, D. (1980). *Wholeness and the implicate order*. Routledge & Kegan Paul.

Bowker, J. (Ed.). (1997). *The Oxford dictionary of world religions*. Oxford University Press.

Casey, E. S. (1993). *Getting back into place: Toward a renewed understanding of the place-world*. Indiana University Press.

Casey, E. S. (1996). How to get from space to place in a fairly short stretch of time: Phenomenological prolegomena. In Feld, S., & Basso, K. H. (Eds.), *Senses of place* (pp. 13–52). School of American Research Press.

Lefebvre, H. (1991). *The production of space*. Blackwell.

Lefebvre, H. (2004). Rhythmanalysis: Space, time and everyday life (S. Elden & G. Moore, Trans.). Continuum.

Malpas, J. (1999). *Place and experience: A philosophical topography*. Cambridge University Press.

Malpas, J. (2008). *Heidegger's topology: Being, place, world*. MIT Press.

National Geographic Society. (n.d.). Concept of place. Retrieved June 26, 2025, from https://education.nationalgeographic.org/resource/resource-library-concept-place/

Piaget, J., & Inhelder, B. (1969). *The psychology of the child*. Basic Books.

Rubin, E. (1915). *Synsoplevede figurer*. Gyldendal.

Schema.org. (n.d.). Place. Retrieved June 27, 2025, from https://schema.org/Place

Seamon, D. (2014). *A geography of the lifeworld*. Routledge.

The Editors of Encyclopedia Britannica. (2016). Epoché. *Encyclopedia Britannica*. Retrieved June 26, 2025, from https://www.britannica.com/topic/epoche

Young, A. T. (n.d.). *Atmospheric refraction: The horizon*. San Diego State University., Retrieved from https://aty.sdsu.edu/explain/atmos_refr/horizon.html

Chapter 3
Earth–Sky Geometries

ABSTRACT

This chapter examines the time-based geometries generated by astronomical cycles: the rotation of the Earth about its axis and its revolution around the Sun. The apparent paths traced by celestial objects and phenomena, as observed from Earth, define geometric configurations of a phenomenological nature, produced by the relational system between an observer in local space and the remote space of the Sky. Within this framework, the architecture of cosmology is understood as a built expression of Earth–Sky Geometries. The chapter also outlines the relationship between astronomy and cosmology as a transition from phenomenological knowledge to theoretical analysis. In doing so, the chapter establishes a conceptual bridge between observational practice and spatial form, situating architectural expression within the larger epistemic framework of cosmic order.

DOI: 10.4018/979-8-3373-3399-1.ch003

Figure 1. Gregor Reisch, Margarita philosophica (Basilea,1535). Woodcut illustration. Public domain. Source: Bayerische Staatsbibliothek, Munich.

SKY FROM EARTH: SPHERICAL ASTRONOMY AND PHENOMENOLOGY

This chapter introduces the astronomical and geometric foundations necessary to articulate Earth–Sky Geometries as perceptual and spatial phenomena. Drawing from spherical astronomy, coordinate systems, and observational practices developed across the history of science, it establishes how the Sky is structured relative to an observer situated on the Earth. Rather than treating celestial space as an abstract or detached domain, the chapter frames astronomical geometry as a relational system in which time, motion, and orientation are inseparable from the observer's location and standpoint. Within this framework, celestial bodies are understood not in terms of physical distance or absolute scale, but through angular relations, reference planes, and apparent motions that organize human perception of the Sky. The emphasis therefore rests not on astrophysical magnitude but on relational geometry as experienced from a terrestrial horizon.

This foundational exposition is not intended as a comprehensive treatise on astronomy, but as a selective and instrumental framework supporting a phenomenological inquiry into how remote space becomes experientially present. By grounding celestial observation in geometric reference systems—particularly those maintaining a direct relationship to bodily orientation—this chapter establishes the conditions under which Earth–Sky Geometries emerge as meaningful configurations linking cosmic motion to lived time and spatial experience. The observer's vertical axis, horizon plane, and cardinal orientation thus function as primary phenomenological references.

The observation of the Sky is approached here s a phenomenological study of remote space. Perceptions of celestial objects are situated within embodied experience, as they appear to human consciousness in first-person perception—centered on the remote space of the Sky and grounded in the local space of the observer on Earth. The focus of this inquiry is to contextualize built constructions within the human-made environment in relation to the perception of remote space, which, although physically out of reach, remains perceptually accessible as phenomenon to a subject situated on Earth. Remote space is therefore not defined by metric distance but by its mode of givenness in perception.

The subject's perception of remote space can be grounded in spherical astronomy, also known as positional astronomy. Observational astronomy, which comprises the earliest astronomical investigations of ancient civilizations, relied primarily on naked-eye observation of the Sky. Spherical astronomy focuses on the apparent positions and motions of celestial bodies—such as the Moon, planets, stars, and galaxies—as well as on cyclical phenomena including solstices, equinoxes, and eclipses. Historically, it has served multiple practical and cultural functions, including timekeeping, navigation, religious practice, and ritual observance.

Spherical astronomy is closely related to astrometry, the science concerned with the measurement of celestial bodies and events. It is based on observational data combined with mathematical methods, principally spherical trigonometry, algebra, and calculus. In the present study, the definition of Earth–Sky Geometries refers specifically to spherical astronomy based on unaided-eye observation, emphasizing phenomenological perception rather than instrumental precision.

Astronomy—from the Greek *astēr* (ἀστήρ, "star") and *nómos* (νόμος, "law") (Liddell et al., 1940)—has been defined as the "oldest of the sciences depending upon observation" (Newcomb, 1890, p. 1). In ancient Greek education, as described in the Pythagorean tradition and attributed to Archytas, the quadrivium comprised four disciplines: geometry, arithmetic, astronomy, and music. As Heath explains, "'sphaeric' means astronomy, being the geometry of the sphere considered solely with reference to the problem of accounting for the motions of the heavenly bodies" (Heath, 1921, pp. 11–12). Spherical astronomy has therefore been studied since antiquity, with origins extending into prehistory.

The central role of the perceiver renders this field of observational astronomy particularly relevant for phenomenological inquiry, as it directly links celestial objects and events in the Sky to the embodied observing subject on Earth. The basic principles of spherical astronomy thus provide a foundation for understanding the architecture of cosmology, both in aesthetic terms and in functional terms—namely, as spatial frameworks enabling the experience and interpretation of celestial cycles and phenomena.

Spherical astronomy provides a coordinate framework for positioning astronomical objects on the celestial sphere, an imaginary sphere representing remote space. The vault of the Sky corresponds to the surface of this sphere, upon which celestial objects and phenomena are mapped as they appear to the observer. Each observation may be described within a four-dimensional framework consisting of three spatial coordinates and time as the fourth coordinate, defined by the moment of observation.

In this sense, astronomical observation aligns conceptually with phenomenological principles such as *Dasein*, being-in-the-world, the Fourfold, and embodiment, as well as with principles derived from Gestalt theory. Dasein (Da-sein, "being-there") emphasizes situated existence; being-in-the-world (In-der-Welt-sein) underscores the inseparability of subject and environment. Each phenomenon on the celestial sphere is perceived as a spatiotemporal event. An event-point in spacetime can be represented by a set of four coordinates (x, y, z, t): the first three identify position in three-dimensional space, and the fourth represents the time at which the event is perceived by an observer on Earth. Every observation of the Sky therefore occurs simultaneously in space and time and constitutes an object of perception for an embodied subject, consistent with phenomenological accounts of lived experience. The conceptual diagram in Figure 2 illustrates a hypothetical observer located at the center of the celestial sphere, with an event-point on its surface identified by four coordinates.

Figure 2. Observing an event identified by four-dimensional coordinates on the celestial sphere.

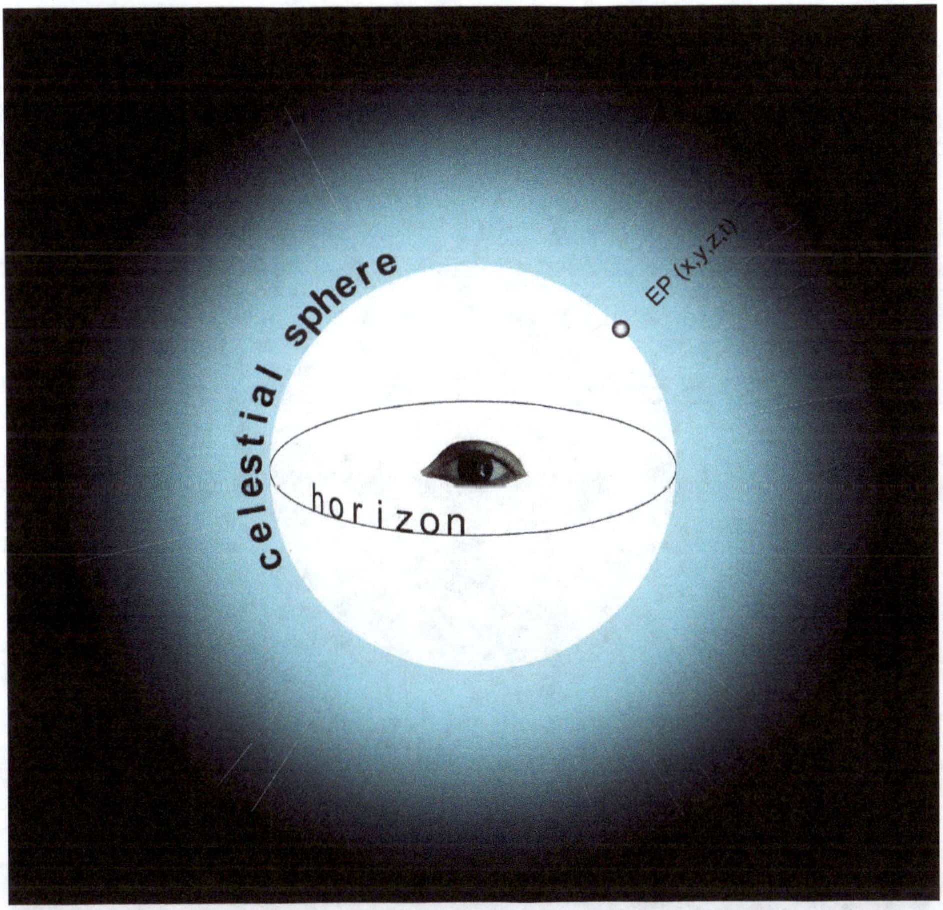

The Celestial Sphere

Spherical astronomy is based on a model of the universe as an imaginary sphere with the Earth at its center, called the celestial sphere: an abstract geometric construct that allows observers to project the apparent motions of celestial bodies onto a unified surface for measurement and interpretation. In this study, the celestial sphere is also considered as a component of the phenomenological system of being-in-the-world. Within this system, the subject–observer is situated at the center of the sphere, in local space on the surface of the Earth, as shown in the conceptual sketch in Figure 3. The sphere functions simultaneously as a geometric abstraction and as a perceptual envelope.

Figure 3. Conceptual representation of an observer and the celestial sphere.

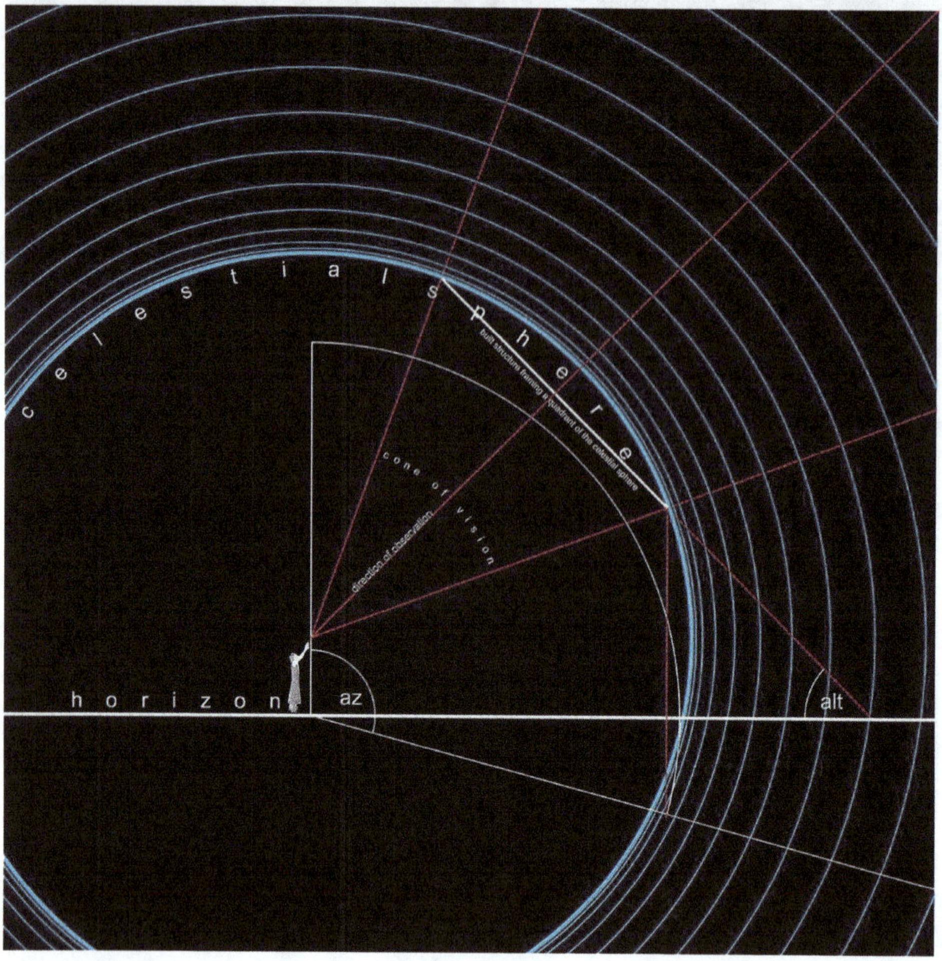

A classical definition of the celestial sphere, found in one of the many treatises and textbooks on astronomy and fully compatible with a phenomenological approach, is provided by Chauvenet:

> FROM whatever point of space an observer be supposed to view the heavenly bodies, they will appear to him as if situated upon the surface of a sphere of which his eye is the centre. If, without changing his position, he directs his eye successively to the several bodies, he may learn their relative directions, but cannot determine either their distances from himself or from each other. (Chauvenet 1874, p. 17)

The celestial sphere is conceptually hollow, and the vault of the sky, as perceived by an observer on Earth, is projected onto its concave inner surface. The positions of celestial objects and phenomena appear projected on this surface and are not fixed: they change in space according to the observer's geographic location and in time according to the daily and annual motions of the Earth, as discussed in later sections of this chapter. More specifically, celestial objects appear to rotate westward (from east to west), opposite to the Earth's rotation around its axis from west to east. This apparent diurnal rotation is a direct phenomenological consequence of terrestrial axial rotation.

Importantly, the center of the celestial sphere does not coincide with the center of the Earth, but with the observer's geolocation. Given astronomical distances, the difference between the Earth's center and the observer's position on its surface is negligible—the Earth's mean radius is approximately 6,371 kilometers, compared to the distances of stars and galaxies, which range from several light-years to millions of light-years. However, the observer's geographic coordinates are essential for determining the apparent positions of celestial objects in terms of direction and angular distance, which constitute measurable and recordable data. The celestial sphere has no measurable radius, and all celestial bodies projected onto its surface appear at the same distance from the observer, although they differ in apparent size and brightness.

Figure 4. Humans on Earth surrounded by the celestial sphere in De sphaera mundi by Sacrobosco. Woodcut illustration. Public domain. Source: https://archive.org/details/3840467.

The concept of the celestial sphere dates back to ancient Greece, when astronomy, as one of the disciplines comprising the quadrivium, was closely associated with geometry. It was known at least as early as the time of the Pythagorean Archytas of Taras (fourth century BCE), when "astronomy was already treated mathematically, the properties of the sphere being studied so far as necessary to explain the movements in the celestial sphere" (Heath, 1921, p. 214). Aristotle also developed a cosmological theory based on nested celestial spheres in the treatise *Metaphysics* (first century BCE), placing the Earth at the center of the universe and surrounding it with multiple interconnected spheres composed of ether. Aristotle's cosmology is fully articulated in *Metaphysics* and *De Caelo* (On the Heavens).

Despite its abstraction and the absence of direct distance measurements, the celestial sphere provided usable observational data for more than two thousand years, prior to the invention of the telescope and modern observational instruments.

These data supported many practical applications, including navigation, the production of star maps, and the compilation of star catalogs. The catalog in Ptolemy's *Almagest* (ca. 150 CE), is one of the best-known ancient examples relying entirely on naked-eye observations of the celestial sphere; originally titled *Mathematikē Syntaxis* (Μαθηματικὴ Σύνταξις), it is known through Arabic transmission as *al-Majisṭī*. Ptolemy lists stars within 48 constellations and provides, for each star, (1) a description of its location within the constellation figure and sometimes of its brightness and color; (2) its longitude; (3) its latitude and position north or south of the ecliptic; and (4) its magnitude (Ptolemy, 1984, p. 14).

The iconography of the celestial sphere illustrates the presence of a wholistic approach to knowledge that persisted even into the early stages of the scientific revolution. Conceptual drawings were used to represent observational methodologies, often incorporating symbolic elements to depict the transition from perception to theoretical interpretation. One of the most influential examples is the treatise *De sphaera mundi* (ca. 1230) by the scholar and astronomer Johannes de Sacrobosco (Figure 4). The Latin title translates as "On the Sphere of the World," where "world" is understood as the sky surrounding the Earth.

The treatise *Cosmographicus Liber* (Apian, 1524) presents further illustrations that can be interpreted as phenomenological diagrams of spherical astronomy. The human observer is frequently depicted, sometimes operating an instrument (*Instrumentum siderale*) to observe celestial bodies (Figure 5), in conjunction with diagrammatic representations of the Earth (Figure 6). These images emphasize the embodied position of the subject within the geometric framework of the cosmos.

Figure 5. Observing stars in the sky. Cosmographicus Liber Cosmographicus liber Petri Apiani mathematici (Apian, 1524). Woodcut illustration. Public domain. Source: https://archive.org/details/Cosmographicusl00Apia.

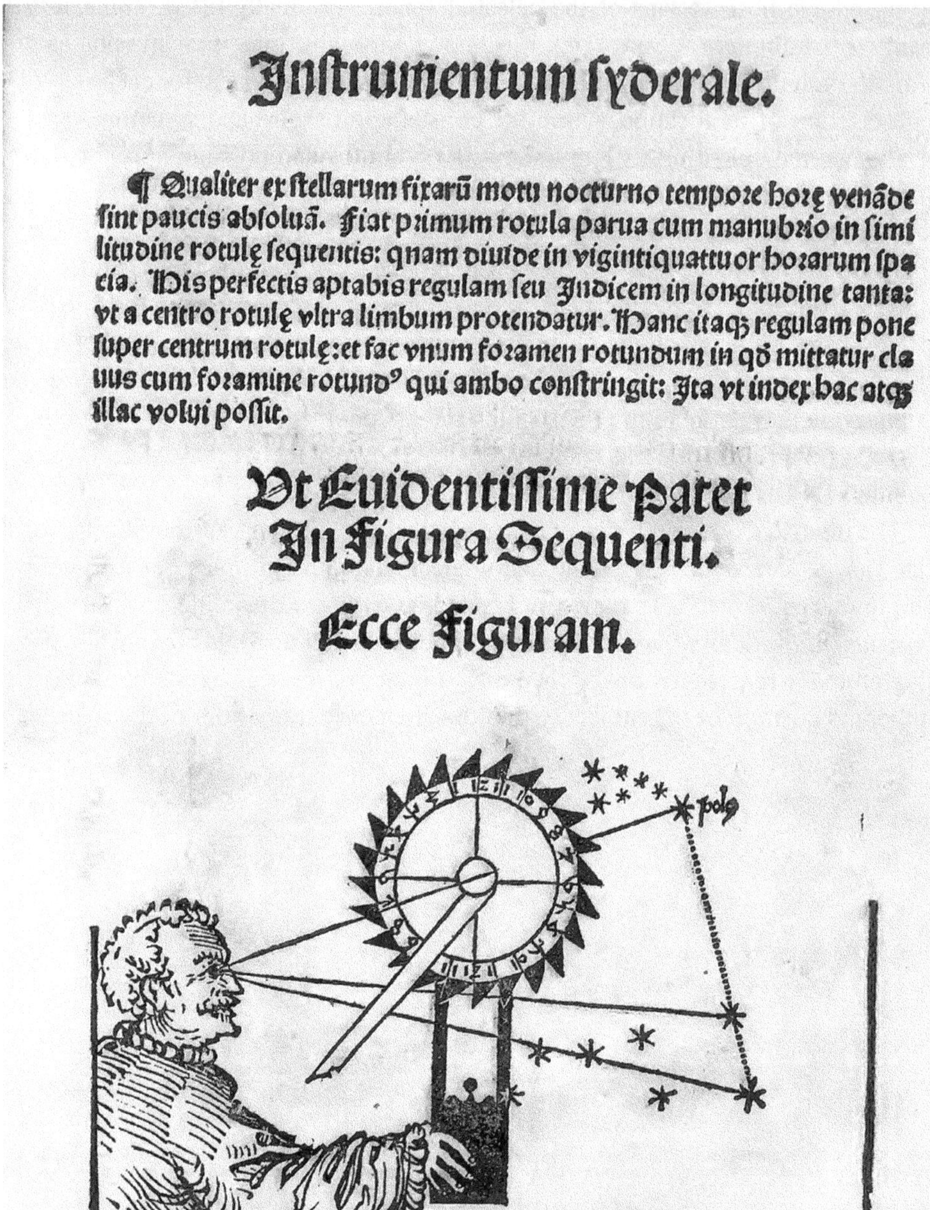

Instrumentum syderale.

¶ Qualiter ex stellarum fixarū motu nocturno tempore horę venãde sint paucis absoluā. Fiat primum rotula parua cum manubrio in similitudine rotulę sequentis: quam diuide in vigintiquattuor horarum spacia. His perfectis aptabis regulam seu Iudicem in longitudine tanta: vt a centro rotulę vltra limbum protendatur. Hanc itaqꝫ regulam pone super centrum rotulę: et fac vnum foramen rotundum in qd mittatur clauus cum foramine rotund⁹ qui ambo constringit: Ita vt index hac atqꝫ illac volui possit.

Vt Euidentissime Patet In Figura Sequenti.

Ecce Figuram.

iij

Figure 6. Celestial sphere in Cosmographicus Liber Cosmographicus liber Petri Apiani mathematici (Apian, 1524). Woodcut illustration. Public domain. Source: https://archive.org/details/Cosmographicusl00Apia.

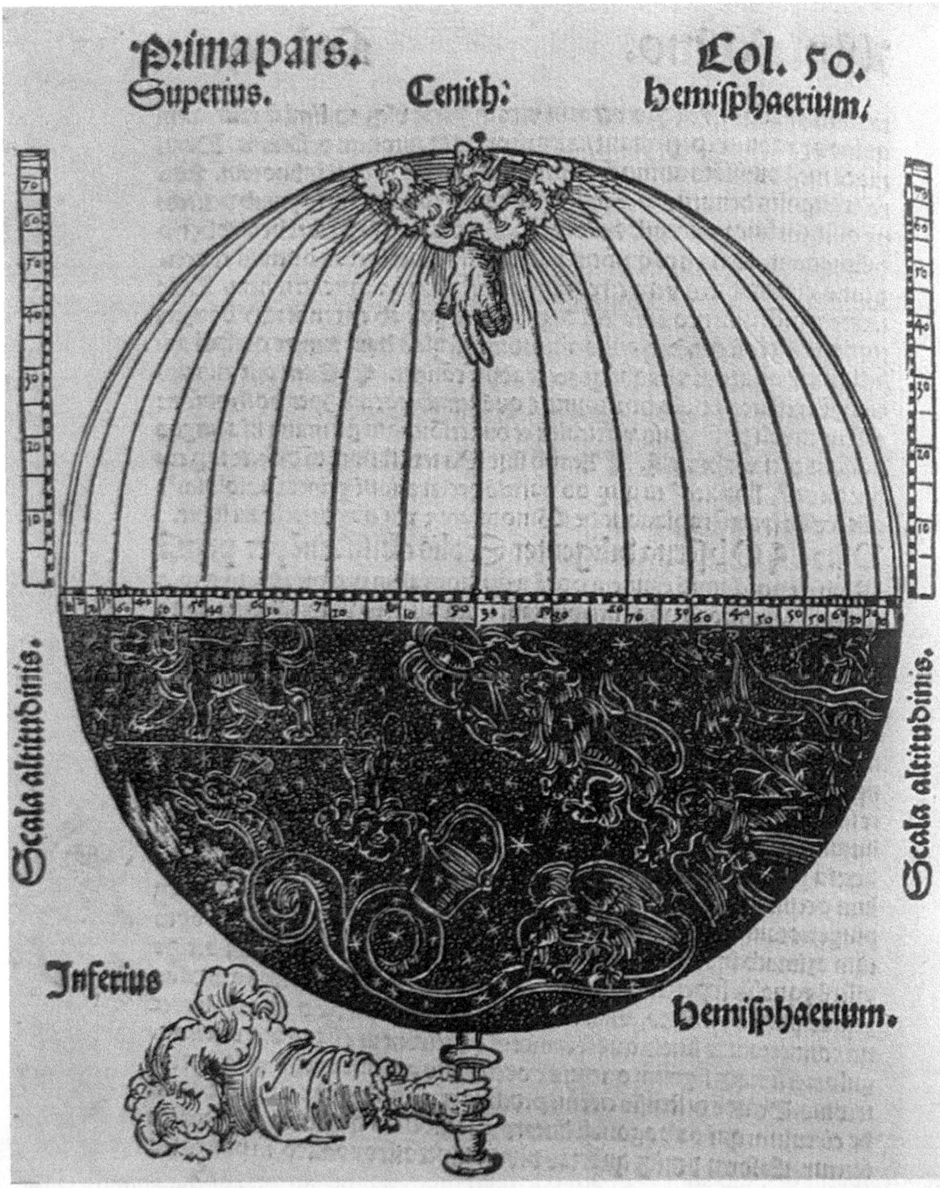

While the celestial sphere originates as a perceptual and phenomenological construct grounded in embodied observation, it also serves as the geometric foundation for formal astronomical description. In order to translate the lived experience of the sky into measurable and communicable knowledge, spherical astronomy introduces coordinate systems and angular relationships that locate celestial objects on the surface of the sphere with mathematical precision. These systems do not replace perception; rather, they formalize it by transforming visual experience into quantifiable spatial relations based on the observer's position on Earth and the time of observation. In this transformation, phenomenological immediacy and mathematical abstraction remain structurally interdependent.

The following sections therefore shift from phenomenological description to geometric representation, outlining the principal reference circles, coordinate systems, and angular measures that structure observational astronomy and enable the analysis of Earth–Sky Geometries as both perceptual and mathematical constructs. This methodological shift prepares the conceptual ground for the subsequent examination of axial rotation, solar revolution, and their architectural implications.

Spherical Geometry

The geometry of the celestial sphere is based on spherical geometry. A hollow sphere is defined as a surface, that is the set of all points in space that are at an equal distance from a fixed point, called the center; this constant distance is the radius of the sphere.

In analytic geometry a sphere is defined in Cartesian coordinates as:

> Let r be a positive real number and $P_o = (X_o, Y_o, Z_o)$ be a fixed point. Then we define a sphere S of radius r and center at P_o to be the locus of all points P whose distance from P_o, is r. It follows that a point P is on S if and only if
>
> $\sqrt{(x - x_o)^2 + (y - y_o)^2 + (z - z_o)^2} = r$ (Albert, 1949, p. 54)

Two points A and B on the surface of a sphere are connected by arcs that are intersections between the spherical surface and planes passing through the two points. If the plane also passes through the center of the sphere O, the resulting circle is called a *great circle*. The arc of a great circle connecting two points represents a *geodesic*, that is, the shortest path between them on the surface of the sphere. If the cutting plane does not pass through the center of the sphere, the resulting circles are called *small circles* (Figure 7a).

Figure 7. Great circles, small circles (a) and spherical triangles (b) in a sphere.

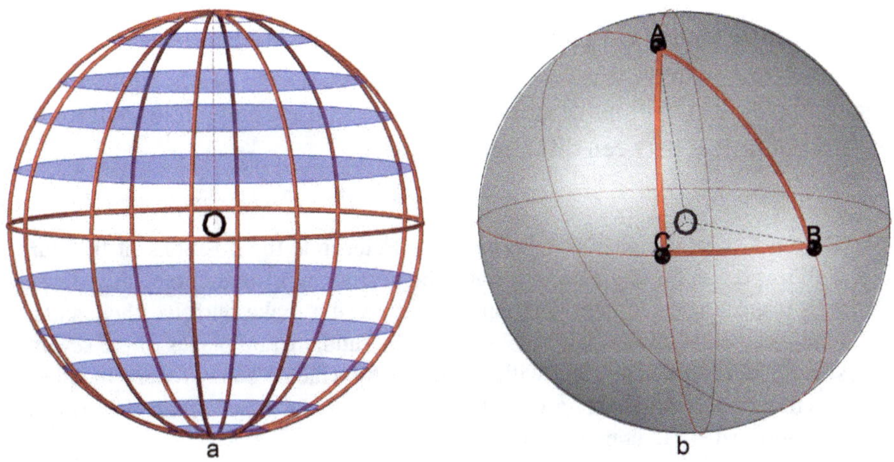

In astronomy, several fundamental great circles are defined on the celestial sphere: the terrestrial equator, the celestial equator (which lies in the same plane as the Earth's equator), and the ecliptic, which is the intersection of the celestial sphere with the plane of the Earth's revolution around the Sun. The equator divides the sphere into two equal hemispheres, the northern and southern hemispheres.

For any two distinct points on a sphere, there is exactly one great circle passing through them, because a plane is uniquely determined by three points—in this case, by points A, B, and the center O. By contrast, there are infinitely many small circles passing through the same two surface points. Three points on a spherical surface define a spherical triangle, whose sides are arcs of great circles representing the minimum-distance paths between the points (Figure 7b). Spherical trigonometry is the branch of spherical geometry that studies the measurable relationships between angles and sides in spherical triangles. Because the radius of the celestial sphere is not measurable, only angular measurements are relevant in observational astronomy.

A sphere can also be defined as the surface generated by rotating a great circle about one of its diameters, which becomes the axis of rotation. The two intersections of the axis with the spherical surface are called poles and are diametrically opposite. On the celestial sphere, these are the North Celestial Pole (NCP) and the South Celestial Pole (SCP), which correspond to the extensions of the Earth's rotational axis to the celestial sphere.

Fundamental Planes

Fundamental planes are another essential concept in spherical geometry and astronomy. As Brünnow explains, fundamental planes serve as reference planes for determining the positions of stars relative to specific great circles and for transforming coordinates between different systems:

> The fundamental planes are the means to determine the positions of the stars with respect to these great circles and the positions of these circles themselves with respect to each other. We must therefore first make ourselves acquainted with these great circles, whose planes are the fundamental planes of the several systems of co-ordinates, and with the means by which we may reduce the place of a heavenly body given for one of these fundamental planes to another system of co-ordinates. (Brünnow, 1865, p. 70).

A fundamental plane is therefore a plane that passes through the center of the sphere and defines a great circle, dividing the sphere into two hemispheres and serving as the basis for a specific astronomical coordinate system.

In astronomy, several fundamental great circles are defined on the celestial sphere: the terrestrial equator, the celestial equator (which lies in the same plane as the Earth's equator), and the ecliptic, which is the intersection of the celestial sphere with the plane of the Earth's revolution around the Sun. The equator divides the sphere into two equal hemispheres, the northern and southern hemispheres.

A sphere can also be defined as the surface generated by rotating a great circle about one of its diameters, which becomes the axis of rotation. The two intersections of the axis with the spherical surface are called poles and are diametrically opposite. On the celestial sphere, these are the North Celestial Pole (NCP) and the South Celestial Pole (SCP), which correspond to the extensions of the Earth's rotational axis to the celestial sphere.

The Earth-Sun Geometry

A phenomenological framework of the architecture of cosmology, articulated through Earth–Sky Geometries, requires an introduction to basic scientific facts concerning the geometry and motion of the Earth. The local space of these geometries is always situated on the planet Earth; therefore, the observer's relationship to the remote space of the celestial sphere must be understood in relation to the Earth's form and motion. Without this physical grounding, celestial orientation would remain abstract rather than embodied. This physical framework underlies both the experience of architectural constructions and the interpretation of the celestial phenomena that historically motivated such constructions.

Geometrically, the Earth is best described as an oblate spheroid—an ellipsoid of revolution flattened at the poles and bulging at the equator. As described in geodetic literature, "the geometrical figure used in geodesy to most nearly approximate the shape of the Earth is an ellipsoid of revolution, obtained by rotating an ellipse about its shorter axis" (United States Defense Mapping Agency, 1983, p. 6). The Earth's equatorial diameter is approximately 12,756 kilometers, while the polar diameter is approximately 12,714 kilometers (National Geographic Society, 2023). For the purposes of naked-eye astronomical observation and the interpretation of celestial alignments in architectural constructions, this difference is negligible; therefore, the Earth may be treated as a sphere with a mean radius of approximately 6,371 kilometers. This approximation is phenomenologically justified: the horizon presents itself as locally spherical, and celestial reference systems presuppose this curvature.

The Earth–Sun geometry must also be understood within the broader context of the Solar System. While heliocentric mechanics explain planetary motion, the architecture of cosmology remains grounded in apparent solar motion as experienced from Earth. The Solar System consists of the Sun (a star), eight planets (Mercury, Venus, Earth, Mars, Jupiter, Saturn, Uranus, and Neptune), their moons, and numerous smaller bodies such as asteroids and comets. These bodies move in elliptical orbits around the Sun, whose mass dominates the gravitational dynamics of the system (NASA Science, 2024). A heliocentric model of planetary motion was already proposed in the third century BCE by the Greek astronomer Aristarchus of Samos, although geocentric models prevailed in Western astronomy until the sixteenth century, when Nicolaus Copernicus formally challenged them in *De revolutionibus orbium coelestium*, first published in 1543 (Figure 8). Copernicus's work is widely regarded as a foundational moment of the scientific revolution, marking a paradigm shift from an Earth-centered to a Sun-centered cosmology.

Figure 8. Page from De revolutionibus orbium coelestium (Copernicus, 1566) showing the Sun surrounded by the orbits of the seven planets.

NICOLAI COPERNICI

net, in quo terram cum orbe lunari tanquam epicyclo contineri diximus. Quinto loco Venus nono menſe reducitur. Sextum deniq; locum Mercurius tenet, octuaginta dierum ſpacio circū currens. In medio uero omnium reſidet Sol. Quis enim in hoc

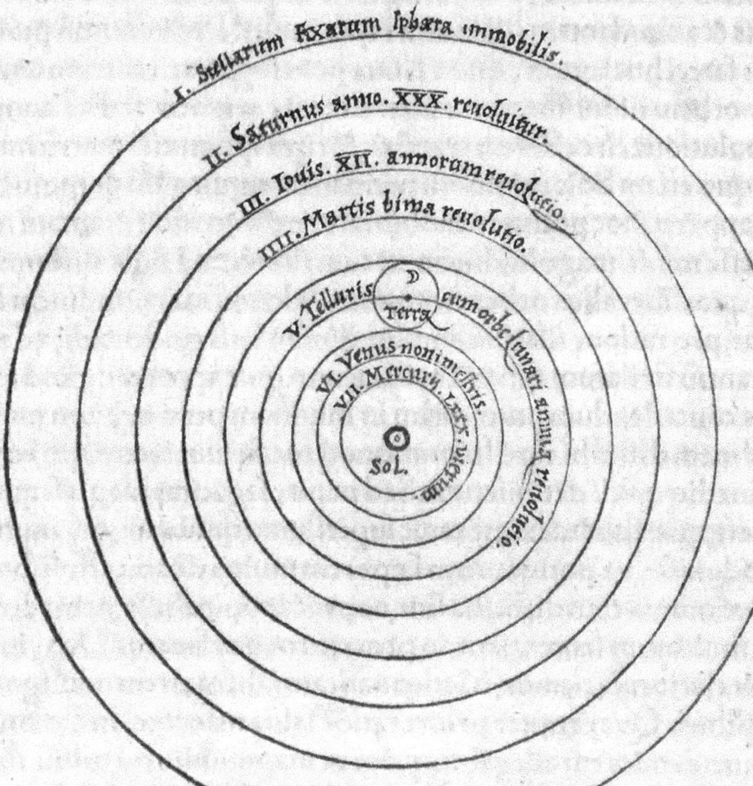

pulcherimo templo lampadem hanc in alio uel meliori loco po neret, quàm unde totum ſimul poſsit illuminare? Siquidem non inepte quidam lucernam mundi, alij mentem, alij rectorem uo= cant. Trimegiſtus uiſibilem Deum, Sophoclis Electra intuentē omnia. Ita profecto tanquam in ſolio re gali Sol reſidens circum

Apparent Motion of the Sun

For Earth–Sky Geometries, however, what is most significant is not heliocentric mechanics but the apparent motion of the Sun as observed from Earth. Phenomenology privileges appearance over physical causation: what matters for orientation is how the Sun traverses the sky vault relative to the observer's horizon. The Sun is the closest star to our planet and the dominant reference for timekeeping, orientation, and seasonal cycles in naked-eye astronomy. Its apparent daily and annual paths across the celestial sphere form the geometric basis of many architectural alignments and spatial markers discussed in this book. While modern astronomy explains these motions in terms of the Earth's rotation and revolution, phenomenological observation registers them as paths traced on the sky vault relative to the observer's horizon and cardinal directions.

The Sun is the closest star to Earth and, as such, has been the object of observation since immemorial times. The following table identifies the planets with their symbols and includes other essential data (National Aeronautics and Space Administration, 2025).

Table 1. Selected physical properties of Solar System bodies.

Name	Symbol	Mass (10^{24} Kg)	Diameter (km)	Gravity (m/s^2)	Distance from Sun (10^6 km)
Sun	⊙	1989100	1,392,000	274	-
Mercury	☿	0.330	4879	3.7	57.9
Venus	♀	4.87	12,104	8.9	108.2
Earth	♁	5.97	12,756	9.8	149.6
Mars	♂	0.642	6792	3.7	228.0
Jupiter	♃	1898	142,984	23.1	778.5
Saturn	♄	568	120,536	9.0	1432.0
Uranus	♅	86.8	51,118	8.7	2867.0
Neptune	♆	102	49,528	11.0	4515.0

The Ecliptic: Geometry of Time

The elliptical orbit of the Earth and the other celestial objects in the Solar System defines a plane called the ecliptic (Figure 9). This plane is fundamental for the observation of phenomena not only in the Solar System but also in relation to the wider celestial background. The term *ecliptic* derives from the Greek ἐκλειπτικός (*ekleiptikós*), meaning "of an eclipse" or "failing to appear" (Liddell, Scott, Jones,

& McKenzie, 1940). The term refers to solar and lunar eclipses, which occur when the Moon crosses the plane of the ecliptic. The ecliptic also coincides with the yearly apparent path of the Sun on the celestial sphere.

Figure 9. The ecliptic, Earth, and celestial sphere.

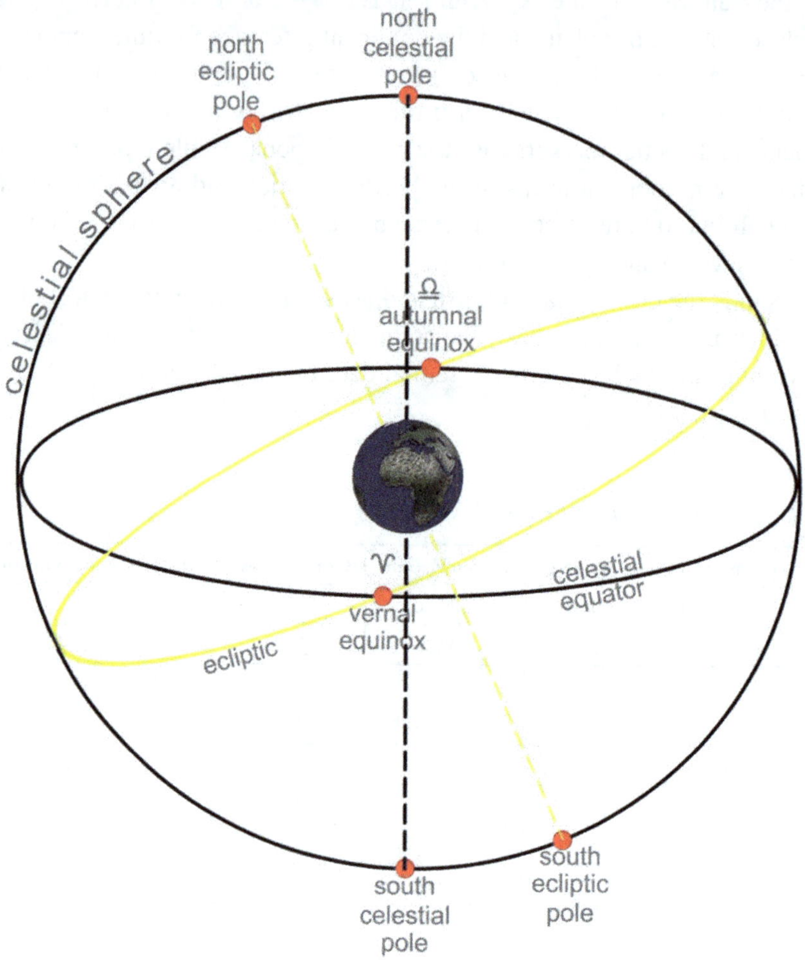

Roy and Clarke provide a comprehensive definition of the ecliptic in relation to other fundamental astronomical elements such as the equator and the equinoxes, while also linking celestial geometry to cultural practices of timekeeping and ritual:

> The Sun's stellar route was called the ecliptic by the ancients. The groups of stars intersected by this path were called the houses of the Zodiac. The ecliptic is found

> to be a great circle inclined at about $23^1/_2$degrees to the equator, the great circle on the sky corresponding to the projection of the Earth's equator, intersecting it at two points, the vernal and autumnal equinoxes, 180 degrees apart.
>
> It was quite natural, then, for the ancients to worship the Sun. Not only did it provide light and warmth by day against the evils of the night but, in addition, its yearly progression was intimately linked to the seasons and so also to seed time and harvest. It was, therefore, necessary to keep track of progress to use it as a clock and a calendar. To this end, the science of sundial-making began, ramifying from simple obelisks that throw shadows on a fan of lines radiating from their bases, to extremely ingenious and complicated erections in stone and metal. Up to the 19th century, these constructions rivalled most pocket-watches in accuracy as timekeepers.
>
> For calendrical purposes, lines of standing stones could be set up, pointing to the midsummer, midwinter and equinoctial rising and setting points of the Sun. In the British Isles, there still remain hundreds of such solar observatories, witnesses to our forefathers' preoccupation with the Sun-god. The observer who watches the nights sky throughout a year, counts about thirteen revolutions of the stellar background by the Moon in that time. Over that period of time, it is not apparent that any simple relationship exists between the sidereal period of revolution of the Moon, the period of its phases and the year (the time it takes the Sun to perform one complete circuit of the ecliptic). That knowledge comes after much more extended observation, certainly measured in decades. (Roy & Clarke, 2003, p. 7)

Geometrically, the ecliptic is a great circle on the celestial sphere, intersecting another great circle, the celestial equator, at two points known as the equinoxes (Chauvenet, 1906, p. 23). The origin of ecliptic longitude is the *First Point of Aries*, denoted by the symbol ♈ (also *T*), historically associated with the constellation Aries. At this point the Sun appears to cross the celestial equator from south to north during its yearly apparent motion. This point also represents the vernal equinox, which formerly occurred when the Sun was located in Aries.[1]

The second intersection occurs at the September (autumnal) equinox, where the Sun crosses from north to south near the constellation Libra; this point is known as the *First Point of Libra* and is denoted by the symbol ♎ (Ball, 1908, p. 84).[2]

In the elliptical orbit of the Earth, the Sun occupies one of the two foci of the ellipse. At *perihelion*, the point closest to the Sun, the Earth is approximately 147.1 million kilometers away; at *aphelion*, the farthest point, the Earth–Sun distance is about 152.1 million kilometers (Williams, 2025).

The Earth's axis is not perpendicular to the plane of the ecliptic but is tilted by an angle ε of approximately 23°26′, or 23.44° (Glossary, 2023). This angle, known as the *obliquity of the ecliptic* (Roy & Clarke, 2003, p. 74), is measured between the Earth's rotational axis and the normal to the plane of the ecliptic. Figure 10 illustrates the geometry of the Earth–ecliptic relationship projected onto a plane passing through the Earth's axis and perpendicular to the ecliptic.

Figure 10. Obliquity ε.

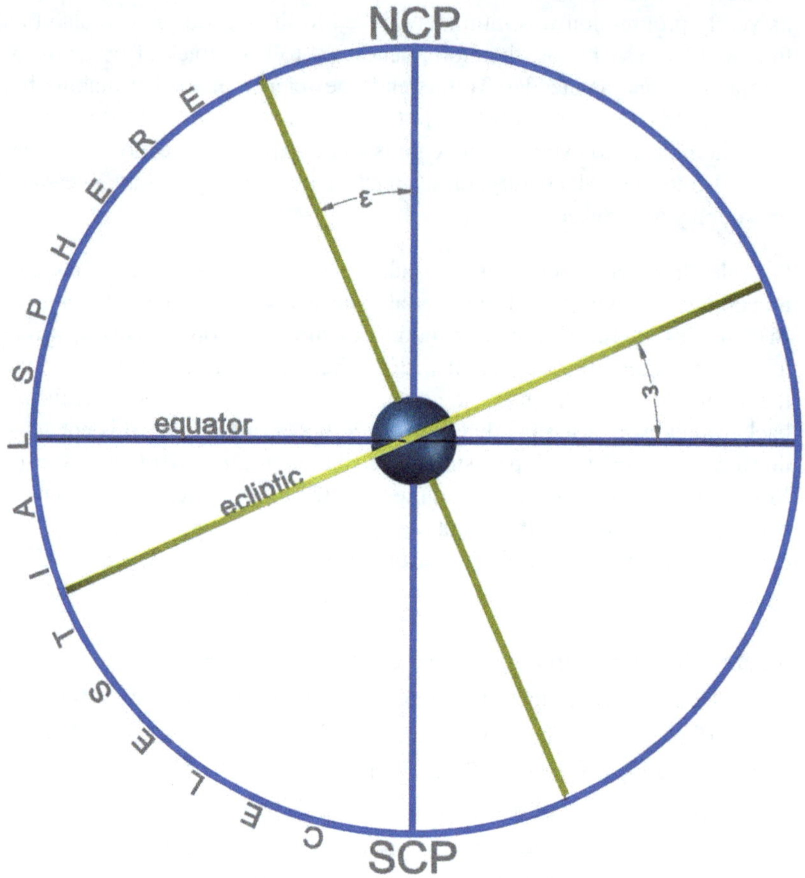

Satellite imagery from Google Earth (Figure 11) with superimposed representations of the equator, meridians, and parallels illustrates the Earth as seen in relation to the ecliptic plane; definitions of these coordinate elements are discussed in the section "Geographic Coordinate System."

Figure 11. Views of Earth from Google Earth.

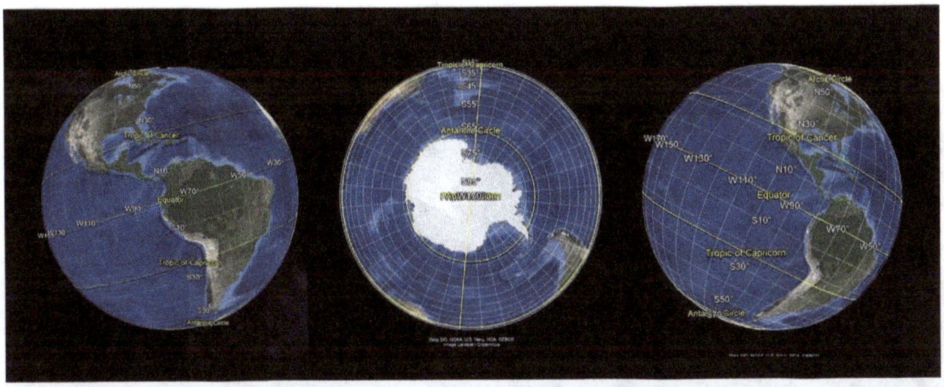

The geometry of the Earth comprises not only its volume and surface but also its movements in space. There are two primary motions: rotation about its axis (also called *spin*) and revolution around the Sun. Human experience of time—day and night, seasonal change, and cyclical rhythms of life—is fundamentally structured by these motions, which directly relate the geometry of the Earth to temporal phenomena.

The Earth rotates about its axis from west to east (eastward), appearing counterclockwise when viewed from above the North Celestial Pole (Figure 12). As a consequence, the apparent daily motion of the Sun and stars is from east to west. The period of Earth's rotation through 360° is approximately 24 hours (Smart, 1936). A day is defined as the interval of time required for a celestial body to cross the same meridian in successive apparent passages. When the observed body is the Sun, the interval defines the mean solar day, whose duration is 24 hours.[3] During each hour, the Earth rotates approximately 15°, an angular measure known as the hour angle. Rotation also produces the alternation of day and night.

Figure 12. Earth spin diagram superimposed on satellite imagery from Google Earth.

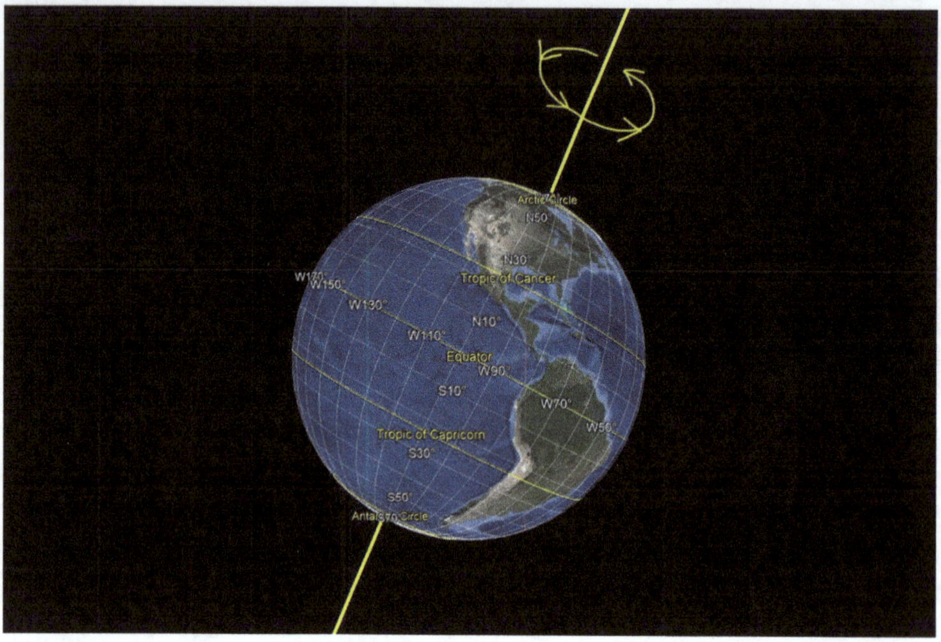

When the interval is measured relative to a star rather than the Sun, the time unit is the sidereal day, which is approximately 23 hours, 56 minutes, and 4 seconds—almost four minutes shorter than the mean solar day (Editors of Encyclopædia Britannica, 2018).

The Earth also revolves around the Sun along the ecliptic. This motion, called revolution, proceeds counterclockwise when viewed from above the North Celestial Pole (Figure 13). One complete revolution takes approximately 365.256 days in sidereal time, and orbital velocity varies between about 29.29 and 30.29 km/s (NASA Science, 2025). The axial tilt remains nearly constant during this motion, producing seasonal variations in solar illumination.

Figure 13. Earth revolution around the Sun.

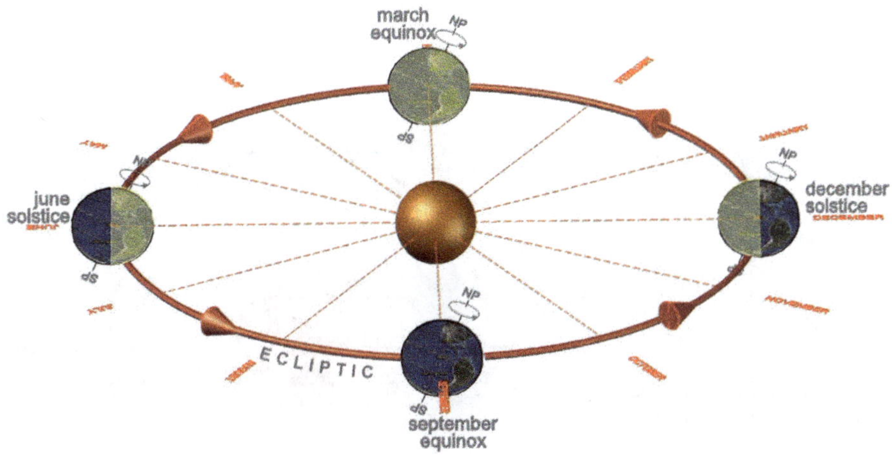

The four seasons—spring, summer, autumn, and winter—result from the combined effect of axial tilt and orbital motion. Seasonal differences manifest in daylight duration, temperature, and environmental change, and occur in opposite phases in the Northern and Southern hemispheres.

Geometrically, four fundamental points are defined on the ecliptic: the two equinoxes and the two solstices. These points are determined by the angle between the Sun–Earth line and the Earth's rotational axis. Considering the plane that contains the Earth's axis and is perpendicular to the equatorial plane (Figure 14), the Sun's declination δ is defined as the angle between the celestial equator and the Sun's apparent position on the celestial sphere.

Figure 14. Angles between the ecliptic plane and the Earth's axis at the equinoxes and solstices.

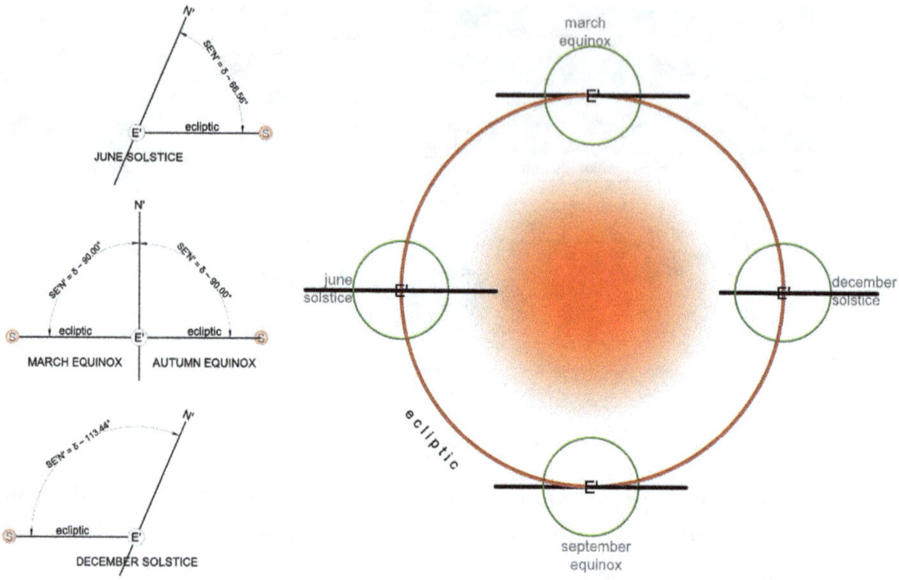

At the equinoxes, the ecliptic intersects the celestial equator, and the Sun's declination is $\delta = 0°$.

Day and night are therefore approximately equal in duration. The term equinox derives from the Latin *aequinoctium* (*aequus* and *nox*), meaning "equal night." The March equinox marks the beginning of spring in the Northern Hemisphere and autumn in the Southern Hemisphere, while the September equinox marks the opposite transition.

At the solstices, the Sun reaches its maximum angular displacement from the celestial equator:

- at the June solstice: $\delta \approx +23.44°$
- at the December solstice: $\delta \approx -23.44°$.

At the June solstice, daylight reaches its maximum duration in the Northern Hemisphere and its minimum in the Southern Hemisphere; the reverse occurs at the December solstice. The term solstice derives from the Latin *solstitium* (*sol* and *sistere*), meaning "Sun standing still," referring to the apparent pause in the Sun's north–south motion before reversing direction.

Observation of the ecliptic is central to the architecture of cosmology, as geometric relationships in remote space are inseparable from temporal experience and biological rhythms. Across cultures, architectural constructions aligned with the ecliptic establish meaningful relationships between built form, seasonal cycles, and social practices of timekeeping.

Forces and the Shape of Space–Time

The Earth–Sun geometry reveals the interrelationship between geometry and force within a spacetime framework. This relationship is articulated in the three laws of planetary motion formulated by Johannes Kepler:

1. All planets move in elliptical orbits, with the Sun at one focus.
2. A line joining a planet to the Sun sweeps out equal areas in equal times.
3. The square of the orbital period of a planet is proportional to the cube of the semi-major axis of its orbit.

These laws, derived from "observations of the movements of the planets and without any reference to the nature of the forces which control these movements" (Ball, 1908, p. 146), were first presented in *Astronomia Nova* (1609) and later in *Harmonices Mundi* (1619). The first law appears as "*Ergo ellipsis est planetæ iter*" ("Therefore, the path of a planet is an ellipse") (Kepler, 1609, p. 285). The third law is stated as "*proportio qua est inter binorum quorumcunque planetarum tempora periodica, sit præcise sesquialtera proportionis mediarum distantiarum*" (Kepler, 1619, p. 189), meaning that orbital periods are in a 3:2 power relation to orbital distances.

Kepler's laws remained empirical until Isaac Newton demonstrated that they follow from the law of universal gravitation. For more than three quarters of a century, they remained isolated facts without explanation, until Newton showed them to be consequences of a universal force governing the motion of every particle of matter. Newton's three laws of motion (Encyclopædia Britannica, 2025) state:

1. A body remains at rest or in uniform straight-line motion unless acted upon by an external force.
2. The force acting on a body equals mass times acceleration: $F = ma$.
3. For every action there is an equal and opposite reaction.

In *Philosophiæ Naturalis Principia Mathematica* (1687), Newton introduced the universal law of gravitation:

$$F = G\frac{m_1 m_2}{r^2}$$

where G is the gravitational constant, approximately 6.6743×10^{-11} $m^3 \cdot kg^{-1} \cdot s^{-2}$ in SI units. This law applies universally, from terrestrial objects to stars and galaxies.

Within a phenomenological interpretation, universal gravitation reinforces the conception of space not as an abstract container but as a relational field structured by forces and motions. For the architecture of cosmology, this supports an understanding of remote space as materially and temporally connected to human experience. In phenomenological terms, force becomes an embodied geometric relation: spatial forms are not inert abstractions but expressions of dynamic interdependence. Within the framework of Earth–Sky Geometries, remote space is not detached from lived experience but participates directly in the structure of being-in-the-world.

Forces and the Structure of Cosmic Order

The geometry described thus far—spherical surfaces, fundamental planes, axial inclinations, orbital paths—does not exist independently of force. Form is inseparable from dynamics. The celestial sphere, though constructed as a surface of projection centered upon the observer, is grounded in gravitational interaction. What appears as stable orientation is sustained by continuous motion and invisible constraint.

The Earth's rotation, which establishes the diurnal cycle and defines the celestial poles, persists through the conservation of angular momentum within a gravitational system. The Earth's revolution around the Sun, which gives rise to the ecliptic and the seasonal cycle, is governed by gravitational attraction balanced against inertial motion. These forces do not merely explain celestial mechanics; they constitute the dynamic substrate upon which geometric order becomes perceptible. The regularity of celestial motion is therefore not static symmetry but sustained equilibrium.

Gravity is not experienced as a visible vector inscribed in space, yet it structures orientation at the most immediate level of embodiment. The vertical axis of architectural construction—up and down—is already a response to gravitational force. Horizon and zenith are not abstract geometric opposites; they arise from the body's alignment within a gravitational field. Weight, balance, and equilibrium are lived correlates of cosmic mechanics. In this sense, the architecture of cosmology is grounded simultaneously in planetary motion and in corporeal orientation.

Newtonian mechanics formalized gravitational attraction as a universal force acting between masses across distance, rendering celestial motion mathematically predictable. Einstein's general theory of relativity, however, reconceptualized gravity not as a force acting within space, but as the curvature of spacetime itself. Mass and energy deform the geometric structure through which motion unfolds; planets follow geodesics within curved spacetime. Geometry and force are thus no longer separable explanatory categories—geometry becomes the manifestation of force. Orbital paths are not imposed upon an empty container called space; they arise from its curvature.

For Earth–Sky Geometries, this shift is decisive. Celestial order is not a rigid vault but a dynamically curved field. The apparent stability of the heavens conceals a relational structure sustained by mass, motion, and curvature. The remote space of the sky is therefore neither symbolic ornament nor metaphysical abstraction; it is the visible trace of invisible structuring forces.

From a phenomenological perspective, this convergence of geometry and force resonates with the structure of lived experience. Husserl's insistence on the primacy of intentional orientation situates perception within a horizon structured by correlation. Merleau-Ponty's account of the body as the pivot of spatiality emphasizes that orientation is not constructed after the fact but arises pre-reflectively from bodily emplacement. Gravitational alignment—the vertical posture of the body, the differentiation of above and below—constitutes one of the most primordial conditions of being-in-the-world. Cosmic mechanics and embodied perception are therefore not disconnected domains; they intersect in the lived structure of orientation.

Architectural alignments with celestial phenomena—solstitial axes, equinoctial orientations, meridian markers—materialize this dynamic order. Built form becomes a local inscription of universal forces. The geometry traced in stone, earth, or landscape corresponds to trajectories governed by gravitational curvature and rotational inertia. Architecture does not stabilize the sky; it stabilizes a relational configuration between planetary motion, embodied perception, and constructed space.

In this sense, the architecture of cosmology does not merely represent the cosmos; it participates in its intelligibility. It situates the human body within a field structured by force, curvature, and motion. Geometry, perception, and gravity converge in a single spatial condition: local space becomes a calibrated response to remote dynamism. Through this convergence, architecture mediates between terrestrial embodiment and cosmic structure, rendering visible the forces that silently sustain orientation, equilibrium, and time.

Reference Systems and Spherical Coordinates

From a Gestalt perspective, the celestial sphere can be understood as a background populated by multiple figures—the celestial bodies—whose perceived configurations continuously change over time. In order to map a celestial object, that is, to measure its position on the celestial sphere at a given time relative to an observer on Earth, a reference system is required. Reference systems constitute the foundation of spherical (positional) astronomy and are based on the geometry of the Earth projected onto the celestial sphere.

Several coordinate systems are used in astronomy to locate objects on the celestial sphere, all grounded in spherical geometry (Green, 1985). Reference systems whose origins are external to the Earth, such as heliocentric or supergalactic coordinate systems, are not introduced here. Since the focus of this inquiry is phenomenological observation and its relationship to built interventions in the human-made environment, only geocentric coordinate systems—centered on the observer on Earth—are considered.

In three-dimensional geometric space, a spherical (also called spherical polar) coordinate system identifies a point P by the coordinates (ρ, θ, ϕ) relative to an origin O (Figure 15a). The point lies on a sphere of radius ρ centered at O and can be related to a Cartesian coordinate system (x, y, z) with the same origin, where:

- ρ is the radial distance from O to P;
- θ is the azimuthal angle measured in the xy-plane from the positive x-axis, counterclockwise;
- ϕ is the polar angle measured from the positive z-axis to the radius vector ρ.

Figure 15. A point in spherical coordinates (a) and corresponding Cartesian coordinates (b).

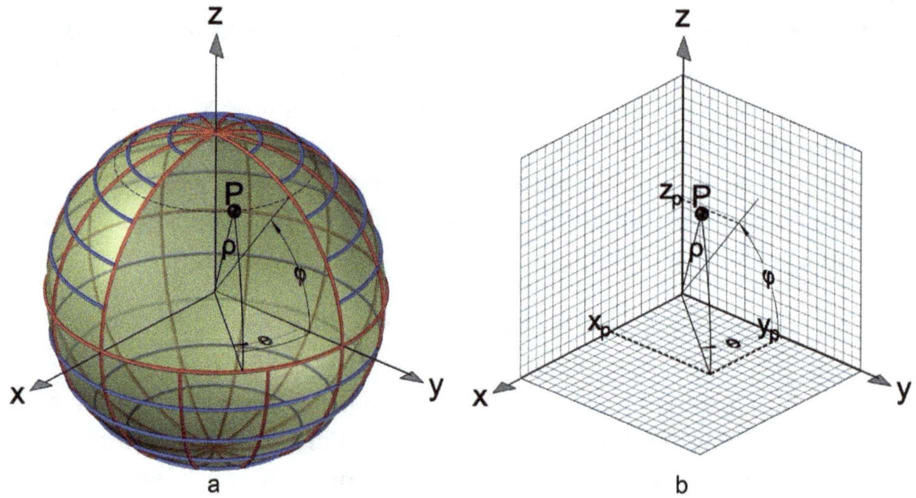

In the astronomical application to the celestial sphere, the radial distance ρ has a constant value equal to the radius of the sphere. The azimuthal angle θ ranges between 0° and 360° (in radians, $0 \leq \theta \leq 2\pi$) and is measured counterclockwise from the reference direction in the fundamental plane. The polar angle ϕ ranges between 0° and 180° (in radians, $0 \leq \phi \leq \pi$) and is measured from the positive *z*-axis toward the equatorial plane (Tanton, 2005, p. 473). At the poles, ϕ equals 0° or 180°. In astronomical usage, the azimuthal angle θ corresponds to longitude, while the polar angle ϕ is related to latitude or, alternatively, to the zenith angle. The polar axis of the spherical system coincides with the *z*-axis of the Cartesian coordinate system.

These relationships provide the mathematical basis for translating angular observations on the celestial sphere into geometric representations.

The diagram in Figure 15a shows a point *P* with spherical coordinates (ρ_P, θ_P, ϕ_P) on a sphere, together with great circles and small circles. Figure 15b illustrates the same point represented in a Cartesian reference system. The transformation from spherical to Cartesian coordinates is given by:

- $x_P = \rho_P \sin \phi_P \cos \theta_P$;
- $y_P = \rho_P \sin \phi_P \sin \theta_P$;
- $z_P = \rho_P \cos \phi_P$

Geographic Coordinates System

The geographic coordinate system (GCS), although it does not locate celestial objects directly, is fundamental to the phenomenological framework of this study because it establishes the position of the observer in local terrestrial space. Moreover, the geometric definitions of the equator, poles, latitude, and longitude used in the GCS are directly transferable to celestial coordinate systems when projected onto the celestial sphere.

The GCS specifies a location *P* on the Earth's surface using two angular coordinates: latitude ϕ and longitude λ (Figure 16a). Latitude is defined as the angle between the equatorial plane and the radius vector extending from the Earth's center to *P*. It ranges from −90° at the South Pole to +90° at the North Pole. The equator, defined as the great circle whose poles are the North and South Poles, divides the Earth into the Northern and Southern Hemispheres (Green, 1985, p. 13). Small circles parallel to the equator are called parallels.

Figure 16. Geographic coordinate system: defined by great and small circles (a) and latitude and longitude superimposed on the terrestrial globe (b).

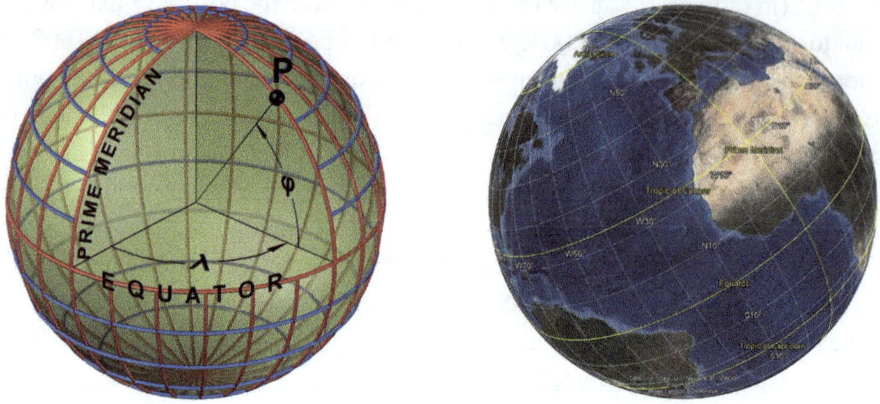

a b

Longitude is measured along the equator as the angle between the Prime Meridian and the meridian passing through point *P*. Meridians are great circles that pass through both geographic poles. By international convention, the Prime Meridian passes through the Royal Observatory in Greenwich, London. Longitude is measured eastward or westward from the Prime Meridian, ranging from 0° to +180° eastward and from 0° to −180° westward.

The principles of the GCS are illustrated in Figure 16b, which shows a view of the Earth from the interactive mapping software Google Earth, with geographic coordinates superimposed on satellite imagery. Digital mapping platforms constitute contemporary applications of the GCS, a system whose conceptual foundations date back to ancient Greek astronomy and geography (Evans, 1998).

Observing the Sky from Earth: North South East West Zenith Nadir & the Horizontal Coordinate System

Among astronomical reference systems, the horizontal coordinate system is the one most directly aligned with embodied perception, because it is defined entirely by the observer's position, local horizon, and vertical axis, rather than by abstract global reference planes. In this system, the observer coincides with the origin of the reference frame and is assumed to be located at the center of the celestial sphere.

In this inquiry, the astronomical observer coincides with the phenomenological self and is also treated geometrically as the center of a sphere. As discussed earlier, the six fundamental directions of embodied perception can be translated into the geometric and astronomical reference systems used in spherical geometry and astronomy, with the observer located at the center of the celestial sphere. In this framework, the origin of the coordinate system and the observer are identical. Geometrically, these six directions correspond to the positive and negative orientations of the three mutually orthogonal axes of a Cartesian coordinate system: the x-, y-, and z-axes (Figure 17).

Figure 17. Directions in a three-dimensional Cartesian coordinate system.

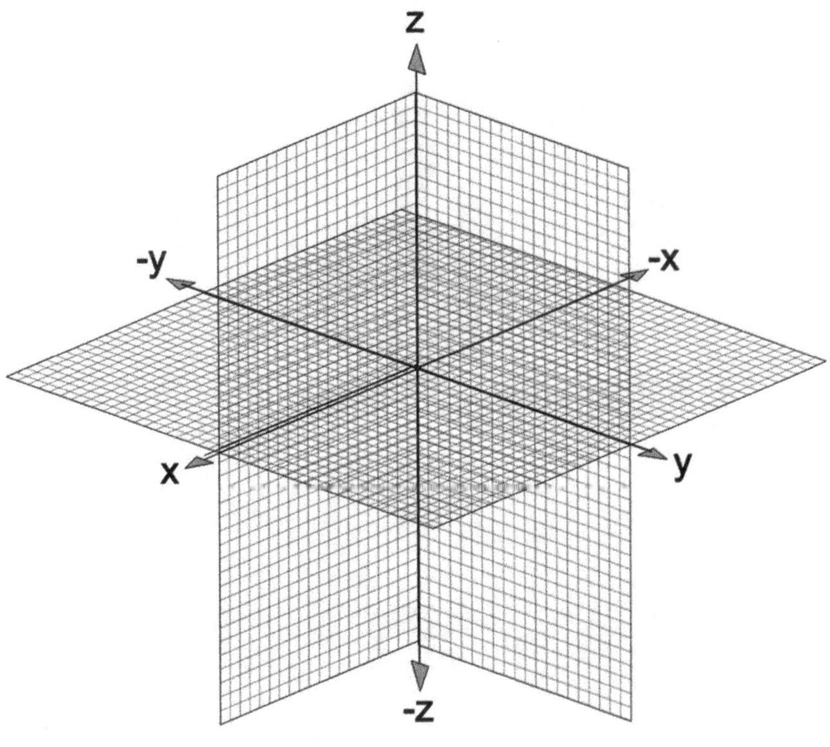

Astronomically, orientation is defined relative to the Earth's surface and the Earth's axis of rotation. The four cardinal directions—north, south, east, and west—derive from the perceived apparent motion of the Sun and the orientation of the Earth's rotational axis. Together, they define two perpendicular horizontal axes: the north–south axis, parallel to the Earth's axis of rotation, and the east–west axis, perpendicular to it. East and west correspond approximately to the directions of sunrise and sunset at the equinoxes. These axes intersect at the observation point, which constitutes the origin of the reference system.

The vertical axis is defined by the zenith, the point directly overhead, and the nadir, the point directly below the observer. Together, these six directions establish the local reference frame for horizontal (topocentric) coordinates (Figure 18).

Figure 18. Zenith, nadir, and cardinal directions in the horizontal reference frame.

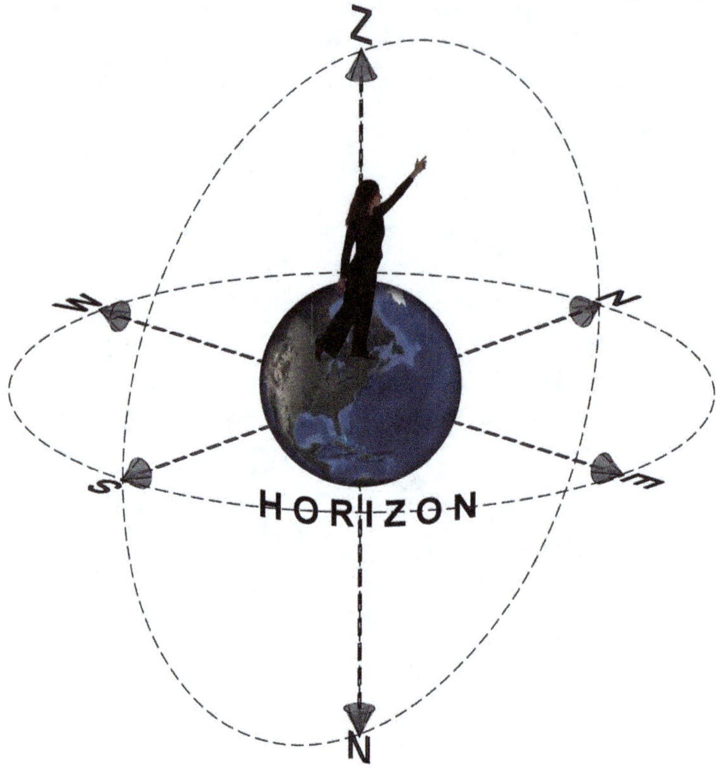

The fundamental plane of the horizontal coordinate system (Figure 19) is the horizon, defined as the great circle on the celestial sphere whose plane is perpendicular to the vertical axis passing through the zenith and nadir. The horizon divides the celestial sphere into an upper hemisphere, which is visible to the observer, and a lower hemisphere, which is not visible (Green, 1985). The term *horizon* derives from the Greek verb ὁρίζω (*horízō*), meaning "to delimit" or "to separate," reflecting both its geometric and perceptual role as a boundary.

Figure 19. Horizontal coordinate system.

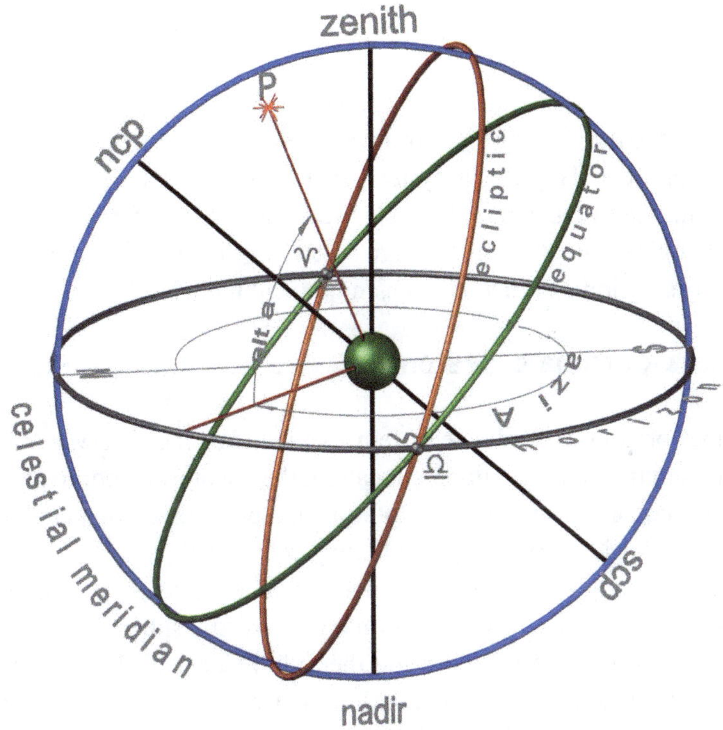

Altitude, denoted by a, is measured along the vertical great circle passing through the object and the zenith, perpendicular to the horizon. It ranges from 0° at the horizon to +90° at the zenith. Objects below the horizon have negative altitude values. The great circle passing through the zenith and the north and south points is called the celestial meridian and serves as the reference plane for altitude measurements.

Because the Earth rotates, the apparent positions of celestial bodies continuously change, and therefore horizontal coordinates depend on both the observer's geographic location and the time of observation. For this reason, the horizontal system is particularly suited to describing immediate perceptual experience of the sky—such as rising, setting, culmination, and visibility—but it is not suitable for cataloging fixed celestial positions.

In this system, the apparent position of celestial bodies continuously changes as the Earth rotates, and therefore the horizontal coordinates of an object depend on both the observer's geographic location and the time of observation. For this reason, the horizontal system is particularly suited to describing immediate perceptual experience of the sky, such as rising and setting, culmination, and visibility, but it is not suitable for cataloging fixed celestial positions.

In professional astronomical practice, observations made in this local, topocentric frame are routinely transformed into other reference systems in order to compare measurements across locations and times. As Seidelmann explains, observed coordinates referred to the rotating terrestrial frame are typically reduced through intermediate geocentric and equatorial systems that are fixed relative to the Earth's axis and the equinox (Seidelmann, 1992). This sequence of transformations highlights the methodological role of the horizontal system as the primary interface between embodied observation and theoretical astronomical models.

Equatorial Coordinate System

While the horizontal coordinate system is centered on the observer and therefore varies with both time and geographic location, the equatorial coordinate system is based on the geometry of the Earth's rotation and provides a comparatively stable reference frame for identifying and cataloging celestial objects.

The fundamental plane of the equatorial system is the celestial equator, defined as the projection of the Earth's equator onto the celestial sphere. The poles of this system are the North Celestial Pole and the South Celestial Pole, determined by extending the Earth's axis of rotation until it intersects the celestial sphere.

In the equatorial system (Figure 20), the position of a celestial object *P* is specified by two angular coordinates: right ascension (RA, or α) and declination (DEC, or δ). Measurements are referenced to the celestial equator and to the great circle perpendicular to it that passes through the intersection of the equator and the ecliptic. This intersection point is known as the *First Point of Aries*, denoted by the symbol ♈ (or T) (see the section "The Ecliptic: Geometry of Time" earlier in this chapter).

Figure 20. Equatorial coordinate system.

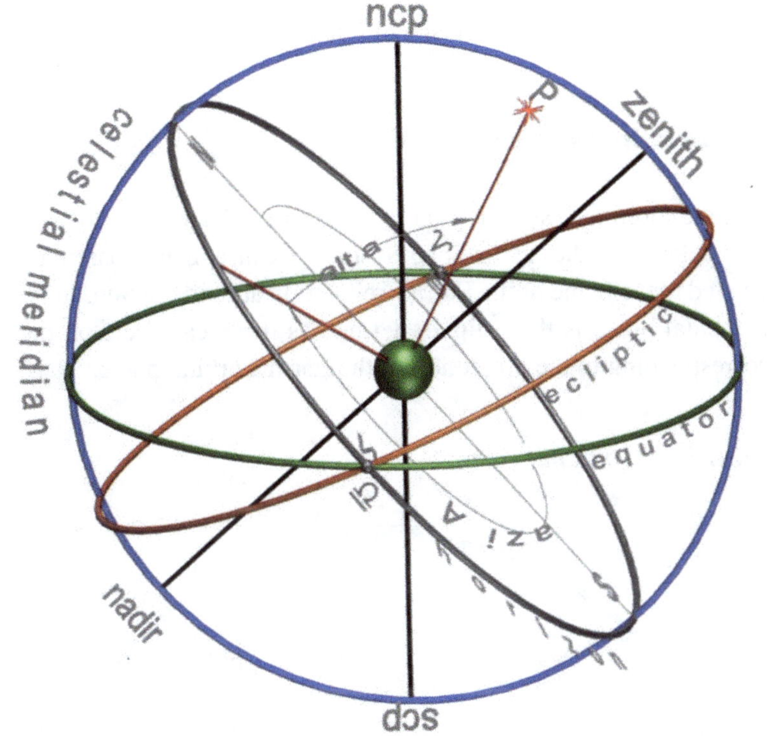

Right ascension is measured eastward along the celestial equator, from the *First Point of Aries* to the point where the great circle passing through *P* and the celestial poles intersects the equator. Although right ascension is an angular quantity, it is conventionally expressed in time units—hours (h), minutes (′), and seconds (″)—because the apparent rotation of the celestial sphere completes a full 360° turn in approximately 24 hours. Thus, one hour of right ascension corresponds to 15° of arc. In this sense, right ascension is analogous to longitude in the geographic coordinate system, though it is referenced to the *First Point of Aries* rather than to a terrestrial prime meridian (Roy & Clarke, 2003).

Declination is the angular distance of *P* north or south of the celestial equator, measured along the great circle passing through *P* and the celestial poles. It ranges from +90° at the North Celestial Pole to −90° at the South Celestial Pole and is directly analogous to latitude in the geographic coordinate system.

For observational purposes, right ascension is often converted into hour angle (HA), which represents the angular distance of a celestial object westward from the observer's local meridian. Hour angle changes continuously with time as the Earth rotates, whereas right ascension remains fixed for a given object (neglecting long-term effects such as precession). This distinction is essential for understanding how stable celestial coordinates are translated into time-dependent, observer-centered measurements.

Ecliptic Coordinate System

The ecliptic coordinate system (Figure 21) is particularly suited to describing the positions of celestial bodies in the Solar System, because their apparent motions are closely related to the plane of the Earth's revolution around the Sun. In this system, the fundamental plane is the ecliptic, defined as the great circle on the celestial sphere corresponding to the projection of the Earth's orbital plane.

Figure 21. Ecliptic coordinate system.

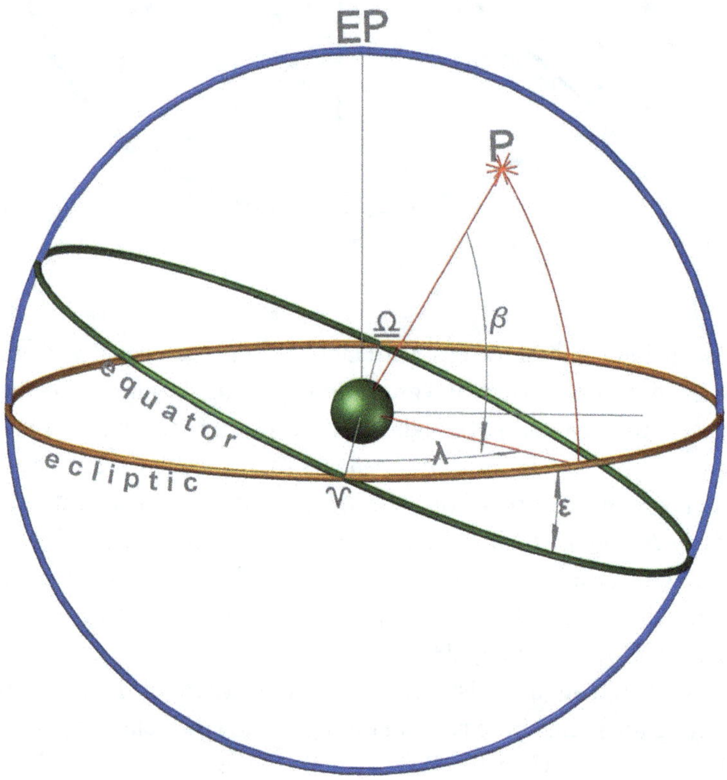

The axis of this system is perpendicular to the ecliptic and passes through the ecliptic poles. The position of a celestial body *P* is specified by two angular coordinates: ecliptic longitude λ and ecliptic latitude β (Green, 1985).

Ecliptic longitude λ is measured eastward along the ecliptic, counterclockwise when viewed from the north ecliptic pole, starting from the *First Point of Aries*, where the Sun crosses the celestial equator from south to north at the March equinox. Longitude ranges from 0° to 360°.

Ecliptic latitude β is measured along the great circle passing through *P* and the ecliptic poles, perpendicular to the ecliptic plane. It represents the angular distance of the object north or south of the ecliptic and ranges from +90° at the north ecliptic pole to −90° at the south ecliptic pole. Objects located exactly on the ecliptic—such as the Sun by definition and approximately the planets—have latitude equal to 0°.

Although the ecliptic coordinate system can be used to describe any celestial object, it is especially effective for analyzing Solar System dynamics and phenomena related to the annual motion of the Sun, the visibility of planets, and the geometry of eclipses. Within the framework of *Earth–Sky Geometries*, the ecliptic system directly links the geometry of remote space to seasonal time and to the apparent solar paths that structure architectural orientation and calendrical construction.

OBJECTS IN THE SKY

The objects of local perception of remote space are celestial bodies, which, in phenomenological observation of the celestial sphere, appear as *figures* against the background of the sky, constituting a classical figure–ground relation in perceptual structure. As shown in the preceding sections, the positions of celestial bodies on the celestial sphere are not defined by distances but by angular relations, which depend on the location of the observer and the time of observation. Nevertheless, physical distances—although not represented in celestial mapping—are crucial for naked-eye perception of stars and other celestial objects, because apparent brightness depends strongly on distance (following the inverse-square law of radiative flux) as well as on intrinsic luminosity.

The perception of celestial objects and phenomena depends on the contrast between their radiation and the background sky luminance (Vollmer, 2025, p. 1). Several environmental conditions interfere with visibility, most notably atmospheric effects and light pollution. Since ancient Greek astronomy, observations of the night sky have been subject to qualitative classification. Most prominently, Hipparchus introduced the concept of stellar magnitude, classifying the brightest stars as first magnitude and progressively decreasing to sixth magnitude for stars barely detectable with the naked eye (Vollmer, 2025, p. 10). Although Hipparchus's system was not quantitative by modern standards, it established the importance of systematic visual assessment. Later studies indicate that well-trained observers can perceive approximately 6,000 stars from Earth over the entire celestial sphere—roughly 3,000 from each hemisphere—under optimal dark-sky conditions (Vollmer, 2025, p. 10).

The Sun is the closest star to Earth and by far the brightest object in the sky, with a luminosity of approximately 3.8×10^{26} W (Fraknoi et al., 2022, p. 525). Its absolute magnitude is $M = +4.83$, meaning that it would appear as a star of apparent magnitude $m = +4.83$ if observed from a distance of 10 parsecs (32.6 light-years) (Vollmer, 2025, p. 15). The Sun has therefore been the most accessible celestial object to observe—not only directly, but also through its effects on Earth—as introduced in the section "The Earth–Sun Geometry."

Observing the Night Sky with the Naked Eye: Moon, Planets, Stars, Constellations, and Galaxies

Besides the Sun, several other celestial objects are visible to the unaided eye. The Moon and the planets, which do not emit their own light, are visible because they reflect sunlight. Under appropriate geometric conditions, the Moon is also visible during daylight hours. The relative geometry of the Sun, Moon, and Earth generates eclipses: a solar eclipse occurs when the Moon passes between the Earth and the Sun, temporarily obscuring the solar disk and casting its shadow on Earth.

The Moon is the most prominent celestial body in the night sky. Its periodic changes in appearance, known as phases, have been observed and predicted since antiquity. During each synodic cycle, lasting approximately 29.5 days, the Moon is seen with varying illuminated portions depending on its angular position relative to the Sun and the observer. The eight conventional lunar phases are: new moon, waxing crescent, first quarter, waxing gibbous, full moon, waning gibbous, third quarter, and waning crescent (NASA, 2025). The diagram in *Cosmographicus Liber* (Figure 22) illustrates the lunar phases in relation to the Sun and as perceived by an observer (Apian, 1524). The geometry of perception is clearly visualized: each ray connects the observer with the Moon, and the illuminated portion is determined by the relative position of the Sun. The lunar phases also have associated symbols, as presented in Table 2.

Figure 22. Moon phases in Cosmographicus Liber (Apian, 1524). Woodcut illustration. Public domain. Source: https://archive.org/details/Cosmographicusl00Apia.

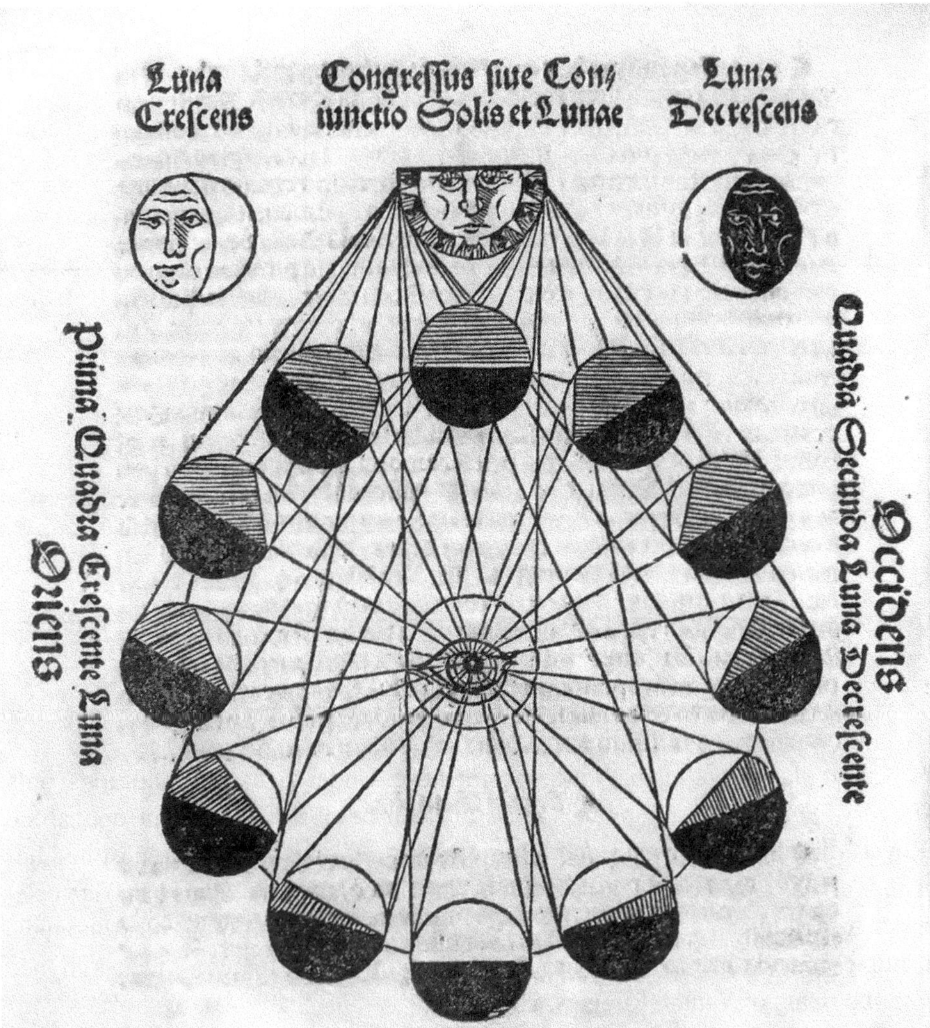

Table 2. Moon phases and their symbols.

Phase	Symbol	Unicode
New Moon	🌑	U+1F311
Waxing Crescent	🌒	U+1F312
First Quarter	🌓	U+1F313
Waxing Gibbous	🌖	U+1F316
Full Moon	🌕	U+1F315
Waning Gibbous	🌖	U+1F316
Last Quarter	🌗	U+1F317
Waning Crescent	🌘	U+1F318

Not all planets are visible to the unaided eye. Under favorable conditions, only Mercury, Venus, Mars, Jupiter, and Saturn can be observed without optical aid (Royal Museums Greenwich, n.d.). Among these, Venus is the brightest and has historically attracted particular attention. The observation of Venus has been celebrated in monuments and buildings since ancient civilizations. Archaeoastronomical examples include the Mayan Governor's Palace at Uxmal (Mexico), which is aligned with the rising of Venus (Ruggles, 2015).

Constellations constitute another category of visible celestial objects in the night sky. They are groupings of stars that form recognizable visual patterns, although the stars themselves are not physically related by proximity or gravitational binding. This distinguishes constellations from star clusters, which are gravitationally bound systems. In ancient traditions, constellations were often associated with mythological figures such as Andromeda, Ursa Major, Scorpius, and Orion. Constellations were described, illustrated, and cataloged in the *Almagest* (Ptolemy, ca. 150 C.E./1515), where stars were listed with their celestial longitude, latitude, magnitude, and hemisphere. In contemporary astronomy, constellations continue to function as practical reference regions for locating stars and deep-sky objects (National Radio Astronomy Observatory, 2024).

Twelve constellations lie along the apparent annual path of the Sun on the celestial sphere, forming the zodiac. The zodiac is a belt-shaped region extending approximately 8° north and south of the ecliptic (Zodiac, n.d.). This belt is divided into twelve equal segments of approximately 30° of celestial longitude in the tropical zodiac system, traditionally associated with the constellations Aries, Taurus, Gemini, Cancer, Leo, Virgo, Libra, Scorpius, Sagittarius, Capricorn, Aquarius, and Pisces, arranged in counterclockwise order (Figure 23). These zodiacal constellations also correspond to astrological signs. The Sun appears to pass through each constellation once per year, beginning with Aries at the March equinox. Due to the precession of the equinoxes, however, the dates traditionally associated with these signs no longer correspond precisely to the current astronomical positions of the constellations, as discussed in the section "The Ecliptic: Geometry of Time."

Figure 23. Constellations and mythological figures in the Almagest (Ptolemy, ca. 150 C.E./1515).

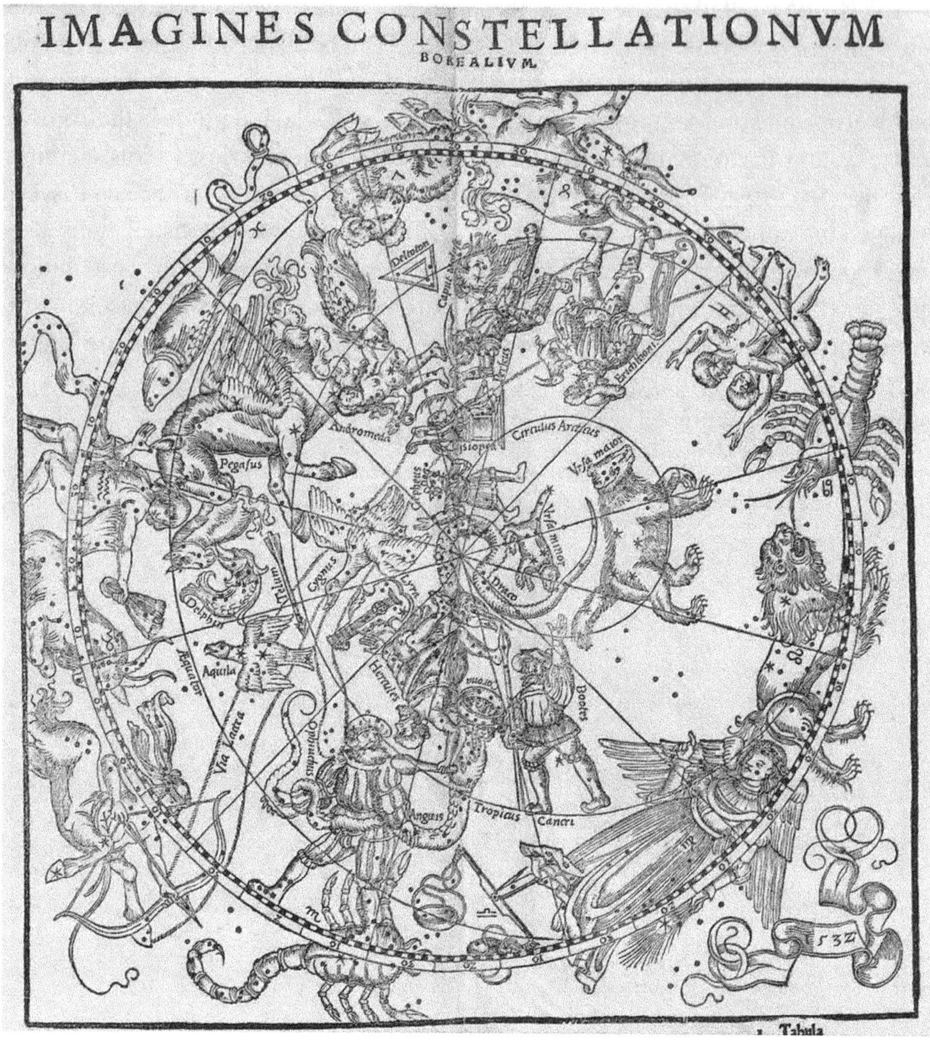

Table 3. The twelve constellations.

Constellation	Symbol	Date
Aries	♈	March 21–April 19
Taurus	♉	April 20–May 20
Gemini	♊	May 21–June20
Cancer	♋	June 22–July 22
Leo	♌	July 23–August22
Virgo	♍	August 23–September 22
Libra	♎	September 23–October 23
Scorpio	♏	October 24–November 21
Sagittarius	♐	November 22–December 21
Capricorn	♑	December 22–January 19
Aquarius	♒	January 20–February 18
Pisces	♓	February 19–March 20

Galaxies are the largest celestial systems observable by the naked eye. They are composed of stars, planets, gas, and dust bound together by gravity. Galaxies range in size from a few thousand to trillions of stars and span distances from thousands to hundreds of thousands of light-years across. Many large galaxies contain supermassive black holes at their centers, with masses millions to billions of times that of the Sun (NASA, 2025). Galaxies exhibit a variety of forms, most commonly spiral and elliptical structures, and their ages typically range from about 10 to 13.6 billion years (NASA, 2025).

The Solar System is part of the Milky Way, a barred spiral galaxy approximately 100,000 light-years in diameter. The Solar System orbits the galactic center once every approximately 240 million years (NASA, 2025). The Milky Way is one of the few galaxies visible to the naked eye from dark locations, appearing as a luminous band across the sky. Other naked-eye galaxies include Andromeda (M31), the Large and Small Magellanic Clouds, and Triangulum (M33) (Masters, 2003).

Star Trails

The continuous motions of the Earth—daily rotation about its axis and annual revolution around the Sun—determine the apparent paths of celestial objects. As discussed in the section "The Ecliptic: Geometry of Time," the ecliptic provides the fundamental reference plane for the annual apparent motion of the Sun. The daily apparent rotation of the sky, however, is governed by the extension of the Earth's rotation axis to the celestial sphere, which defines the North and South Celestial Poles.

Stars near the celestial poles trace small circular paths and appear to rotate around fixed points in the sky. Such stars are called circumpolar stars. Polaris, in the constellation Ursa Minor, currently lies within less than one degree of the North Celestial Pole and therefore appears nearly stationary. With an apparent visual magnitude of approximately 2.0, it serves as a pole star in the Northern Hemisphere. In the Southern Hemisphere, there is no comparably bright star located near the South Celestial Pole.

Because the Earth's rotation axis slowly precesses, completing a full cycle in approximately 26,000 years, the identity of the pole star changes over long timescales. As NASA explains, the celestial pole traces a slow circular path among the stars, passing near different bright stars over millennia; about 14,000 years ago it pointed near Vega, and it will do so again in approximately 12,000 years (NASA, 2021).

In the night sky, celestial objects appear to move along circular arcs centered on the celestial poles—time-dependent geometric configurations whose angular extent corresponds directly to the duration of observation. Since the Earth rotates at approximately 15° per hour (360° in 23h 56m of sidereal time)s, an exposure time of 1 hour, 30 minutes, and 30 seconds corresponds to an angular arc of 22.625° on the celestial sphere. These apparent paths are known as star trails and can be recorded using long-exposure photography. While stars are perceived by the naked eye as point sources, star-trail images render visible the time-dependent geometric structure of the rotating celestial sphere.

Figure 24 shows a long-exposure photograph recording star trails at Mandala Pyramid in S.U.N.F.A.R.M., New York. Each trail subtends an angle of approximately 25.083°, corresponding to an exposure time of about 1 hour, 40 minutes, and 20 seconds.

Figure 24. Star trail at Mandala Pyramid, S.U.N.F.A.R.M., New York, 2011.09.06.

Star trails need not be centered on the pole star. Depending on the direction of observation, different curvatures appear. For example, when observing toward the eastern or western horizon, star trails appear as inclined arcs. Figure 25 shows star trails observed while looking east at Sunrise Trellis on September 13, 2007; the arc spans approximately 9°, corresponding to an exposure time of about 36 minutes.

Observable configurations of celestial objects vary with the Earth's position along its orbit around the Sun. Consequently, star trails—like all time-based astronomical images—cannot be exactly repeated. Their geometry functions as a temporal signature of a specific observation of remote space from a specific local place.

Figure 25. Star trail looking east at Sunrise Trellis, S.U.N.F.A.R.M., New York, 2007.09.13.

CONCLUDING THOUGHTS: THE PHENOMENOLOGY OF EARTH-SKY GEOMETRIES

This chapter has emphasized the central role of geometry in astronomy and how geometric relations articulate the interconnectedness of time, space, movement, and light. This introductory presentation of spherical astronomy and sky-related reference systems provides the scientific context for the definition of Earth–Sky Geometries (ESGs): the geometric configurations created by unaided-eye observations of the sky by a subject situated on Earth. The term *geometries* is used in the plural because the number of possible configurations is unlimited, varying with location, time, and the celestial phenomena involved.

From the foregoing analysis, Earth–Sky Geometries may be characterized as follows:

1. ESGs involve alignments between the observer and a celestial body or phenomenon, such as the Sun rising due east at the March equinox or the geometric configuration producing a lunar eclipse.
2. ESGs include static configurations formed by groups of celestial objects during a fixed moment of observation, such as a constellation observed at a specific date and time.
3. ESGs also include dynamic configurations generated by the apparent motion of celestial bodies and recorded over a given duration from the same location, such as the observed solar path on July 7, 2025, from 7:40 a.m. to 10:15 a.m. at latitude 42.1729° and longitude −73.6755°.
4. ESGs require a reference system; in this framework, the horizontal coordinate system is adopted because it is directly grounded in the observer's bodily position, local horizon, and vertical axis.
5. In ESGs, the celestial object functions as figure, while the celestial sphere constitutes the perceptual *background*, constituting a phenomenological figure–ground structure analogous to terrestrial perception.
6. ESGs are associated with astronomical time as defined by the Earth's rotation and revolution, producing the sidereal day, the alternation of day and night, the seasons, and the sidereal year.
7. ESGs establish a correspondence between astronomical time and existential time, as experienced by the observer as embodied consciousness.
8. ESGs provide a phenomenological interpretation of remote space as lived experience, in which the body of the observer participates in the perception of celestial figures against the sky.
9. ESGs may function as a method of scientific and spatial analysis, recording observations as four-dimensional events within a spacetime framework defined by the observer's terrestrial location, the time of observation, and the celestial coordinates of the observed object.

From a phenomenological perspective, inquiry begins with lived experience. Phenomenology, as Merleau-Ponty states, does not seek understanding from any starting point other than the facticity of human existence, offering an account of space, time, and the world as they are lived rather than abstractly conceived (Merleau-Ponty, 1962, p. vii). Within this framework, ESGs operate as a methodological bridge between scientific description and existential meaning. Although celestial phenomena remain physically remote and largely beyond direct interaction, they exert continuous influence on terrestrial life and human temporal structures and are encountered through embodied perception rather than detached measurement alone.

These geometries are fundamental to what may be termed the architecture of cosmology: built constructions, spatial orientations, and material instruments that have historically mediated human understanding of the universe as manifested in spacetime. From prehistoric alignments and ritual landscapes to observatories and contemporary land-based installations, architectural form has repeatedly engaged celestial order as a structuring principle.

The broader aim of this inquiry is therefore to examine how Earth–Sky Geometries shape architectural space as a mode of being-in-the-world, emerging from the interaction between self and universe, which Heidegger describes as a unitary phenomenon rather than a subject–object duality (Heidegger, 1962, p. 68). In this sense, architecture becomes not merely a container for experience but an active participant in orienting the body within cosmic rhythms.

By bringing phenomenology, geometry, and astronomy into a single observational framework, Earth–Sky Geometries are translated into architectural practices that align perception, movement, and construction with solar, lunar, and stellar dynamics across seasonal and planetary timescales. Earth–Sky Geometries establish a shared ground where scientific description and lived spatial experience are not opposed but structurally intertwined. This conceptual framework prepares the ground for subsequent volumes, in which Earth–Sky Geometries are explored not only as observational constructs but as generators of spatial organization, material systems, and landscape interventions. Later chapters extend the discussion from celestial reference systems to site-specific design strategies, embodied movement, and seasonal inhabitation, integrating astronomical orientation with ecological cycles and human vulnerability. Projects developed within the S.U.N.F.A.R.M. research platform and related installations are presented as experimental fields in which Earth–Sky Geometries are translated into architectural practices that align perception, movement, and construction with solar, lunar, and stellar dynamics. In this way, ESGs evolve from descriptive geometries of observation into operative frameworks for design, enabling architecture to function as an interface between terrestrial habitation and cosmic order.

This chapter has established Earth–Sky Geometries as observational and conceptual frameworks grounded in spherical astronomy and phenomenological perception. The following chapter shifts from celestial reference systems to terrestrial situations, examining how these geometries are encountered through bodily movement, site conditions, and material constraints. If Earth–Sky Geometries describe how the sky is structured relative to an observer, the next step is to investigate how the observer, in turn, is structured by place—by slope, horizon, vegetation, built form, and seasonal change—and how architectural interventions may register, amplify, or transform these relations into spatial experience.

REFERENCES

Albert, A. A. (1949). *Solid analytic geometry* (1st ed.). McGraw-Hill.

Apian, P. (1524). *Cosmographicus liber Petri Apiani mathematici studiose collectus*.

Ball, R. S. (1908). *A treatise on spherical astronomy*. Cambridge University Press.

Brünnow, F. (1865). *Spherical astronomy*. Ferd. Dümmler.

Chauvenet, W. (1906). *A manual of spherical and practical astronomy*. J. B. Lippincott.

Copernicus, N. (1566). De revolutionibus orbium coelestium []. Basileae] [Basel] [, Switzerland: Ex officina Henricpetrina.]. *Libri*, VI.

Evans, J. (1998). *The history and practice of ancient astronomy*. Oxford University Press.

Fraknoi, A., Morrison, D., & Wolff, S. C. (2022). *Astronomy* (2nd ed.). OpenStax.

Galilei, G. (1623). *Il saggiatore*. Giacomo Mascardi.

Glossary. In (2023). *Astronomical Almanac Online*. United States Naval Observatory.

Green, R. M. (1985). *Spherical astronomy*. Cambridge University Press.

Heath, T. L. (1921). *A history of Greek mathematics*. Clarendon Press.

Heidegger, M. (1962). *Being and time* (Macquarrie, J., & Robinson, E., Trans.). Harper & Row. (Original work published 1927)

Kepler, J. (1609). *Astronomia nova*. G. Voegelinus.

Kepler, J. (1619). *Harmonices mundi*. Johann Planck.

Liddell, H. G., Scott, R., Jones, H. S., & McKenzie, R. (1940). *A Greek–English lexicon*. Clarendon Press.

Merleau-Ponty, M. (1962). *Phenomenology of perception* (Smith, C., Trans.). Routledge & Kegan Paul. (Original work published 1945)

NASA. (2021, July 28). *What is the North Star and how do you find it?* NASA Science. Retrieved August 25, 2025, from https://science.nasa.gov/solar-system/what-is-the-north-star-and-how-do-you-find-it/

National Geographic Society. (2023, October 19). *Equator*. Retrieved August 25, 2025, from https://education.nationalgeographic.org/resource/equator/

National Radio Astronomy Observatory. (2024, October 29). *What are constellations?* Retrieved August 24, 2025, from https://public.nrao.edu/ask/what-are-constellations/

Newcomb, S., & Holden, E. S. (1890). *Astronomy*. Henry Holt.

Newton, I. (1687). *Philosophiæ naturalis principia mathematica.*

Norton, W. A. (1872). *A treatise on astronomy, spherical and physical.* John Wiley & Son.

Ptolemy, . (1984). *Ptolemy's Almagest* (Toomer, G. J., Trans.). Princeton University Press.

Reisch, G. (1525). *Margarita philosophica.* Johann Schott.

Roy, A. E., & Clarke, D. (2003). *Astronomy: Principles and practice* (4th ed.). Institute of Physics Publishing.

Royal Museums Greenwich. (n.d.). *Astronomy with the naked eye.* Retrieved August 20, 2025, from https://www.rmg.co.uk/stories/space-astronomy/astronomy-naked-eye

Ruggles, C. L. N. (2015). *Handbook of archaeoastronomy and ethnoastronomy.* Springer.

Sacrobosco, J. (1531). *Sphaera tractatus.* Lucantonio Giunta.

Science, N. A. S. A. (2024, October 22) *Our Solar System.* Retrieved August 25, 2025, from https://science.nasa.gov/resource/our-solar-system-2

Science, N. A. S. A. (2025, January 16). *Chapter 2: Reference systems. Basics of Spaceflight.* Retrieved August 25, 2025, from https://science.nasa.gov/learn/basics-of-space-flight/chapter2-1

Seidelmann, P. K. (Ed.). (1992). *Explanatory supplement to the Astronomical Almanac* (Rev. ed.). University Science Books.

Smart, W. A. M. (1936). *Text-book on spherical astronomy.* Cambridge University Press.

Tanton, J. S. (2005). *Encyclopedia of mathematics.* Infobase Publishing.

Vollmer, M. (2025). Naked-eye celestial objects and phenomena: How far can we see at night? *European Journal of Physics*, *46*(3), 035301.

Zodiac. (n.d.). In *Encyclopedia.pub.* Retrieved August 25, 2025, from https://encyclopedia.pub/entry/31827

ADDITIONAL READING

Masters, K. (2003). *Curious about astronomy: Can any galaxies be seen with the naked eye? Ask an Astronomer*. Cornell University.

Smart, W. M. (1942). *Foundations of astronomy*. Oxford University Press.

Williams, D. R. (2025). *Planetary fact sheet – metric*. NASA National Space Science Data Center.

ENDNOTES

1 Due to the precession of the equinoxes, the Sun currently appears in the constellation Pisces at the March equinox, although the symbolic designation ♈ for the First Point of Aries remains in use.

2 The terms *vernal* and *autumnal* are hemisphere-specific; *March* and *September equinox* are globally accurate.

3 The term *mean* reflects small variations in apparent solar day length caused by Earth's elliptical orbit and axial tilt (equation of time).

Chapter 4
The Sky as Universe and Culture

ABSTRACT

Spherical astronomy, grounded in the geometry of unaided-eye observation, constitutes a foundational framework for cosmological inquiry when approached from a phenomenological perspective. This chapter examines how the geometry implicit in human perception of the sky has informed cosmological theories and, in turn, shaped cultural expressions, including the built environment. Tracing a conceptual trajectory from cosmology to cosmogony, cosmography, and mapping practices, the chapter shows how scientific models of the universe are translated into systems of meaning, belief, and spatial order, contributing to the formation of the human-made world.

DOI: 10.4018/979-8-3373-3399-1.ch004

Figure 1. Camille Flammarion, L'atmosphère: météorologie populaire (Paris: Hachette, 1888), 163. Wood engraving illustrating the layered structure of the atmosphere and the cosmos as a cultural and scientific construct. Public domain.

Un missionnaire du moyen âge raconte qu'il avait trouvé le point où le ciel et la Terre se touchent...

FROM ASTRONOMY TO COSMOLOGY

The previous chapter established geometry as a fundamental framework for the perception of remote space. The sky, geometrically conceptualized as the celestial sphere, is populated by stars whose apparent configurations continuously change in spacetime. Since time immemorial, human beings have observed the sky and its celestial objects and phenomena—most prominently the Sun. These observations have progressively developed into theories of the universe, constituting the domain of cosmology.

While astronomy is grounded in empirically acquired data, cosmological theories often extend beyond direct experience and may also be understood as cultural constructs embedded within philosophical, spiritual, and religious frameworks. Figure 2, an illustration from the early sixteenth-century treatise *Cosmographicus Liber*, visually expresses the translation of astronomical observation into a cosmological model structured around a geocentric universe.

Figure 2. Cosmological model: geocentric representation translating astronomical observation into a cosmological order, from Cosmographicus Liber Cosmographicus liber Petri Apiani mathematici (Apian, 1524). Woodcut illustration. Public domain. Source: https://archive.org/details/Cosmographicusl00Apia.

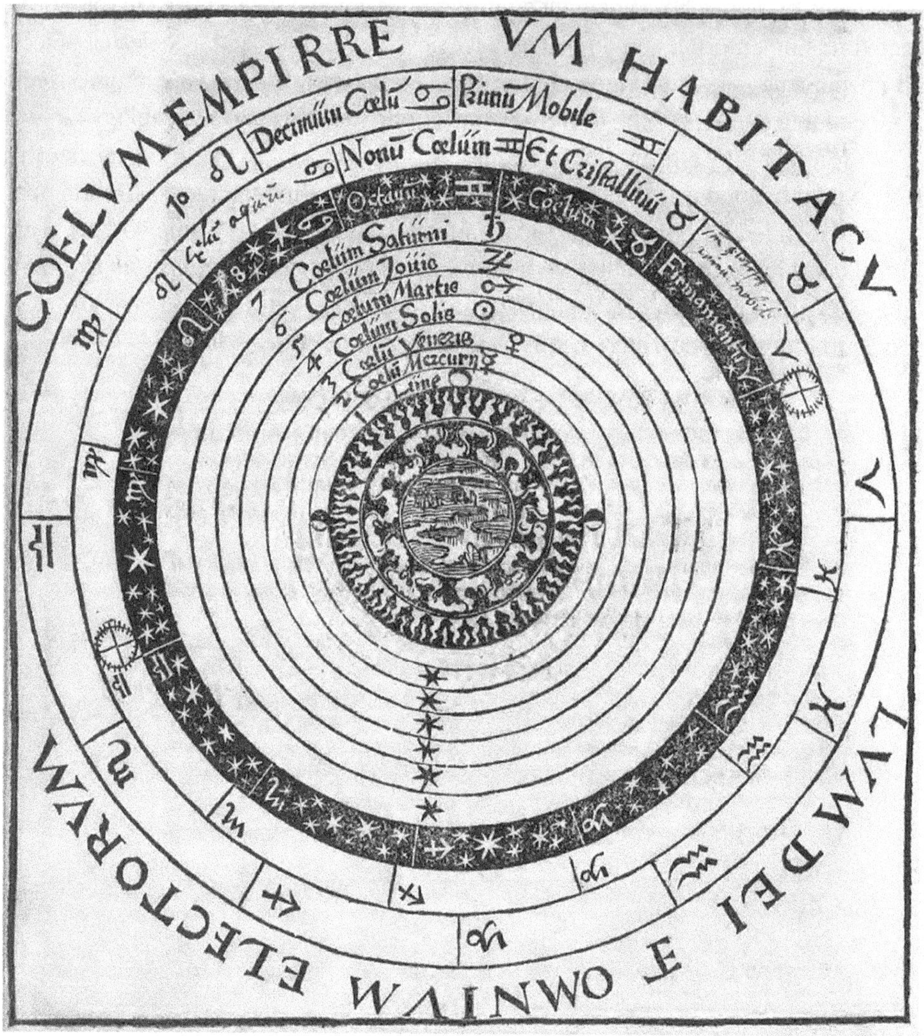

The etymology of *cosmology* already suggests its scope. The term derives from the Greek κόσμος (*kosmos*: universe, world, order) and λογία (*logia*: study), and thus literally means the "study of the universe." In contemporary scientific usage, cosmology is a field within astronomy and astrophysics concerned with the structure, origin, and evolution of the physical universe, including celestial systems such as the Solar System, stars, galaxies and large-scale phenomena governed by physical laws and understood within a spacetime framework.

Cosmology extends beyond the physical data and quantitative methodologies of these sciences. Geometry and other branches of mathematics are foundational to cosmology, often legitimizing models that are counterintuitive from the standpoint of ordinary experience. In its attempt to account for the universe as a totality, cosmology develops theoretical models that address conditions which are incommensurable or not directly accessible to measurement, such as the origin of the universe and the emergence of time itself. In the theory of relativity, geometry is not merely descriptive but constitutive of physical reality. In special relativity, Minkowski space provides a four-dimensional geometric framework integrating space and time; in general relativity, non-Euclidean Riemannian geometry is employed to describe spacetime curvature generated by the distribution of mass and energy (Einstein, 1920). Geometry thus defines the interaction between space, time, and force, and general relativity is widely regarded as the theoretical foundation of modern cosmology (Fazio, 2018). In doing so, cosmological inquiry frequently crosses the boundaries of empirical science, intersecting with philosophical and metaphysical perspectives. Many cosmological models therefore articulate realities that appear to contradict, or at least exceed, the experiential and perceptible world of everyday life, revealing a tension between scientific abstraction and lived experience that is central to the interpretation of cosmology within a phenomenological framework. This tension does not diminish cosmology's cultural significance; rather, it intensifies its role as a mediating structure between measurable reality and existential orientation. Cosmological models, whether geocentric or relativistic, therefore function simultaneously as scientific explanations and as frameworks through which human beings situate themselves within the totality of existence.

Current cosmological theories assert that the universe is expanding and that this expansion began approximately 13.8 billion years ago in an event known as the Big Bang (Center for Astrophysics, 2023). Evidence for cosmic expansion derives from observations of galaxies receding from one another, inferred through redshift measurements obtained by ground-based and space-based telescopes. The initial discovery of this expansion was made in 1929 by Edwin Hubble, after whom the Hubble Space Telescope—launched in 1990—was named. Observations further suggest that the rate of expansion is accelerating, a phenomenon commonly expressed through the Hubble constant (National Aeronautics and Space Administration, 2025). While debates concerning the precise value of this constant lie beyond the scope of this study, the broader implications of expansion are central to contemporary cosmological models.

The Big Bang framework does not arise from a single theory but from the convergence of multiple lines of evidence and theoretical developments. These include solutions of general relativity implying expansion, systematic observations of galactic redshift, measurements of cosmic microwave background radiation (Figure 3), and hypotheses concerning dark matter and dark energy as formalized in the ΛCDM (Lambda Cold Dark Matter) model (Disney, 2007). Together, these elements constitute the dominant explanatory structure of contemporary cosmology.

Figure 3. Cosmic microwave background radiation, WMAP nine-year data set (2012). NASA / WMAP Science Team (WMAP #121238). Public domain.

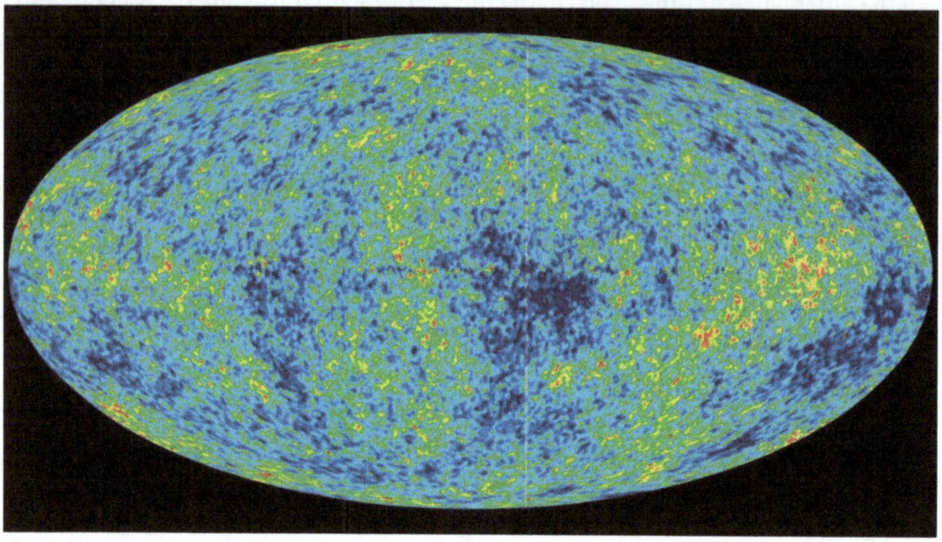

Despite its explanatory power, modern cosmology appears increasingly distant from the phenomenological and intuitive character of spherical astronomy, whose geometric foundations emerged from naked-eye observation. Cosmology is not solely a scientific and mathematical construction; it is also shaped by the historical, cultural, and intellectual contexts in which it is produced. As such, it constitutes a defining component of a civilization's *Weltanschauung*. Understanding a cosmological model therefore offers insight not only into scientific knowledge but also into the broader cultural frameworks through which time, space, force, and geometry are interpreted and lived.

From a humanistic perspective, cosmology may be understood not only as the study of the physical universe but also as a conception of the world encompassing human existence, belief systems, and place-making practices. The historian of religion Mircea Eliade addresses cosmology in this expanded sense, arguing that "in extremely varied cultural contexts, we constantly find the same cosmological schema and the same ritual scenario: settling in a territory is equivalent to founding a world" (Eliade, 1961, p. 47). For Eliade, dwellings at all scales—from the house to the city—symbolically reproduce a cosmos, embedding cosmological order in everyday life (Eliade, 1961, pp. 56–58). Similarly, geographer Yi-Fu Tuan discusses cosmology as a worldview deeply rooted in culture and fundamental to the human experience of place (Tuan, 1974; 1977).

In this expanded sense, cosmology becomes operative not only in theoretical discourse but in spatial practice. The ordering of territory, settlement, and architecture may thus be understood as the material crystallization of a cosmological worldview. The concept of the architecture of cosmology, as defined in this book, arises from this convergence of scientific, cultural, and experiential dimensions. Built constructions oriented toward celestial phenomena—such as solar or stellar alignments—do more than register observable events; they materialize cosmological beliefs and encode systems of meaning. In this sense, architecture becomes a spatial embodiment of cosmology, transforming geometric relations between Earth and sky into cultural identity, ritual practice, and enduring form.

ABOUT COSMOGONY, COSMOGRAPHY, AND COSMOGRAMS

The Greek term κόσμος (*kosmos*) signifies not only "universe" or "world," but also "order" and "ornament." This semantic richness already suggests that cosmology concerns not merely the physical structure of the universe, but its intelligibility as an ordered whole. From its earliest formulations, cosmological thought has been intertwined with the search for harmony—an attempt to discern proportional relationships among celestial bodies, temporal cycles, and spatial forms.

This pursuit of cosmic harmony is clearly articulated in Pythagorean philosophy through the concept of the "harmony of the spheres" (*musica universalis*), which expresses the idea that celestial motions correspond to mathematical ratios analogous to musical intervals. As Heath explains, "as astronomy is the motion of bodies and appeals to the eye, so there is a harmonious motion, a motion according to the laws of harmony, which appeals to the ear" (Heath, 1991, p. 13). Geometric harmony thus becomes both a conceptual and a representational principle in cosmological thought.

A paradigmatic example is Johannes Kepler's *Mysterium Cosmographicum* (1596), in which the orbits of the known planets are associated with the Platonic solids (Figure 4). Another influential visual synthesis of cosmic order is found in Andreas Cellarius's *Harmonia Macrocosmica* (first published 1660; later editions 1708), where astronomical knowledge is rendered through elaborate cartographic imagery that integrates geometry, myth, and measurement.

The same Greek root κόσμος (*kosmos*) underlies several related terms that are central to this inquiry—cosmogony, cosmography, and cosmogram—each of which extends cosmological thinking beyond scientific description into symbolic, cultural, and spatial domains.

Figure 4. Planetary orbits and Platonic solids from Johannes Kepler, Mysterium Cosmographicum (1596). ETH-Bibliothek Zürich, RAR 1367:1.

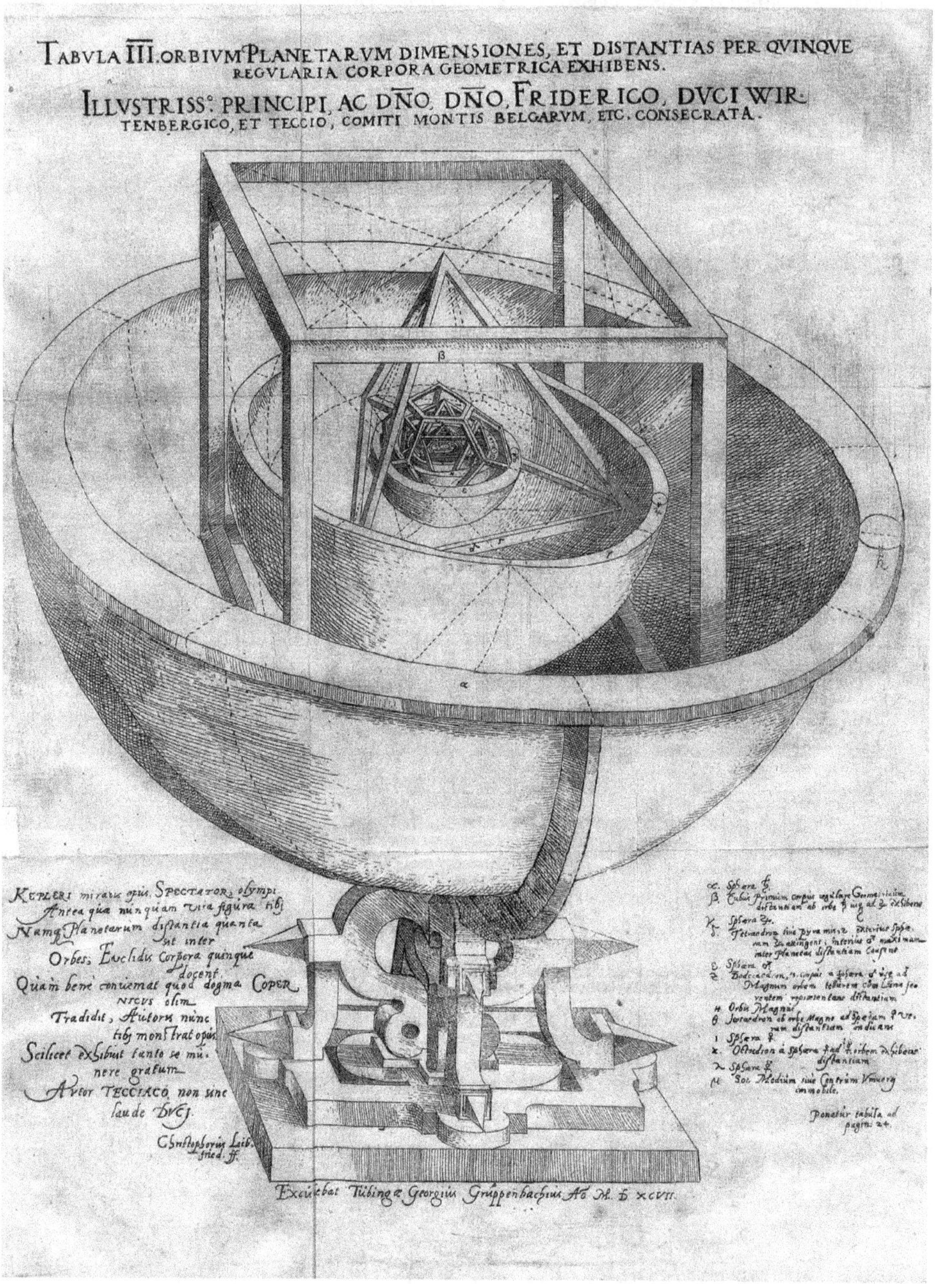

Figure 5. Planetary orbits and Platonic solids from Andreas Cellarius, Harmonia Macrocosmica (1708 edition). Public domain.

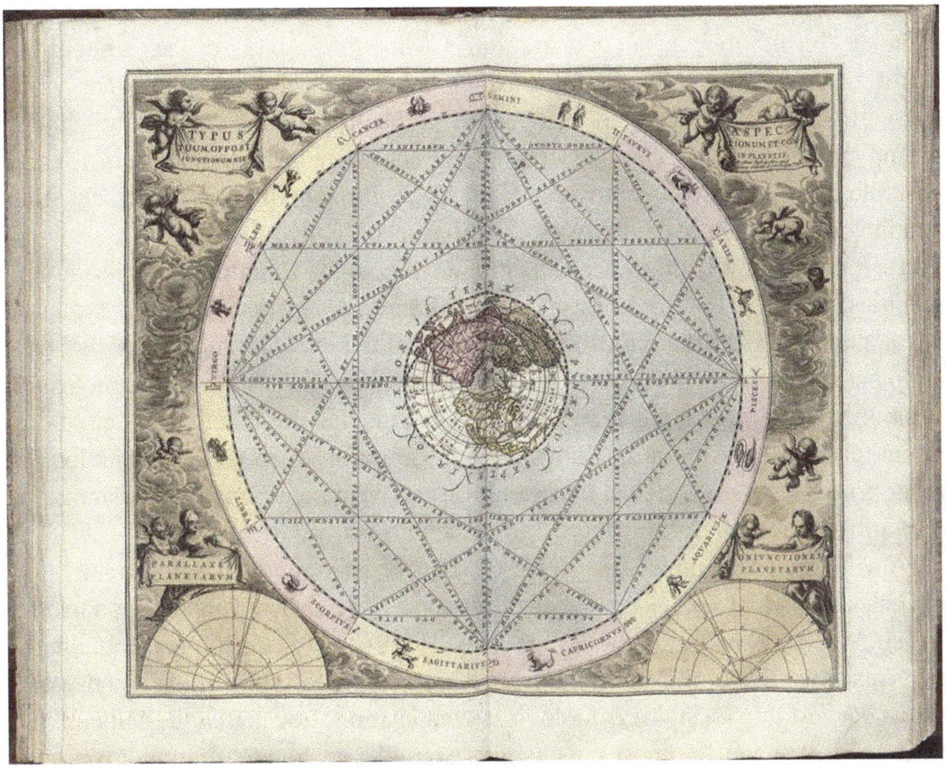

In this study, cosmogony, cosmography, and cosmograms are not approached merely as historical or lexical distinctions, but as analytical modalities through which Earth–Sky relations become intelligible and spatially operative. Cosmogony articulates the origin of the world and establishes an ontological framework; cosmography describes and maps the structural order of the cosmos; cosmograms materialize these conceptions in diagrams, symbols, and built configurations. Together, they trace the progressive translation of cosmological thought into embodied orientation and spatial form. This tripartite structure therefore serves as a methodological framework for examining how Earth–Sky Geometries move from narrative and metaphysical formulation to representational systems and, ultimately, to architectural and landscape practice.

Cosmogony

The term *cosmogony* derives from the Greek κόσμος (*kosmos*) and γόνος (*gonos*), meaning "birth," "origin," or "generation." Cosmogony addresses the origin of the world and has distinct yet overlapping meanings across disciplines. In astronomy and astrophysics, cosmogony refers to the study of the origin and early development of the universe and its characteristic structures (Encyclopædia Britannica, n.d.). While cosmology investigates the universe from its beginnings through its present state and into possible futures, cosmogony focuses specifically on origins.

Beyond the physical sciences, cosmogony occupies a central place in the humanities, where it is closely associated with creation myths. As Long explains, cosmogony refers to the philosophical and theological elaboration of primal creation narratives within a religious or cultural community; here, "myth" denotes the symbolic narration of what is apprehended as fundamental reality, and "creation" refers to the beginning of all things, whether through divine will, emanation, or other principles (Long, 2025). Every culture and religion articulates a cosmogony, providing an account of how the world came into being.

A well-known example is the Judeo-Christian cosmogony described in the Book of Genesis, in which the universe is created over six days. Michelangelo Buonarroti's fresco cycle on the ceiling of the Sistine Chapel (1508–1512) constitutes a monumental visual narrative of this cosmogony, in which Earth, Sky, humanity, and divinity are integrated into a single cosmological composition. Located in the Vatican, the frescoes are among the most outstanding examples of how cosmogony becomes spatially and architecturally embedded at the symbolic center of a religious world.

The Sistine Chapel may also be interpreted through the lens of Heidegger's fourfold—Earth, Sky, Divinities, and Mortals—where terrestrial existence, celestial order, human presence, and the divine are gathered into a single spatial event. In this sense, the chapel does not merely depict creation; it stages the ontological unity of the world as a structured relation between Earth and Sky.

Cosmogony is directly relevant to the *architecture of cosmology* because it links the creation of space with the creation of meaning. As Mircea Eliade argues, the discovery or projection of a center is equivalent to the creation of the world itself: "the discovery or projection of a fixed point—the center—is equivalent to the creation of the world" (Eliade, 1961, p. 22). In traditional societies, acts of construction replicate the primordial act of creation; "every construction or fabrication has the cosmogony as paradigmatic model" (Eliade, 1961, p. 45). Sacred spaces arise through *hierophany*—from the Greek ἱερός (*hieros*), "sacred," and φαίνειν (*phainein*), "to reveal"—and "every spatial hierophany or consecration of a space is equivalent to a cosmogony" (Eliade, 1961, p. 63).

Cosmography

The term *cosmography* derives from the Greek κόσμος (*kosmos*) and γραφία (*graphia*), meaning "description" or "writing." Cosmography has historically encompassed multiple meanings and disciplinary contexts. In classical and early modern usage, it referred to the descriptive study of the structure of the heavens and the Earth, including *uranography*—the mapping of celestial bodies based on spherical astronomy (Treccani, n.d.).

During the Renaissance, cosmography became closely associated with geography and cartography, forming an integrative field that combined astronomical knowledge, terrestrial mapping, and visual representation (Figure 6). As Baldasso notes, Renaissance cosmography sought to describe heaven and Earth through both textual and graphic means, producing representations such as *imago mundi*, *mappaemundi*, and regional maps (Baldasso, 2022). In this context, iconography played a central methodological role, not merely illustrating knowledge but actively shaping it (Cosgrove, 2007).

Figure 6. Terrestrial and celestial globes, from Cosmographicus liber Petri Apiani mathematici (Apian, 1524). Woodcut illustration. Public domain. Source: Internet Archive.

In contemporary physical cosmology, *cosmography* refers to the mapping of the large-scale structure of the universe through computational methods and data visualization. Here, cosmography is defined as "the study and making of maps of the universe or cosmos" (Pomarède et al., 2017, p. 1), relying on extensive datasets derived from observational astronomy and astrophysics.

In this study, cosmogony, cosmography, and cosmograms are not treated merely as historical categories, but as analytical lenses through which cosmological thought becomes spatially operative. Together, they trace a movement from origin (cosmogony), to description and representation (cosmography), to material and symbolic inscription (cosmogram). This tripartite structure provides a methodological framework for examining how conceptions of the universe are progressively translated into spatial orientations, geometric systems, and built environments. By distinguishing yet interrelating these modes, the analysis clarifies how cosmology shifts from narrative and metaphysical speculation to embodied practice and architectural form.

Cosmography, in this light, is not merely descriptive but performative: by mapping the world, it establishes hierarchies, centers, and axes that condition how space is inhabited and interpreted. Representation becomes an act of ordering.

Cosmograms

The term *cosmogram* derives from the Greek κόσμος (*kosmos*) and γράμμα (*gramma*), meaning "drawing," "mark," or "inscription." Cosmograms are objects, images, or spatial configurations created to express and make intelligible a particular understanding of the universe. As Tresch explains, cosmograms "define and express the universe" and may take many forms, ranging from creation narratives and ritual practices to diagrams, symbols, and scientific visualizations (Tresch, 2005).

Cosmograms function as material condensations of a worldview or *Weltanschauung*. They are not limited to diagrams but appear across media and scales, from drawings and artifacts to buildings, landscapes, and urban plans. Although their application to city planning has been debated (Smith, 2005), cosmograms have historically played a role in structuring both sacred and civic spaces.

Yantras and mandalas provide well-known examples of cosmograms associated with contemplative practices. Figure 7 presents two canonical instances: the *yin–yang* symbol, originating in Taoist philosophy in China (ca. third century BCE), and the *Śrī Yantra*, associated with Hindu traditions. Yantras (from the Sanskrit यन्त्र, meaning "instrument" or "diagram") and mandalas (from Sanskrit ⌧⌧⌧⌧⌧, meaning "circle") function as visual aids in meditation, facilitating concentration and the apprehension of abstract principles. As Khanna defines it, a yantra is a geometrical diagram composed of abstract symbols, often employed in tantric practices (Khanna, 2003). Eliade describes the yantra as a diagrammatic and geometrically simplified form of the mandala, whose structure "may be regarded as the linear paradigm of the mandala" (Eliade, 2009, p. 219).

Figure 7. Yin–yang symbol (left) and Śrī Yantra (right).

Another influential example is the Bakongo cosmogram (Figure 8), also known as the *dikenga* or *yowa* cross. This diagram consists of a circle intersected by a cross and an arrow indicating cyclical motion. It represents the interrelation of the physical world (*Ku Nseke*) and the spiritual world (*Ku Mpémba*), structured around the "four moments of the sun": dawn, noon, sunset, and midnight. These solar moments correspond to the four stages of human life—conception, birth, maturity, and death (Thompson & Cornet, 1981, p. 27). Here, cosmological structure and temporal experience coincide: the daily solar cycle becomes a diagram of human existence, revealing how cosmograms compress cosmic time into lived biography.

Figure 8. Diagram of the Bakongo cosmogram (dikenga or yowa cross).

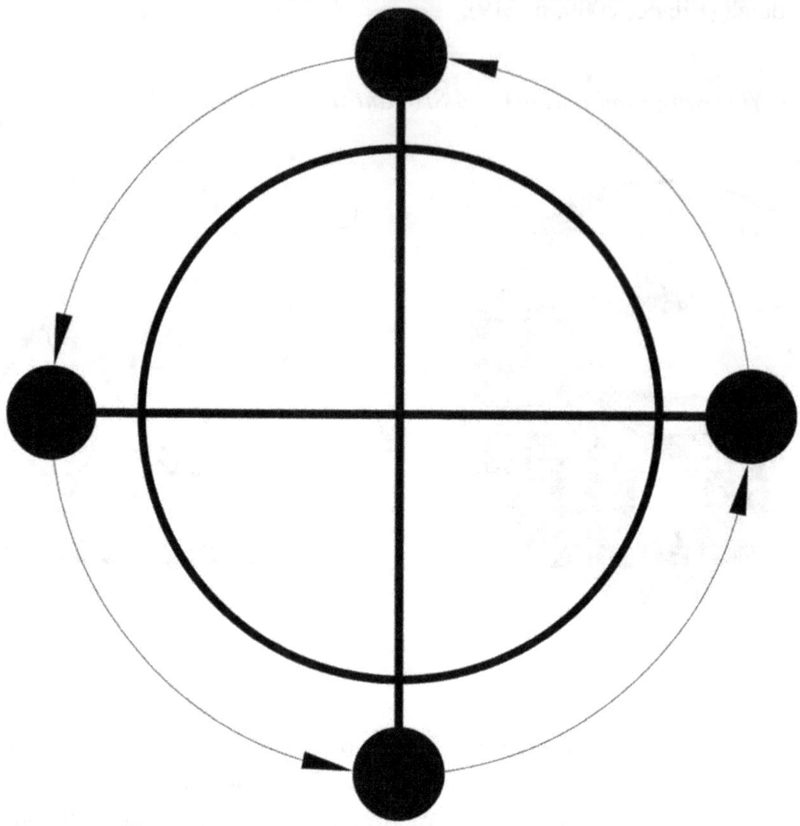

Cosmograms also operate as expressions of place identity and genius loci. The top of Figure 9 presents the cosmogram developed for S.U.N.F.A.R.M., based on its guiding vision and site-specific spatial logic. Two interlocking spirals symbolize expansion and contraction as metaphors for both cosmic processes and human life cycles, extending from birth toward death (Bertol, 2006; 2008; 2015). This cosmogram is inscribed not only in diagrams but also in built form, including a sculptural gate marking the entrance to the site (Figure 9, bottom).

Figure 9. S.U.N.F.A.R.M. cosmogram (top) becomes a sculptural entrance (bottom).

From Cosmological Thought to Spatial Practice

By following cosmology through its related expressions in cosmogony, cosmography, and cosmograms, this section has traced how conceptions of the universe are progressively translated into structured forms of orientation and representation. What begins as an account of origins (cosmogony) becomes a descriptive ordering of the heavens and the Earth (cosmography), and is ultimately inscribed in diagrams, artifacts, and spatial configurations (cosmograms).

Through this progression, Earth–Sky relations shift from metaphysical narrative to geometric articulation and finally to material enactment. Cosmological thought does not remain confined to theoretical speculation or visual imagery; it becomes operative through orientation, mapping, alignment, and construction. In this sense, built environments may be understood as material mediations of Earth–Sky Geometries—spaces in which celestial order, terrestrial position, and human perception are brought into relation.

The sections that follow examine how these symbolic and geometric structures inform concrete interventions on Earth. Through built alignments, mapped correspondences, and spatial frameworks, cosmological thought is transformed into experiential environments that situate human existence within a broader and measurable order of Earth–Sky relations. At this threshold, cosmology ceases to be solely an explanatory model of the universe and becomes an operative framework for spatial meaning. The built environment emerges as a site where cosmological abstraction, geometric order, and embodied perception converge—transforming the sky from distant object into structuring presence.

ORIENTING OURSELVES

The online *Etymology Dictionary* defines orientation as the "arrangement (of an object) to face east or any other specified direction," a noun of action derived from the verb *orient*. The sense "process of determining the points of the compass" is attested by 1868, followed shortly thereafter by the extended meaning of "determining one's mental bearings" (1870). The verb *orient* derives from the Latin *oriens*, meaning "east" or "rising," referring to the rising of the Sun (Harper, n.d.).

Although the term has acquired metaphorical meanings, it is used here in its literal and spatial sense: orientation as the act of establishing directional reference within a structured environment. Orientation therefore presupposes structure: a field of relations within which direction becomes meaningful. Without a structured Earth–Sky framework, direction would remain indeterminate and spatial experience unanchored.

Orientation is phenomenologically grounded in being-in-the-world: it is the act by which the observer situates the self in relation to space and to the universe, within Earth–Sky relations. Since antiquity, the Sun has provided the primary reference for orientation through its apparent daily and annual motions as observed from Earth. The cardinal directions derive from this solar geometry. East and west correspond to the Sun's rising and setting at the equinoxes; south in the northern hemisphere corresponds to the Sun's highest daily culmination (and conversely in the southern hemisphere); north is defined as the opposite direction. These directions are not arbitrary abstractions but arise from systematic observation of solar motion. The geometry of orientation is thus grounded in cyclical celestial phenomena translated into terrestrial reference.

In astronomy, the cardinal directions—east, west, north, and south—structure the plane of the celestial horizon. In geography, they define a horizontal plane tangent to the Earth's surface at the observer's location. Two additional reference points complete the local spatial framework: the zenith, directly above the observer along the vertical axis, and the nadir, directly below (Figure 10).

Figure 10. Observer-centered orientation system showing cardinal directions, zenith, and nadir within the local Earth–Sky framework.

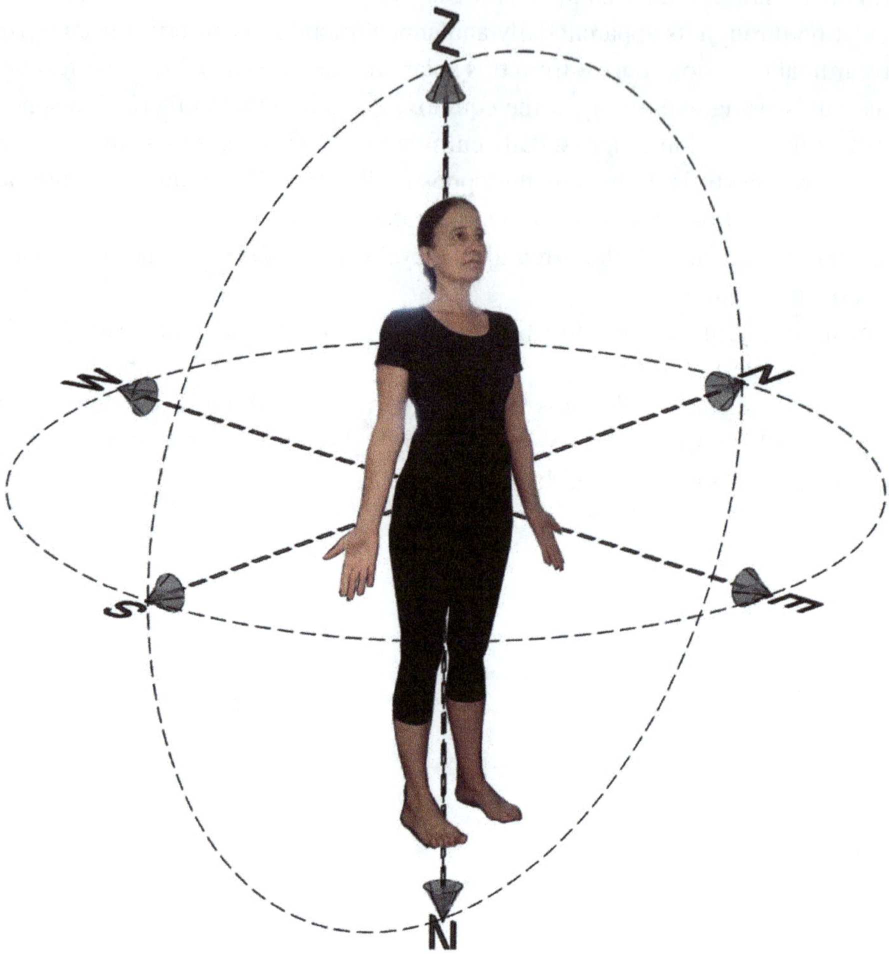

Geometrically, north and south define a meridional axis orthogonal to the east–west line. The resulting 360° circle of a polar diagram is conventionally measured clockwise from north. These geometric relations are embodied in orientation instruments such as the wind rose and the magnetic compass, which translate Earth–Sky alignments into practical systems of navigation and spatial organization. Such instruments do not create direction; they stabilize and externalize preexisting Earth–Sky relations. Orientation devices formalize what observation already reveals: that spatial order is anchored in celestial regularity.

MAPPING THE EARTH AND THE SKY

Maps are intrinsically linked to orientation, since they externalize directional knowledge into structured representation. A definition of map is therefore central to this inquiry. Within the architecture of cosmology, constructions may be understood as local, physical, and tangible representations of remote and intangible celestial phenomena—analogous, in an expanded sense, to acts of mapping. The aim of this section is to establish a conceptual framework in which buildings can operate as maps, not merely as graphic depictions or technical plans, but as spatial correspondences between terrestrial and celestial domains. In this expanded definition, mapping becomes an ontological act: it establishes correspondences between domains that differ in scale, materiality, and perceptibility. Architecture, when aligned to celestial phenomena, performs this mediating function spatially rather than graphically.

Maps are the products of mapping. In mathematics, mapping denotes "any prescribed way of assigning to each object in one set a particular object in another (or the same) set" (Encyclopædia Britannica, 2025). Understood in this formal sense, mapping establishes a relation between two domains through defined correspondence. Any geometric entity may be mapped onto another if structural relationships are preserved. From this perspective, the geometry of the celestial sphere—the apparent vault of the sky—may be mapped onto constructions on Earth, provided that a coherent system of correspondence is established between Earth and Sky. The validity of such mapping does not depend on visual resemblance but on structural congruence. What matters is the preservation of relational geometry rather than surface similarity.

The word *map* derives from the Latin *mappa*, meaning "napkin" or "cloth," referring to the material upon which early maps were drawn (Harper, n.d.). Maps are defined as "symbolic representations of selected characteristics of a place, usually drawn on a flat surface" (National Geographic Society, 2025). While this definition traditionally refers to geographic maps of the Earth, it extends conceptually to representations of extra-terrestrial space. Celestial maps—also known as uranography (from the Greek ο ρανός (*ouranos*), "sky," and γράφειν (*graphein*), "to write" or "to describe")—render the celestial sphere through systematic projection and symbolic notation (Figure 11).

Figure 11. A map of the celestial sphere: the constellation Ursa Minor (Bayer, 1603). Public domain.

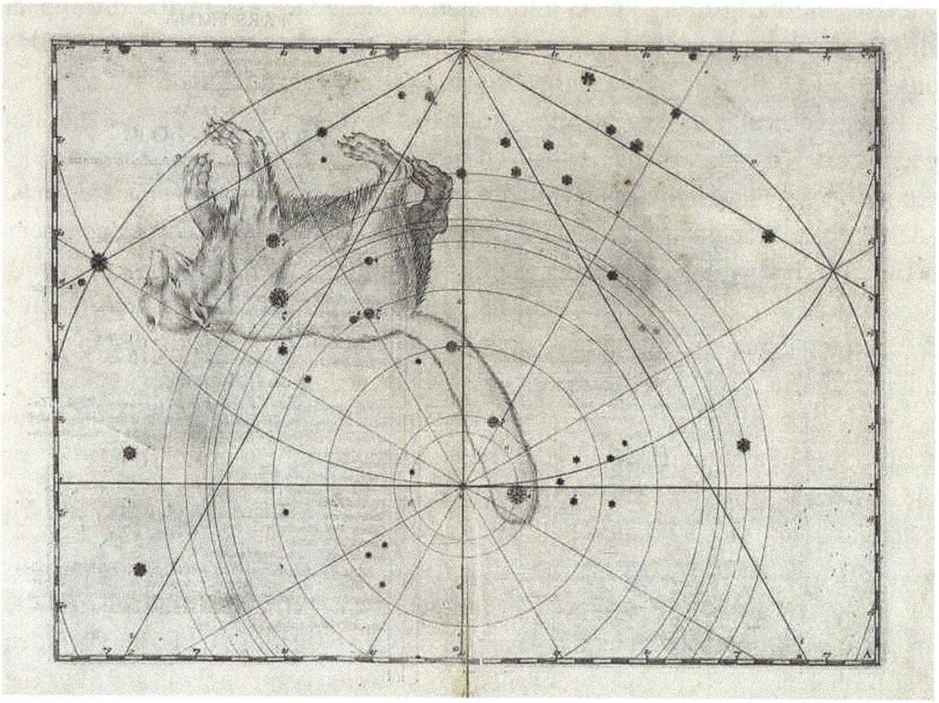

When mapping is understood as structured correspondence rather than graphic depiction alone, architecture itself may be interpreted as a spatial map. Built forms can encode directional systems, solar paths, stellar alignments, and cosmological diagrams within their geometry and orientation. In this sense, architectural space becomes a three-dimensional cartography of Earth–Sky relations: a constructed interface through which celestial order is materially registered and experientially apprehended. If architecture may operate as a spatial map, it must also confront the fundamental problem faced by cartography: how to translate one geometric domain into another. The Earth is approximately spherical, while architectural space is constructed within planar and rectilinear systems. Mapping therefore requires projection—an operation that transforms curved surfaces into planar representations through defined geometric rules. The study of projection is not merely technical; it formalizes the very act of correspondence between Earth and Sky, between celestial geometry and built form. For this reason, an examination of cartographic projection provides a critical conceptual bridge between mapping and architectural practice.

Symbolic representation and place are therefore defining characteristics of maps. Importantly, the "place" represented by a map need not be strictly physical. Mind maps, for example, depict conceptual or cognitive domains through diagrammatic structures, extending cartographic logic into abstract spatial fields.

Maps convey information about places and are traditionally characterized by scale, symbology, and orientation. In some cases, scale may be implicit rather than explicitly stated, provided the visual organization clearly communicates relational structure. For example, the 1843 topographic map of Rome by Venanzio Monaldini (Figure 12), derived from Giambattista Nolli's 1748 *Nuova Pianta di Roma*, employs a strong figure–ground representation to articulate urban morphology and orientation.

Figure 12. Venanzio Monaldini, Nuova Pianta di Roma Moderna (1843), derived from Giambattista Nolli's 1748 plan. Public domain.

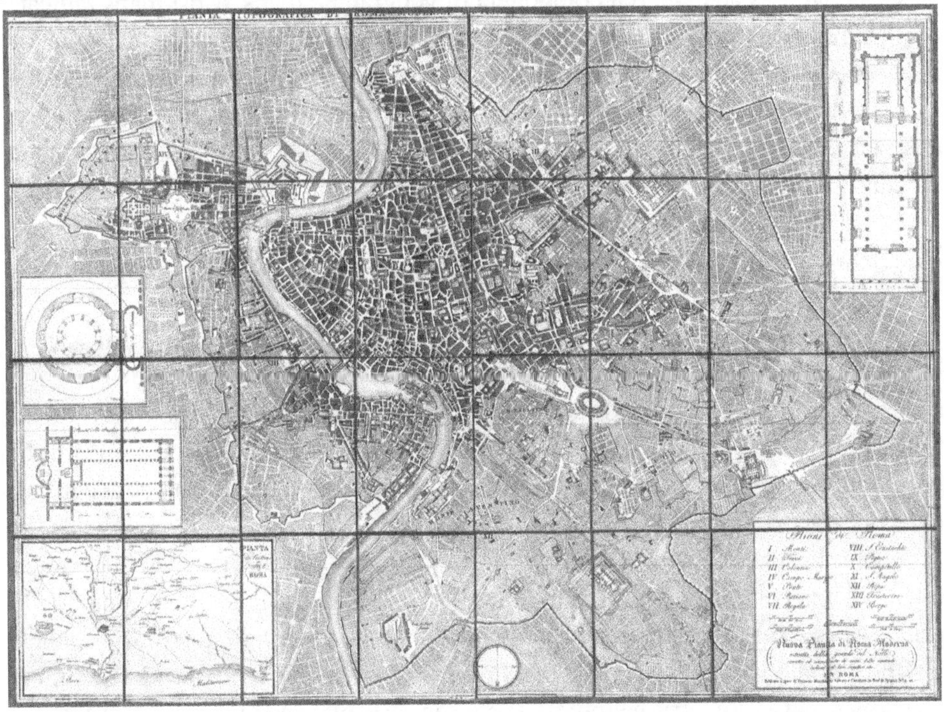

The full title reads: *Nuova Pianta di Roma Moderna estratta dalla grande del Nolli, corretta ed accresciuta de' nomi delle contrade indicati al loro rispettivo sito in Roma.* The map includes enlarged plans of significant monuments located beyond the main frame, such as the Mausoleum of Santa Costanza and the basilicas of San Paolo and San Lorenzo fuori le mura, thereby expanding the representational field beyond strict cartographic boundaries.

Scale is defined as the "relationship between the distances on the map and the corresponding distances on Earth" (National Geographic Society, 2025). It may be expressed graphically, verbally, or numerically. The representative fraction—for example, 1:1,000,000—is unitless and indicates proportional correspondence between map and territory. As the denominator increases, the level of visible detail decreases; thus, scale and detail are inversely related. This inverse relationship reveals an epistemic condition of mapping: as spatial reach expands, experiential immediacy diminishes. Scale mediates between abstraction and perceptual engagement.

Maps rely on symbols and legends to encode information. Cities, boundaries, roads, rivers, and topographic features are represented through standardized graphic conventions and color systems, typically explained in a legend or key (National Geographic Society, 2025). Through this symbolic system, maps translate spatial complexity into legible structure.

Orientation is an essential component of geographic maps. Most maps indicate north through a compass rose or directional arrow. A distinction must be made between true north (geographic north), aligned with the Earth's rotational axis, and magnetic north, toward which a magnetic compass needle points. The angular difference between them—known as magnetic declination—varies by geographic location and epoch and is sometimes noted on maps to ensure navigational accuracy.

Mapping Projections

Geographic maps rely on projection to represent the three-dimensional surface of the Earth on a two-dimensional medium. Locations on Earth are defined through spherical coordinates—latitude and longitude—which must be geometrically transformed into planar representations. This transformation involves projecting arcs of great and small circles from the sphere onto a developable surface or plane according to defined geometric rules.

Figure 13 schematically illustrates the projection of curved arcs onto orthogonal axes, where angular intervals are transformed into linear segments. A comparable principle governs the projection of geographic latitude and longitude from the spherical Earth onto a planar map surface. Projection is therefore not a mere technical adjustment but a geometric negotiation between curvature and planarity. It formalizes the translation of spherical order into rectilinear representation.

Figure 13. Schematic diagram illustrating projection of spherical arcs onto orthogonal planar axes.

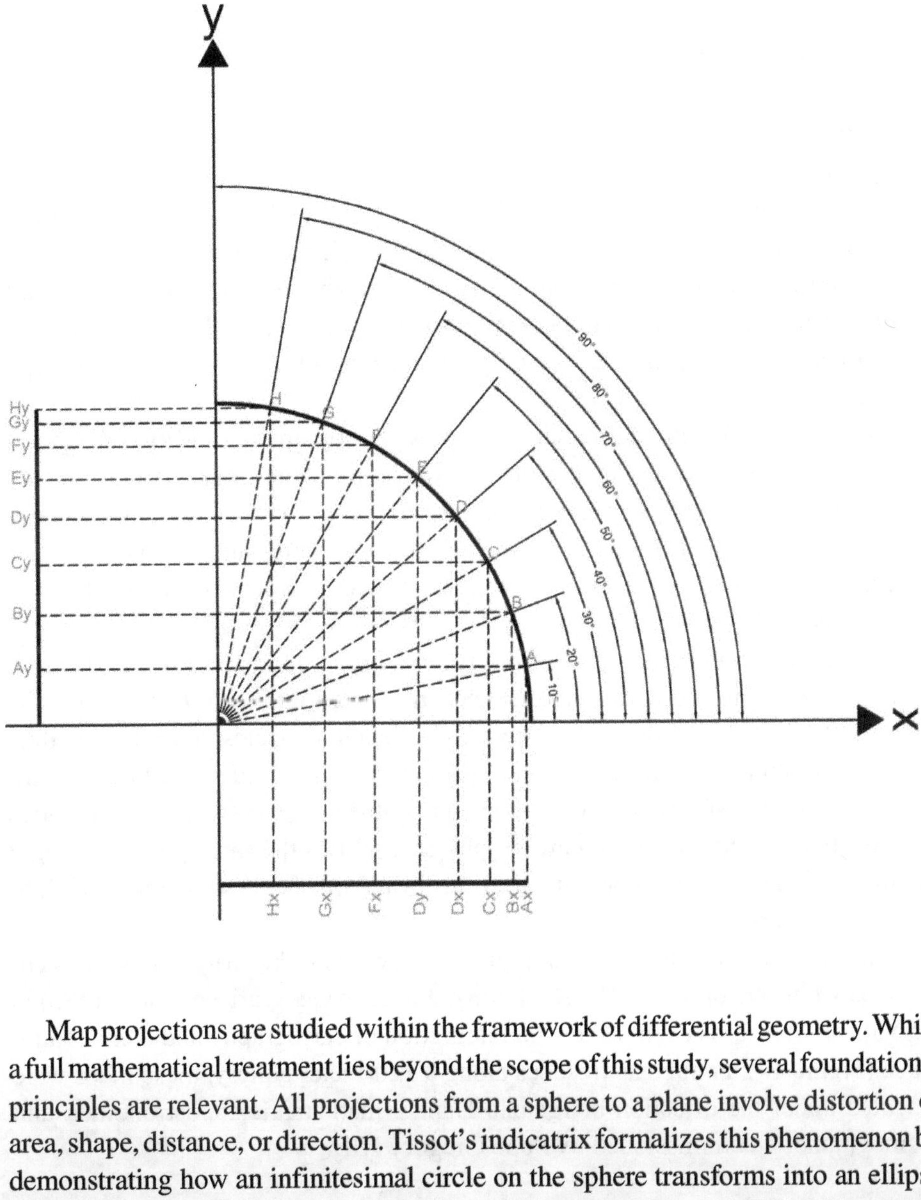

Map projections are studied within the framework of differential geometry. While a full mathematical treatment lies beyond the scope of this study, several foundational principles are relevant. All projections from a sphere to a plane involve distortion of area, shape, distance, or direction. Tissot's indicatrix formalizes this phenomenon by demonstrating how an infinitesimal circle on the sphere transforms into an ellipse on the map, thereby visualizing local distortion (Robinson et al., 1985). Distortion is not an error but an inevitability rooted in differential curvature. The cartographer's task is thus to determine which properties—area, angle, distance, or direction—are to be preserved and which may be sacrificed.

Projections are commonly classified according to the developable surface employed—conic, cylindrical, or azimuthal—or according to preserved properties, such as equal-area, conformal, equidistant, or azimuthal (Snyder, 1987; Lapaine & Frančula, 2022). Widely used projections include the stereographic (conformal), Mercator (cylindrical conformal), and Albers equal-area conic projections.

Celestial Pojections

The problem of projection is not limited to terrestrial cartography. Celestial mapping likewise requires the transformation of a spherical surface—the celestial sphere—into planar representation. Just as geographic coordinates (latitude and longitude) are projected from the Earth's surface onto a map, celestial coordinates (right ascension and declination, or altitude and azimuth) must be projected from the celestial sphere onto a plane.

Classical celestial charts frequently employ stereographic projection, in which points on the sphere are projected from one pole onto a plane tangent to the opposite pole. This projection preserves angles (conformal property), making it particularly suited for astronomical applications, including planispheres and astrolabes. Other projections, such as gnomonic or orthographic projections, serve different observational purposes by preserving great-circle paths or visual appearance.

Celestial projection thus formalizes the geometric mediation between Earth and Sky. It translates observed spherical configurations into measurable planar diagrams, enabling calculation, prediction, and orientation. Celestial projection thus reveals a structural symmetry between astronomy and architecture: both require the disciplined transformation of spherical phenomena into constructed geometric systems. In this sense, celestial cartography exemplifies the same projective logic that underlies architectural mapping: the transformation of curved cosmic order into constructed spatial systems.

Before the advent of digital technologies, maps were produced manually on parchment, paper, wood, or plaster. The Galleria delle Carte Geografiche (1580–1585) in the Vatican Museums exemplifies the integration of cartography and architectural space (Figure 14). Painted as frescoes along a 120-meter gallery, these maps were executed under the direction of Ignazio Danti—mathematician, astronomer, and cosmographer—commissioned by Pope Gregory XIII.

Figure 14. Galleria delle Carte Geografiche (1580–1585), Vatican Museums, Rome. Fresco cycle directed by Ignazio Danti.

Here, cartography transcends portable representation and becomes architectural environment: the map is no longer held in the hand but inhabited through bodily movement along the gallery axis. The architectural frame thus transforms geographic projection into spatial experience, prefiguring the notion of architecture as immersive cosmography. In this environment, the body becomes the moving reference point within a projected world. The gallery transforms cartographic abstraction into sequential spatial experience.

Digital Maps

The information age has fundamentally transformed both maps as products and mapmaking as a process. Digital cartography relies on computational methodologies, Geographic Information Systems (GIS), and Global Positioning Systems (GPS) to collect, integrate, and visualize spatial data (Figure 22). Satellite imagery, aerial photography, geodetic measurements, and non-visual datasets are organized within spatial databases and rendered through interactive interfaces.

Equally significant is the shift in medium. Whereas traditional maps were fixed representations printed on paper or other material supports, digital maps are dynamic, continuously updated, and accessible through computers, tablets, and smartphones. Global mapping platforms—including Google Maps, Microsoft Bing, Waze, Apple Maps, and OpenStreetMap (OSM)—provide near-instant access to multi-scalar geographic information. Unlike proprietary platforms, OpenStreetMap operates as a collaborative, open-source system based on user-contributed data.

Interactivity distinguishes digital maps from their analog predecessors. Users may zoom seamlessly across scales, shifting from continental to urban to architectural resolution. While the mathematical definition of scale remains unchanged—the proportional relationship between map distance and terrestrial distance—digital implementation is achieved through continuous zoom operations rather than discrete printed scales. The sequence of Figures 15–20 illustrates this multi-scalar capacity, presenting digital maps centered on the Campidoglio (City Hall) in Rome across successive orders of magnitude. The increasing zoom level follows successive powers of ten.

Figure 15. Digital map centered on Rome at approximately continental scale (~1,000 miles). Google My Maps interface. Screenshot by author.

Figure 16. Digital map centered on Rome at regional scale (~100 miles). Google My Maps interface. Screenshot by author.

Figure 17. Digital map centered on Rome at metropolitan scale (~10 miles). Google My Maps interface. Screenshot by author.

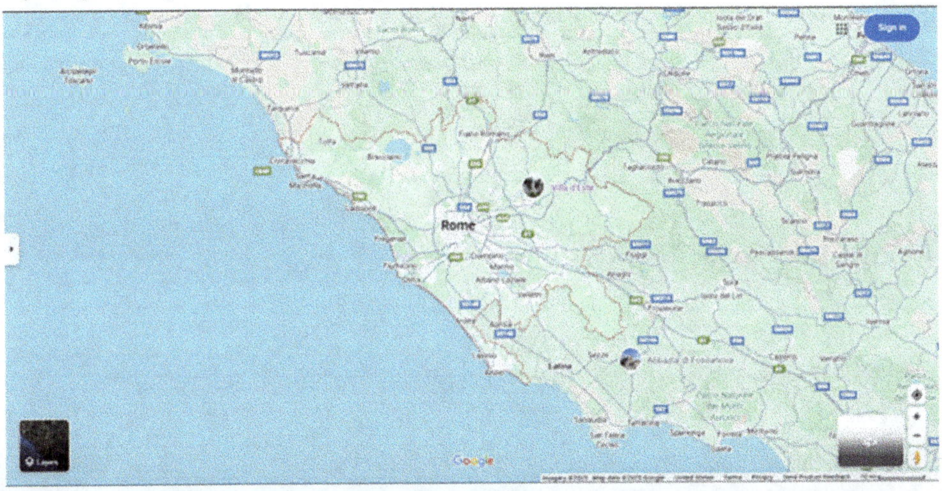

Figure 18. Digital map centered on Rome at urban scale (~1 mile). Google My Maps interface. Screenshot by author.

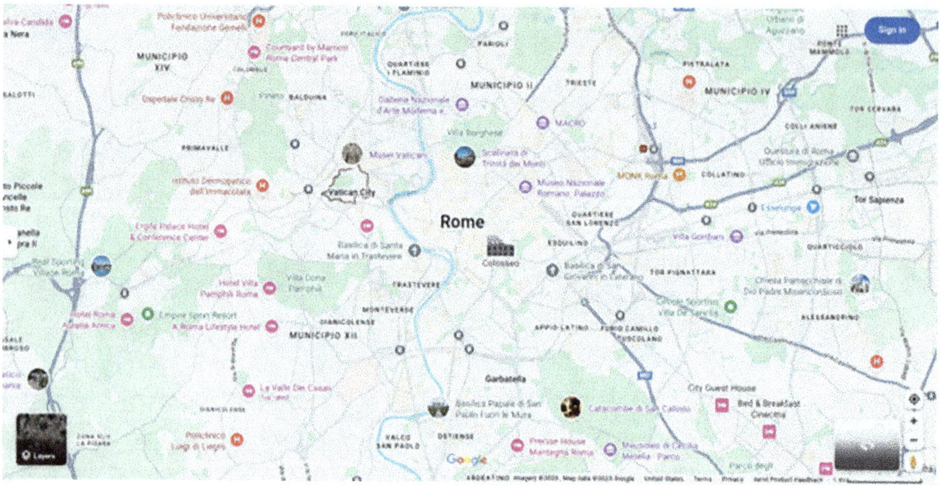

Figure 19. Digital map centered on Rome at neighborhood scale (~1,000 feet). Google My Maps interface. Screenshot by author.

Figure 20. Digital map centered on Rome at architectural scale (~100 feet). Google My Maps interface. Screenshot by author.

Beyond scale, digital cartography introduces real-time positional awareness. The integration of GPS allows the user's geographic coordinates to be embedded directly within the map interface. Figure 23 shows the author's location at the time of writing (August 30, 2025, 9:00 pm EDT). This feature—now commonplace in everyday navigation—represents a technologically mediated form of spatial self-location. From a phenomenological perspective, digital geolocation literalizes being-in-the-world: the user appears simultaneously as observer and mapped entity, situated "here" in space and "now" in time. The user is no longer merely orienting within a mapped world but is algorithmically inscribed within it. Spatial self-location becomes continuous, automated, and globally referenced.

Figure 21. Digital map centered on Rome (~100 feet), displaying aerial imagery. Google My Maps interface. Screenshot by author.

Figure 22. Digital map centered on Rome (~1,000 feet), displaying transit overlay. Google My Maps interface. Screenshot by author.

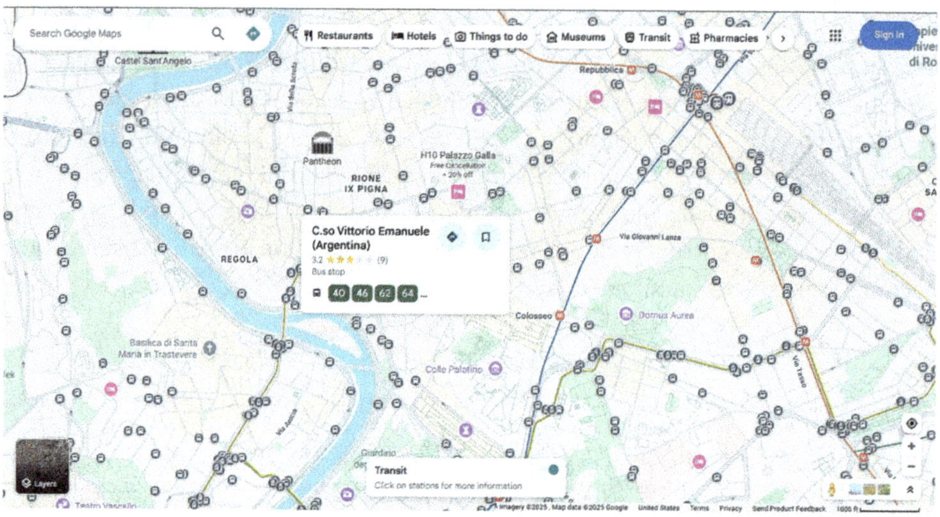

Figure 23. Digital map displaying the author's geolocated position at the time of writing, 2025.08.30 9:00 pm EDT. Google My Maps interface. Screenshot by author.

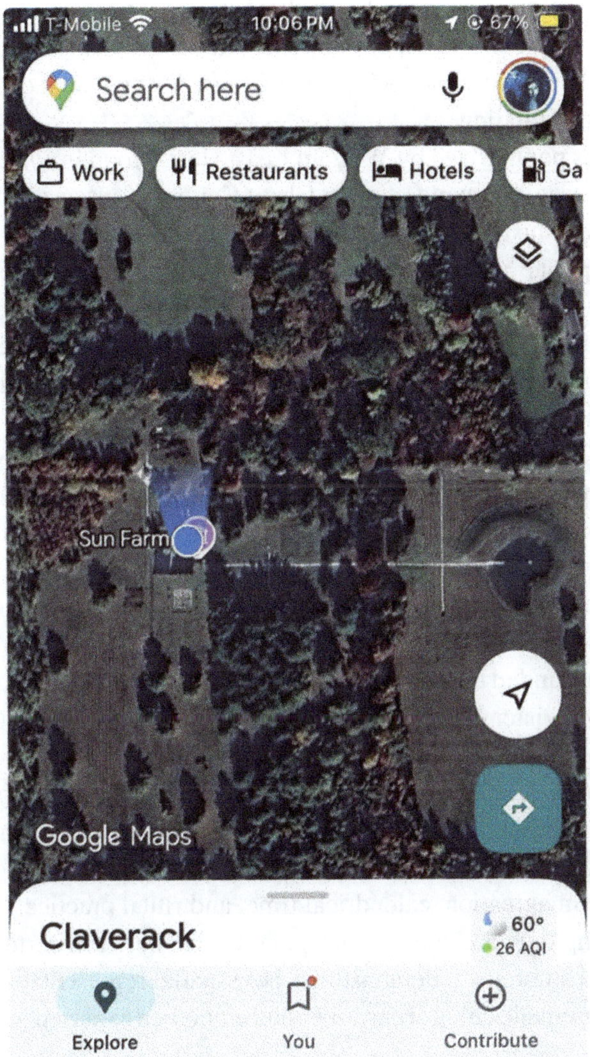

Digital mapping thus extends the logic of cartography from representation to participation. The map is no longer a static object external to the observer but an interactive environment in which orientation, measurement, and embodied presence converge. Digital cartography thus extends the projective logic of mapping into real-time feedback systems, collapsing the temporal distance between representation and experience.

Deeps Maps

Maps represent regions of the Earth through layered forms of information, including not only physical and geographic characteristics but also cultural, historical, and experiential dimensions associated with place. The concept of the "deep map" extends this understanding.

The term was popularized by William Least Heat-Moon in *PrairyErth (A Deep Map)* (1991), a sustained exploration of Chase County, Kansas, in which geography is interwoven with history, ecology, social memory, and narrative. In this expanded sense, a deep map is not merely a spatial depiction but a thickened account of place—one that integrates quantitative data with qualitative, lived experience. Deep mapping resists reduction to coordinate systems alone. It acknowledges that place is constituted not only by measurable geometry but by memory, narrative, and lived duration.

As Roberts (2016, p. 5) notes, a deep map may evolve into a complex dataset "in which a near limitless range and quantity of sources can be included, interrogated, manipulated, archived, analyzed, and read," potentially becoming "a big data-driven, totalizing model." Yet deep mapping also retains anthropological, embodied, and performative dimensions. It is not static but accumulative and iterative, developing as places themselves evolve.

Within this expanded framework, celestial maps may be understood as deep maps of the sky. They register not only spatial positions of celestial objects but also temporal coordinates, observational location, cultural interpretation, and cosmological meaning. A star chart, like a terrestrial deep map, encodes layers of astronomical measurement, mythological association, calendrical structure, and navigational practice. The sky, like the Earth, is experienced as stratified: empirical observation overlays mythic imagination, calendrical time, and ritual practice.

Deep mapping thus provides a conceptual bridge between cartography and the architecture of cosmology. It demonstrates how spatial representation can integrate empirical measurement, cultural narrative, and embodied experience within a unified system of place.

MAPPING ARCHITECTURE OF COSMOLOGY: URANIBORG

Maps and buildings converge in one of the most compelling historical examples of the architecture of cosmology: Uraniborg, the astronomical observatory established by Tycho Brahe (1546–1601). Constructed between 1576 and 1580 on the island of Hven (now Ven, Sweden), Uraniborg represents the last major observatory designed to operate exclusively with naked-eye instruments.

Brahe documented the instruments and architectural arrangements in *Astronomiæ Instauratæ Mechanica* (1602). The complex was later illustrated in detail in Joan Blaeu's *Atlas Maior* (1662–1665), which includes both cartographic material and architectural engravings of Uraniborg (Figures 24-29) and its subterranean companion, Stjerneborg (Figure 30).

Figure 24. Uraniborg site plan (Blaeu, 1665). From Atlas Maior, Vol. 1. Reproduced with the permission of the National Library of Scotland.

Figure 25. Uraniborg elevation view (Blaeu, 1665). From Atlas Maior, Vol. 1. Reproduced with the permission of the National Library of Scotland.

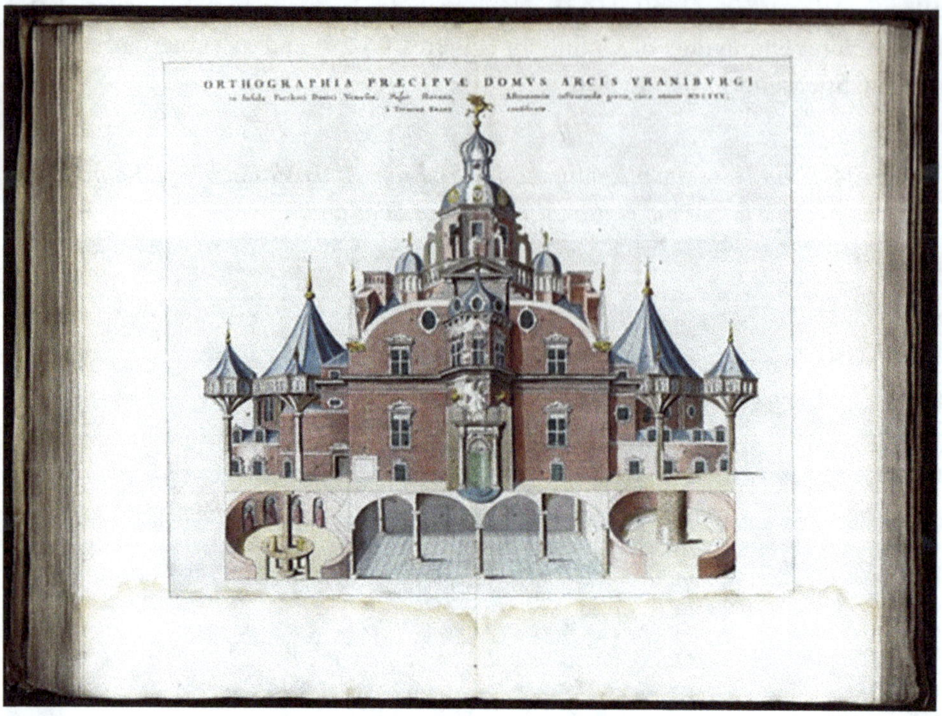

Figure 26. Uraniborg floor plan (Blaeu, 1665). From Atlas Maior, Vol. 1. Reproduced with the permission of the National Library of Scotland.

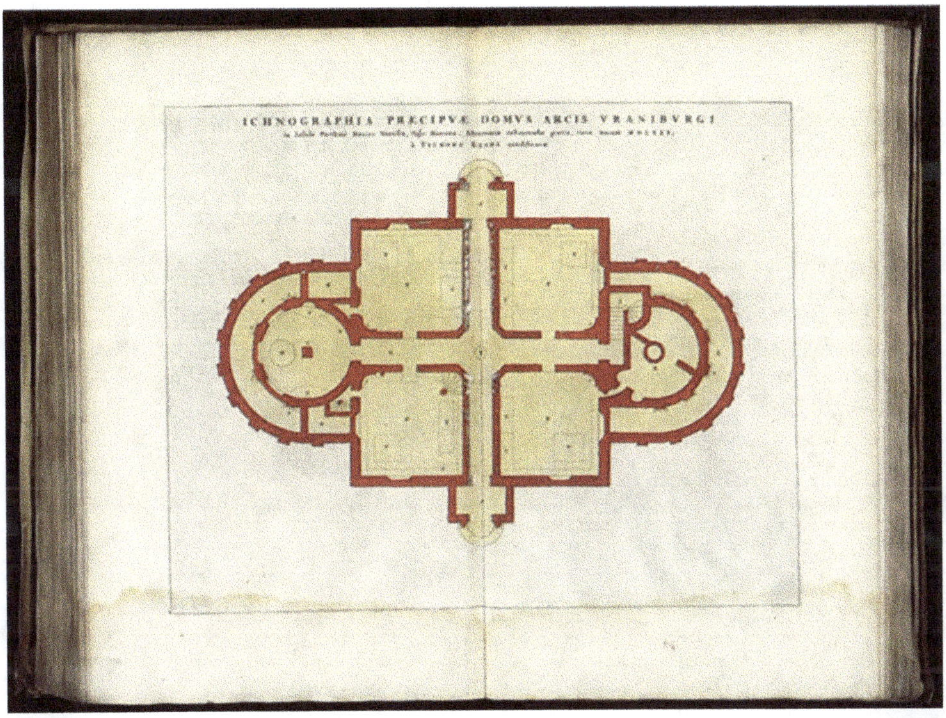

Figure 27. Uraniborg instruments: Parallacticum (Blaeu, 1665). From Atlas Maior, Vol. 1. Reproduced with the permission of the National Library of Scotland.

Figure 28. Uraniborg instruments: Sextans trigonicus (Blaeu, 1665). From Atlas Maior, Vol. 1. Reproduced with the permission of the National Library of Scotland.

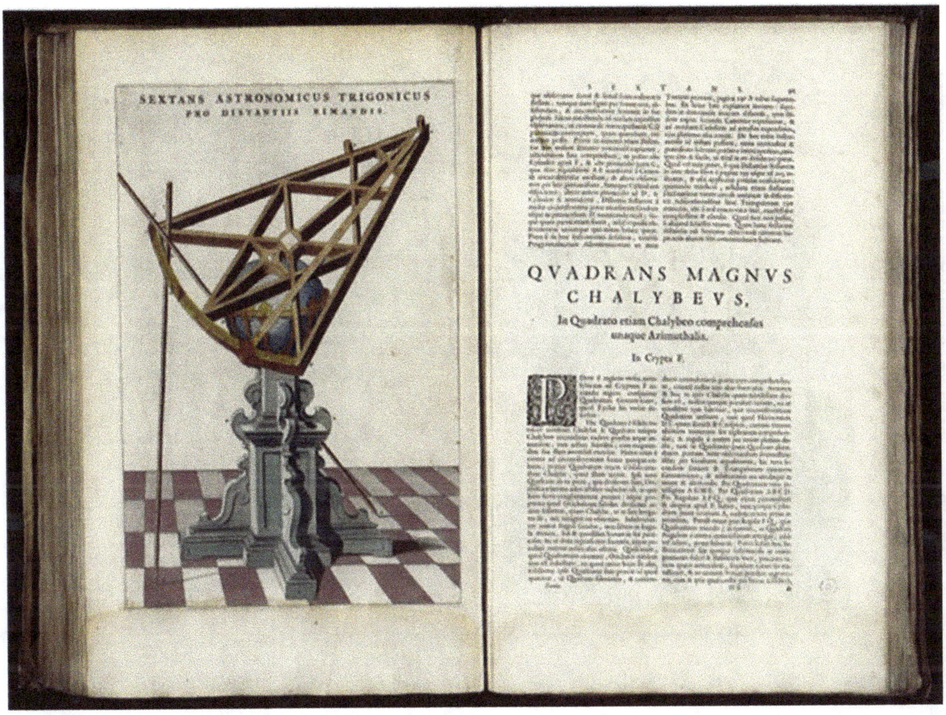

Figure 29. Stjerneborg, perspective view (Blaeu, 1665). From Atlas Maior, Vol. 1. Reproduced with the permission of the National Library of Scotland.

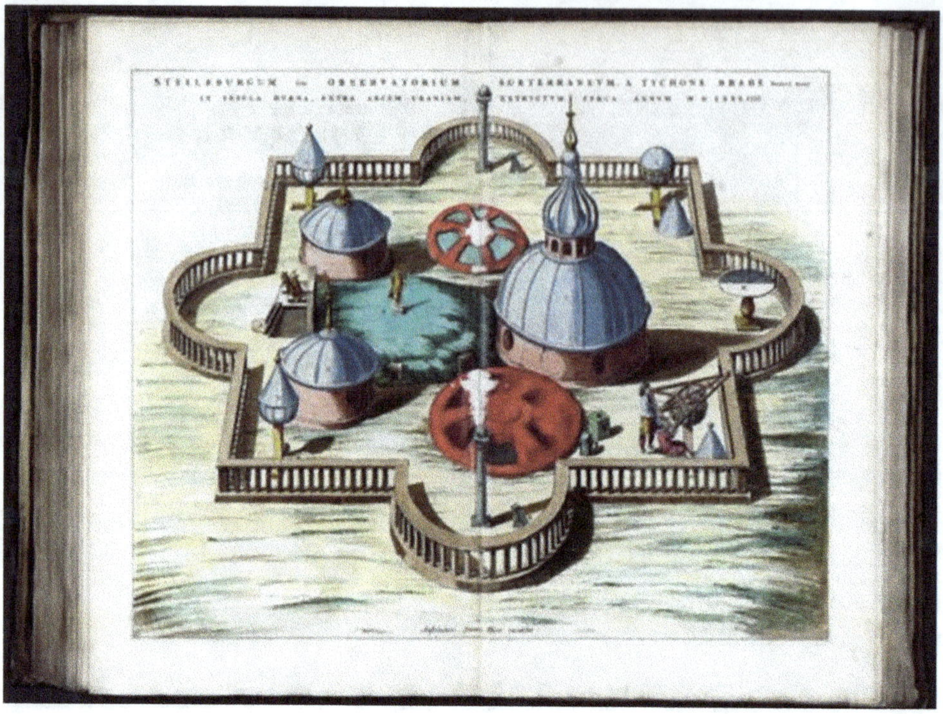

Figure 30. Stjerneborg floor plan (Blaeu, 1665). From Atlas Maior, Vol. 1. Reproduced with the permission of the National Library of Scotland.

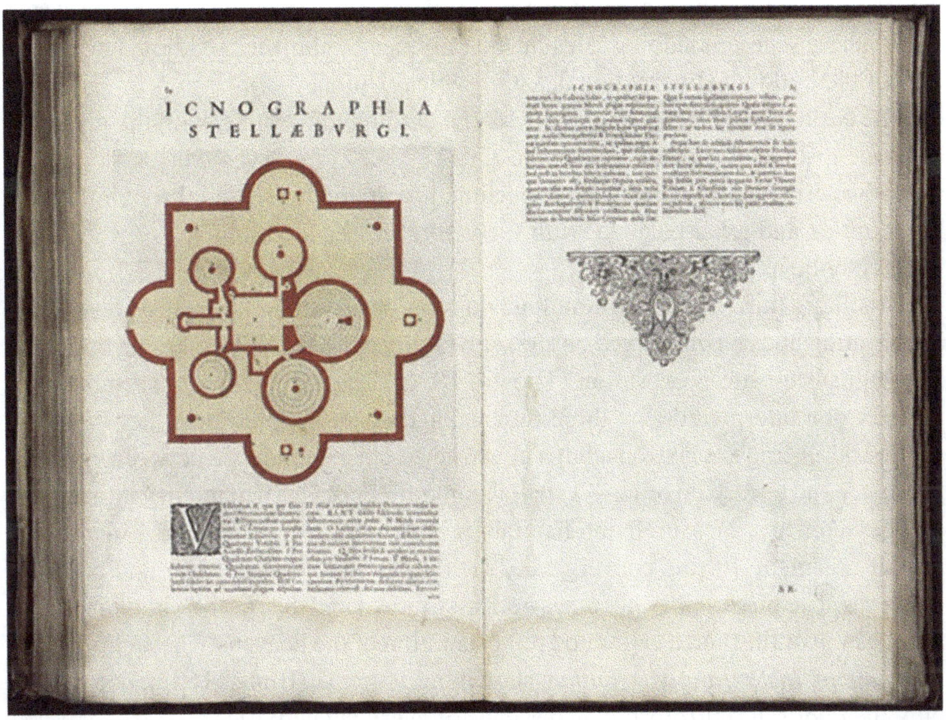

Located at approximately 55°54′28″ N, 12°41′48″ E, Uraniborg measured approximately 16 m × 16 m in plan, with a central tower rising to 19 m and two smaller round towers (6 m diameter) to the north and south (International Astronomical Union, n.d.). The building stood within a rigorously geometrical Renaissance garden organized according to axial symmetry. The site also included a printing press and facilities for paper production, enabling the integration of observation, calculation, and publication within a single spatial system. Knowledge production was therefore spatially consolidated: measurement, computation, inscription, and dissemination were architecturally unified. Uraniborg functioned not merely as an observatory but as an epistemic machine embedded in built form.

In 1584, Brahe constructed Stjerneborg (*Stellæburgus*), approximately 80 meters south of the main structure. Unlike Uraniborg, Stjerneborg housed partially subterranean observing chambers designed to reduce wind interference and improve measurement precision. Together, the two complexes supported research in astronomy, alchemy, chemistry, and meteorology. The coexistence of these disciplines within one site reflects the late Renaissance conception of cosmic inquiry as unified rather than fragmented. The architecture materialized an integrated cosmological worldview.

Several of Brahe's instruments—including the triquetrum, sextant, armillary sphere, and mural quadrant—were directly integrated into the architectural fabric. The large mural quadrant affixed to a north–south wall measured stellar altitudes at meridian passage, embedding celestial measurement within the meridional structural axis of the building (International Astronomical Union, n.d.).

The entire complex was aligned with the cardinal directions, with the principal entrance oriented eastward. Westman (2011, p. 237) observes that its floor plan embodied Vitruvian *symmetria*: proportional harmony expressed through Pythagorean ratios and axial balance, with the central tower functioning as a geometric and symbolic axis.

Uraniborg thus functioned simultaneously as observatory, laboratory, residence, and printing house. Brahe lived on site with family and assistants, merging domestic life with systematic observation (Thoren, 1990). The architectural ensemble may therefore be interpreted as a three-dimensional cosmographic map: not merely a site for observing the heavens, but a constructed correspondence between celestial geometry and terrestrial order—a spatial embodiment of human inquiry into remote space. Geometry, measurement, habitation, and publication converged in a single spatial organism. In this convergence, the distinction between map and building dissolves. The observatory becomes both instrument and diagram—simultaneously a place of dwelling and a calibrated representation of the heavens. The etching from *Astronomiæ Instauratæ Mechanica*, shown in Figure 31 (Brahe, 1602), depicts not only the *quadrans muralis* but also scenes of everyday activities.

Figure 31. Uraniborg mural quadrant. From Astronomiæ instauratæ mechanica (Brahe, 1602).

Interpreted through the framework of Earth–Sky Geometries, Uraniborg formalizes the observer-dependent relation between celestial motion and terrestrial construction. The cardinal alignment of the plan, the meridional wall supporting the mural quadrant, and the calibrated placement of instruments transform astronomical coordinates into architectural axes. Observation is not abstracted from space but embedded within it. The observatory thus becomes a constructed interface in which spherical celestial geometry is translated into orthogonal terrestrial structure—an operational synthesis of perception, measurement, and built form. Uraniborg demonstrates that architecture can function as a projective device: it receives celestial curvature through observation and rearticulates it through rectilinear geometry. Projection here is not graphic but tectonic.

Brahe departed Hven in 1597 after losing royal patronage under King Christian IV. The observatory gradually fell into disrepair during the seventeenth century. Archaeological excavations and partial reconstructions of the garden began in the twentieth century, with major restoration efforts initiated in 1985. Contemporary satellite imagery (Figures 32) reveals the restored axial symmetry and cardinal alignment of the site. Even when reduced to archaeological traces, the geometric logic remains legible. The persistence of axial order underscores that cosmological alignment is not ornamental but structural.

Figure 32. Site of Uraniborg and Stjerneborg, satellite view, 2024.09.26. Google Maps imagery.

CONCLUDING THOUGHTS

The inquiries presented in this chapter began with the phenomenological condition of being here and now in astronomical observation and traced its development into cosmological theory and spatial practice. Geometry emerged as the fundamental framework through which the universe is rendered intelligible. Observations of remote celestial space give rise to structured systems of harmony, proportion, and order that extend beyond empirical measurement into cultural and symbolic domains.

Earth–Sky Geometries transcend observable data to embody social, cultural, and often spiritual constructs associated with humanity's search for meaning within cosmic order. Geometry functions not only as analytic tool but as aesthetic and symbolic medium, transforming diagrams into cosmograms—articulations of the relationship between culture and cosmos. The sky, with its continuously shifting configurations, becomes both observational field and narrative stage. Its apparent motions generate measurable cycles, while its symbolic interpretations generate cultural meaning. Cosmology operates at the intersection of these two dimensions.

Mapping establishes correspondences between domains. Whether expressed in two-dimensional charts, immersive galleries, or architectural complexes such as Uraniborg, mapping translates remote and intangible realities into coherent local spatial systems. Architecture, in this expanded sense, becomes a three-dimensional cartography: a material interface through which celestial order is registered, measured, and inhabited. The act of inhabiting such spaces transforms abstract geometry into lived orientation. Measurement becomes experience.

The trajectory traced in this chapter—from astronomical observation to cosmological theory, from symbolic representation to spatial embodiment—demonstrates that cosmology is not solely speculative or scientific. It is also a spatial practice enacted through geometry, ritual, and construction. Through cosmogonies, cosmographies, cosmograms, maps, and mapped architectures, cosmological thought becomes operative in the human-made world, shaping orientation, settlement, and the construction of meaning in place. Cosmology, therefore, is not imposed upon space from without; it emerges through the reciprocal calibration of perception, geometry, and construction.

If Chapter 3 established the geometric structures of Earth–Sky relations and the present chapter examined their cartographic and architectural embodiment, the following chapter turns to their temporal dimension. By shifting focus from representational frameworks to architectural implementation, Chapter 5 concretizes the geometry of space and time through specific sites and alignments to celestial events. Spatial orientation alone does not constitute cosmology; it must be integrated with the measurement of duration, recurrence, and celestial cycles. The architectural cosmogram therefore operates not only as a spatial map but also as a temporal instrument—calibrating built form to the rhythms of the heavens..

REFERENCES

Blaeu, J. (1665). *Atlas Maior, sive cosmographia Blaviana* (Vol. 1). Joan Blaeu.

Brahe, T. (1602). *Astronomiæ instauratæ mechanica*. Noribergae [Nuremberg], Germany: Levinum Hulsium.

Cosgrove, D. E. (2007). Images of Renaissance cosmography, 1450–1650. In Woodward, D. (Ed.), *The history of cartography* (Vol. 3, pp. 55–98). University of Chicago Press.

Heat-Moon, W. L. (1991). *PrairyErth (A Deep Map)*. Houghton Mifflin.

Lapaine, M., & Frančula, N. (2022). Map projections classification. *Geographies*, *2*(2), 274–285. DOI: 10.3390/geographies2020019

National Library of Scotland. (n.d.). *Atlas Maior, sive cosmographia Blaviana* (Vol. 1) [Digital reproduction of Blaeu, 1665]. Edinburgh, UK: Author. Retrieved from https://maps.nls.uk/atlas/blaeu-maior/vol/1

Roberts, L. (2016). Deep mapping and spatial anthropology. *Humanities (Washington)*, *5*(1), 5.

Robinson, A. H., Sale, R., Morrison, J., & Muehrcke, P. (1985). *Elements of cartography* (5th ed.). John Wiley & Sons.

Smith, M. E. (2005). Did the Maya build architectural cosmograms? *Latin American Antiquity*, *16*(2), 217–224.

Snyder, J. P. (1987). *Map projections: A working manual*. Washington, DC: United States Government Printing Office.

Thompson, R. F., & Cornet, J. (1981). *The four moments of the sun*. National Gallery of Art.

Thoren, V. E. (1990). *The lord of Uraniborg: A biography of Tycho Brahe*. Cambridge University Press.

Tresch, J. (2005). Cosmogram. In Ohanian, M., & Royoux, J. C. (Eds.), *Cosmograms* (pp. 67–76). Lukas & Sternberg.

ADDITIONAL READING

Apian, P. (1524). *Cosmographicus liber Petri Apiani mathematici studiose collectus*.

Baldasso, R. (2022). Cartography in the Renaissance. In Sgarbi, M. (Ed.), *Encyclopedia of Renaissance philosophy*. Springer., DOI: 10.1007/978-3-319-02848-4_921-1

Bayer, J. (1603). *Uranometria*. Christophorus Mangus.

Bertol, D. (2006). Framing the land and sky: Art meets cosmology in a sustainable environment. *Leonardo*, *39*(2), 125–130.

Bertol, D. (2008). Architettura e cosmologia: Percezioni del cielo sulla terra. In *Matematica e cultura 2008* (pp. 285–294). Springer Milan.

Bertol, D. (2015). Sun Farm: Body, self, universe. *PAJ a Journal of Performance and Art*, *37*(1), 36–47.

Bertol, D. (2017). *Form mind body space time – The geometry of human movement* (Doctoral dissertation, Deakin University). Geelong, Australia: Deakin University.

Brown, N. (2006). *Engaging the cosmos: Astronomy, philosophy and faith*. Liverpool University Press.

Cellarius, A. (1708). *Harmonia macrocosmica*. Gerardum Valk & Petrum Schenk.

Copernicus, N. (1566). De revolutionibus orbium coelestium []. Basileae] [Basel] [, Switzerland: Henricpetrina.]. *Libri*, VI.

Disney, M. J. (2007). Modern cosmology: Science or folktale? *American Scientist*, *95*(5), 383–387.

Einstein, A. (1920). *Relativity: The special and general theory* (Lawson, R. W., Trans.). Henry Holt.

Eliade, M. (1951). *Images and symbols*. Harvill Press.

Eliade, M. (1961). *The sacred and the profane*. Harper & Brothers.

Evans, J. (1998). *The history and practice of ancient astronomy*. Oxford University Press.

Fazio, G. G. (2018). *The encyclopedia of cosmology*. World Scientific.

Flammarion, C. (1888). *L'atmosphère: Météorologie populaire*. Librairie Hachette.

Galilei, G. (1623). *Il saggiatore*. Giacomo Mascardi.

Greenstein, G. (2013). *Understanding the universe*. Cambridge University Press.

Hawking, S. (2009). *A brief history of time*. Random House.

Heath, T. L. (1991). *Greek astronomy*. Courier Corporation.

Hilbert, D., & Cohn-Vossen, S. (1952). *Geometry and the imagination*. Chelsea Publishing.

Kepler, J. (1596). *Mysterium cosmographicum*. Georg Gruppenbach.

Khanna, M. (2003). *Yantra: The tantric symbol of cosmic unity*. Inner Traditions.

Krupp, E. C. (1983). *Echoes of the ancient skies*. Dover Publications.

Lefebvre, H. (1991). *The production of space*. Blackwell.

Penprase, B. E. (2011). *The power of stars*. Springer.

Pomarède, D., Courtois, H. M., Hoffman, Y., & Tully, R. B. (2017). Cosmography and data visualization. *Publications of the Astronomical Society of the Pacific, 129*(975), 058002.

Rajawat, D. S. (1980). *Astronomical observatory of Jaipur*. Delta Publications.

Roos, M. (2003). *Introduction to cosmology* (3rd ed.). John Wiley & Sons.

Tresch, J. (2007). Technological world-pictures: Cosmic things and cosmograms. *Isis, 98*(1), 84–99.

Tresch, J. (2010). Even the tools will be free. In Aubin, D. (Eds.), *The heavens on earth* (pp. 253–285). Duke University Press.

Tuan, Y. F. (1974). *Topophilia*. Prentice-Hall.

Tuan, Y. F. (1977). *Space and place*. University of Minnesota Press.

Westman, R. S. (2011). *The Copernican question*. University of California Press.

Chapter 5
Time Architecture

ABSTRACT

This chapter reflects on time as embodied through Earth–sky geometries and expands upon the introduction of astronomical time presented in Chapter 3. It also introduces an alternative approach to time measurement drawn from an Eastern tradition, specifically Hindu astronomy. The chapter discusses how the Sun has served, for millennia, as the primary reference for measuring time and how solar cycles have been materially embodied in the architecture of cosmology. The Jantar Mantar observatories in India are presented as a case study, and gnomonics is introduced through Vitruvius' treatise De Architectura.

DOI: 10.4018/979-8-3373-3399-1.ch005

Figure 1. Astronomical time represented as a spiral, symbolizing cyclical recurrence within linear progression.

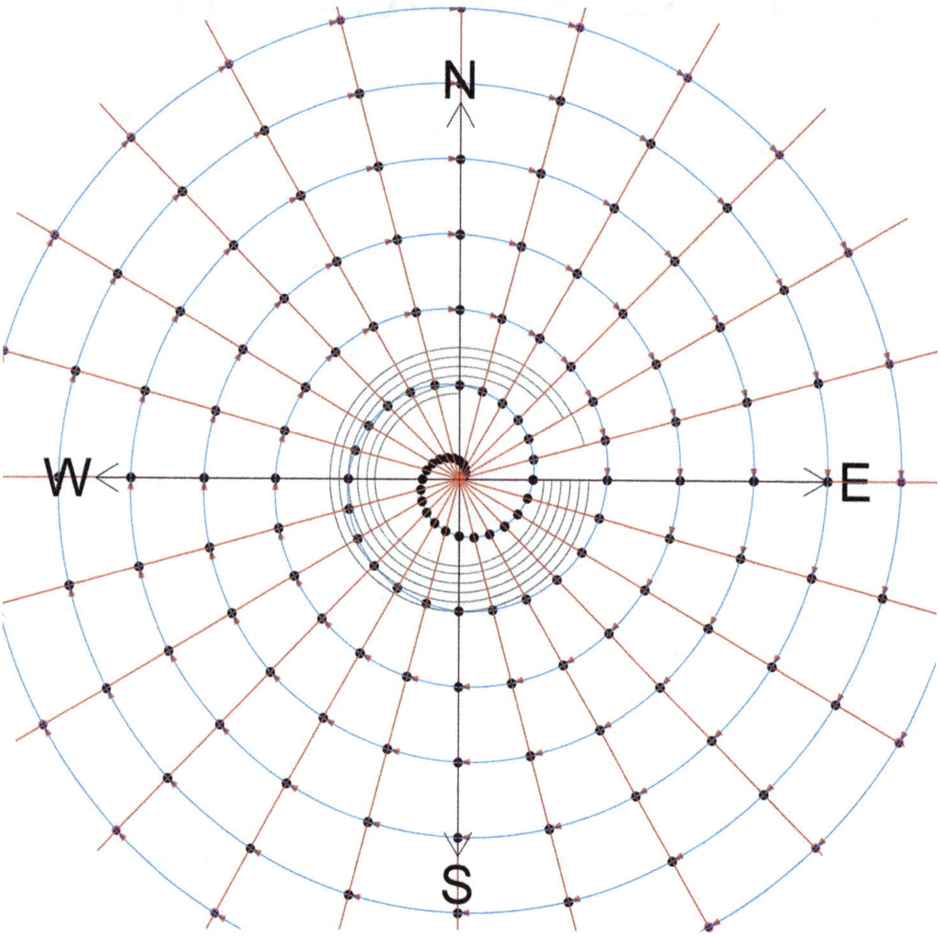

"WHAT IS TIME?" AND "WHAT TIME IS IT?"

The title of this section outlines two fundamental and complementary inquiries into time: its nature and its measurement. The question *"What is time?"* concerns the essence of time as it relates to reality and the human condition and has been the object of philosophical investigation for thousands of years. The question *"What time is it?"* refers instead to time as measurement, a practical necessity embedded in everyday human life.

Responses to these questions draw from fields of knowledge that have traditionally been considered separate or even antithetical, yet which can be integrated within a holistic framework. Without entering into an extended philosophical discussion—which is not the aim of this inquiry—an intuitive and comprehensive definition is provided by the *Oxford English Dictionary*, which defines time as "the indefinite continued progress of existence and events in the past, present, and future regarded as a whole" (Oxford Dictionaries, 2012).

Within the phenomenological framework adopted in this study, *"What is time?"* can be approached as lived experience. Time became a central topic in phenomenology with Edmund Husserl, who devoted a series of lectures and manuscripts to its exploration, later collected in *On the Phenomenology of the Consciousness of Internal Time* (1893–1917) (Husserl, 1991). As the title suggests, for Husserl time is not separate from consciousness but coincident with awareness itself. In contrast to the Newtonian conception of time as a neutral container, phenomenological time is continuous and inseparable from conscious experience (Brough et al., n.d.). Time, in this view, is not an external parameter through which events pass, but the very condition through which experience unfolds. Consciousness does not occur "in" time; it is temporally structured.

Maurice Merleau-Ponty further reinforces the inseparability of subject and time:

> To analyse time is not to follow out the consequences of a pre-established conception of subjectivity, it is to gain access, through time, to its concrete structure. If we succeed in understanding the subject, it will not be in its pure form, but by seeking it at the intersection of its dimensions. We need, therefore, to consider time itself, and it is by following through its internal dialectic that we shall be led to revise our idea of the subject. (Merleau-Ponty, 1962, p. 77)

Time as lived experience is not antithetical to scientific time. Rather, the integrated system of a perceiving subject and an observed phenomenon connects time and space within a phenomenological continuum. In this sense, the question *"What time is it?"*—time as measurement—finds its answer in the coincidence of existential and astronomical time. The act of looking at the sky at a given moment fuses the subjective present with an objectively measurable celestial configuration. Measurement and awareness converge in a single event.

EXISTENTIAL AND ASTRONOMICAL TIME

A concise and essential account of astronomical time is provided in the 1675 treatise *An Introduction to Astronomy and Geography* by the mathematician and surveyor William Leybourne:

> Whatſoever is to be known of the Order, Situation and Motion of the great Bodies of the World, is all to be found out, by the Observation of Celestial Appearances; amongst which the first, chiefeſt and moſt known, is that by which we see the Sun, Moon and all the Stars, in every 24 Hours, to be mov'd about the Earth from Eaſt to West ; and this by Astronomers is call'd the firſt or Diurnal Motion, to distinguish it from the ſecond or Annual Motion, which the Sun, Moon, and all the Planets are obſerved differently to perform from E. to W. (Leybourne, 1675, pp. 1–2)

Described in its title as "plain and easy," the treatise exemplifies an early modern approach to knowledge grounded in observation and multidisciplinary practice. Leybourne's concise statements—written in archaic English—identify key components of astronomical time: the apparent motion of celestial bodies, cyclic repetition, and directional movement relative to the cardinal points.

These elements—the real motion of the observer on the rotating Earth and the apparent paths of celestial bodies—together define astronomical time. As outlined in the introduction to spherical astronomy in Chapter 2, the position P of a celestial body or phenomenon, understood as a spacetime event, can be defined by four spacetime coordinates (x, y, z, t), where t represents astronomical time. The perception of this celestial event is linked to an observer O located on the surface of the Earth, which is also assumed as the center of the celestial sphere. This assumption is phenomenological rather than cosmological: it reflects the observer's experiential horizon rather than a physical claim about the universe.

From a phenomenological perspective, existential time is identified with lived experience, awareness, and consciousness. When the subject is also the observer of a celestial event, astronomical time coincides with existential time during the act of observation. The perception of the remote space of the sky—the surface of the celestial sphere—is thus inseparable from the astronomical time in which celestial events unfold.

This coincidence of astronomical and existential time further demonstrates how the systemic relationship between self and universe, perceiving subject and observed phenomenon, leads to an integrated understanding of mind and body, overcoming Cartesian dualism. The observer is neither detached spectator nor passive recorder; the observer is structurally implicated in the temporal unfolding of the observed. The experience of *"I am now here"* situates the subject within the astronomical spacetime continuum, which becomes not only a scientific framework but also a condition of existential awareness and lived experience.

TIME MEASUREMENT AS CYCLES

Human life is structured by the measurement of time, from daily biological rhythms and social routines to the broader arc of existence from birth to death. Time measurement can be defined as "the science and practice of counting the repetitions of recurring phenomena and subdividing the intervals between each repetition into smaller units capable of being measured by various devices, including mechanical and atomic clocks" (EBSCO, n.d.). An interval thus defines units of time that can be subdivided into smaller measurable units.

Cycles—series of phenomena that repeat at regular intervals—are fundamental to time measurement. In astronomical time, two primary cycles arise from the movements of the Earth: its daily rotation around its axis (spin) and its annual revolution around the Sun. As discussed in Chapter 3, these motions are perceived by an observer on Earth as apparent movements of the Sun; for this reason, the term *apparent solar path* is used.

Each cycle defines a principal time unit. The *solar day* is the interval required for the Sun to return to the same azimuth following one complete rotation of the Earth. The *solar year* is the interval required for the Sun to return to the same apparent position relative to the celestial equator and the ecliptic. In addition to solar reference systems, stellar reference systems are used: the *sidereal day* is defined as the interval required for the Earth to rotate such that the same reference star returns to the same position relative to the local meridian. The sidereal day is approximately four minutes shorter than the solar day because, during one Earth rotation, the Earth also advances nearly one degree along its orbit around the Sun.

The subdivisions of the solar day—hours, minutes, and seconds—are directly related to angular measurements of the Earth's rotation. Time measurement is therefore angular at its foundation. Chronometry is inseparable from rotational geometry. One hour corresponds to a 15-degree rotation, one minute to 0.25 degrees, and one second of time to approximately 0.00417 degrees.

These astronomical cycles are essential elements in the architecture of cosmology. Built constructions—monuments, ritual sites, and sundials—often mark geometric alignments with the cyclic positions of celestial bodies, particularly the Sun. Across cultures, materials, and historical periods, such structures share a common function: by situating the observer within space and time, they enable both orientation and the measurement of time through embodied experience. Architecture thus stabilizes recurrence. It anchors ephemeral celestial motion within durable material form.

HINDU ASTRONOMY: SŪRYA SIDDHĀNTA

The *Sūrya Siddhānta* (from the Sanskrit *sūrya*, "Sun," and *siddhānta*, "established doctrine" or "treatise") is a foundational text of Hindu astronomy, written in Sanskrit by an anonymous author and generally dated to the late fourth or early fifth century CE (Encyclopædia Britannica, n.d.). Like much of classical Sanskrit literature, the treatise is composed in *śloka* verse: stanzas of two lines, each line consisting of two halves (*pādas*) of eight syllables (Burgess, 1935, p. iv).

Although the cosmological model presented in the *Sūrya Siddhānta* is geocentric and therefore scientifically outdated, many of its mathematical and astronomical computations remain remarkably accurate, including trigonometric sine tables and the calculation of the obliquity of the Earth's axis (Burgess, 1935). More significant for the present inquiry, however, is the treatise's conception and measurement of time, which integrates astronomical cycles with embodied human experience.

The following verses, quoted from Burgess's translation, outline the hierarchical structure of time units defined in the *Sūrya Siddhānta*:

11. That which begins with respirations (*prāṇa*) is called real; that which begins with atoms (*truṭi*) is called unreal. Six respirations make a *vināḍī*, sixty of these a *nāḍī*.
12. And sixty *nāḍīs* make a sidereal day and night. Of thirty of these sidereal days is composed a month; a civil (*sāvana*) month consists of as many sunrises.
13. A lunar month [consists] of as many lunar days (*tithi*); a solar (*saura*) month is determined by the entrance of the Sun into a sign of the zodiac: twelve months make a year … (Burgess, 1935, pp. 5–7)

These relations are summarized in Table 1

Table 1. Hierarchical system of time measurement in the Sūrya Siddhānta, relating respiration (prāṇa) to astronomical cycles and modern time units.

Measure (Hindu Unit)	Definition	Seconds	Prāṇa
10 long syllables (*gurvaṣṭakṣara*)	1 respiration (*prāṇa*)	4	1
6 respirations	1 vināḍi	4 x 6 = 24	6
60 vināḍis	1 nāḍi	4 x 6 x 60 = 1,440	360
60 nāḍis	1 day	4 x 6 x 60 x 60 = 86,400	21,600

According to the *Sūrya Siddhānta*, time measurement is grounded in breathing—one of the most fundamental biological processes. The interval between two respirations corresponds to four seconds, and the duration of a day is defined by the number of breaths taken. This approach, without precedent in Western astronomy, establishes a direct equivalence between cosmic cycles and bodily rhythms. The macrocosm is calibrated through the microcosm. The human body becomes the primary chronometric instrument. Time, space, and motion are not abstract quantities but are integrated into a lived, physiological experience in which the self and the universe form a single, interrelated system. Breath, rotation, and celestial recurrence become homologous structures. Time is simultaneously cosmic, angular, and respiratory.

This embodied conception of astronomical time provides a conceptual foundation for understanding how cosmological knowledge is translated into spatial practices and built form, particularly within the Indian architectural tradition.

VĀSTU ŚĀSTRA AND *ŚILPA ŚĀSTRA*: INDIAN COSMOLOGY, ART, AND URBAN PLANNING

Hindu cosmology finds concrete expression in art, architecture, and urban planning through the design systems of *vāstu śāstra* and *śilpa śāstra* (Shukla, 1993). The Sanskrit term *vāstu* (वास्तु) derives from a root meaning "to dwell" and is commonly translated as "architecture" or "habitation," while *śāstra* (शास्त्र) denotes "science," "treatise," or "systematic teaching." The term *śilpa* (शिल्प) refers to "skill," "craft," or "art" (Puranavedas, 2021). These two bodies of knowledge are closely interrelated and together inform the creation of the human-made environment in harmony with cosmic order. Design is therefore cosmological in intent. Settlement is not merely functional occupation of land but calibrated participation in universal structure.

Often described as the "Indian art of placement," *vāstu śāstra* "reaches out to the universe even as it simultaneously imposes structure upon it," forming "a strange amalgam of science and religion, of design and instinct" (Arya, 2000, p. 2). The system shares fundamental principles with geomancy and practices such as *feng shui*, insofar as all are concerned with energy flows generated by the relationship between human beings, natural forces, and the built environment, ultimately aiming at forms aligned with the laws of the cosmos (Arya, 2000, p. 2).

The design prescriptions of *vāstu śāstra* extend from town planning and infrastructure to residential architecture and temple construction (Shukla, 1993). Spatial layouts are governed by geometric configurations aligned to the cardinal directions and structured by the *vāstu-puruṣa maṇḍala*, a cosmological grid that organizes space symbolically and functionally. The establishment of this grid was traditionally accompanied by ritual procedures, including *śaṅkusthāpana* (literally, "fixing the conch shell"), used to locate the navel of the *Vāstu Puruṣa* on the site—typically slightly offset from the geometric center (Arya, 2000, p. 31). Ritual action formalizes geometric intention. Measurement and consecration occur simultaneously.

The relationship between architecture and cosmology in *vāstu śāstra* reflects the holistic Hindu approach to knowledge, in which theory is inseparable from practice. Yoga offers a parallel example: philosophical interpretations of reality are embodied in postures, breath, and movement. As Shukla notes, "a science without a philosophy or religious background is not a complete science, according to the Hindu view of life" (Shukla, 1993, p. 35).

This integration is articulated with particular clarity in the concept of the *Vāstu-puruṣa maṇḍala*, described by Shukla as the metaphysical plan of a Hindu building:

> … the surface of the earth, in traditional Indian cosmology, is regarded as demarcated by sunrise and sunset, by the points where the sun apparently emerges above and sinks below the horizon; by the East and West and also by the North and South points. It is therefore represented by the ideogram or maṇḍala of a square. … The earth is therefore called *caturbhṛṣṭi*, four-cornered, and is symbolically shown as *Pṛthivī-maṇḍala*, whereas, considered in itself, the shape of the earth is circular. (Shukla, 1993, p. 184)

Within this framework, the orientation of buildings and their alignment with the cardinal directions constitute a form of applied astronomy embedded in everyday life. Following the principles of *vāstu śāstra*, human settlements become spatial instruments through which cosmic order is made present and perceptible. Orientation, proportion, and alignment are not symbolic overlays but operative principles embedded in construction. In this sense, Indian architecture offers a paradigmatic example of architecture of cosmology, where astronomical knowledge, embodied time, and spatial practice converge.

This convergence reaches a monumental and explicitly astronomical expression in the observatories of early modern India, most notably the Jantar Mantar complexes. There, architecture ceases to symbolize time and instead becomes a calibrated instrument for its measurement.

SUN ARCHITECTURE: GNOMONICS & SUNDIALS

An introduction to gnomonics is central to this study: the apparent path of the Sun, as perceived by the naked eye, defines the geometric logic underlying the majority of examples of the architecture of cosmology.

The term *gnomonics* derives from the Greek γνώμων (*gnṓmōn*), meaning "indicator," "interpreter," or "discerner" (Liddell et al., 1940), and denotes the study and practice of understanding, designing, and constructing solar clocks, commonly known as sundials. Gnomonics is based on the gnomon, a vertical element—sometimes inclined—whose shadow, cast by the Sun, measures the time of day according to the Sun's position in the sky.

Gnomonics originated in China, with evidence dating back to the Shang dynasty (ca. 18th–12th century BCE) (Shi, 2015). The measurement of gnomon shadows constituted one of the primary astronomical observations and was essential for timekeeping and calendar making. Other sources date the presence of the gnomon to even earlier periods: a painted stick discovered at the prehistoric astronomical site of Taosi (ca. 2300 BCE) is considered the oldest known gnomon in China. From the second century BCE onward, gnomon shadow measurements became an essential component of calendrical practice (Li, 2015, p. 2095). Early Chinese gnomons comprised a post measuring approximately 8-feet "and a template for measuring its shadow lengths at noon during the year. The earliest records of ancient Chinese gnomon shadow measurements are found in the Rituals of the Zhou Dynasty *Zhou li* (周礼) and the *Zhou bi suan jing* (周髀算经)" (Li, 2015, p. 2096).

In the eighth century CE, gnomons in China were also employed in one of the most ambitious geodetic and astronomical survey projects ever undertaken. During the Tang dynasty, the Buddhist monk Yi Xing (683–727 CE, also spelled I-Hsing), assisted by the astronomer Nankung Yüeh, oversaw the installation of eight-foot gnomons at four stations aligned along a meridian line extending approximately 200 km along a meridian across eastern China. One objective of this project was the measurement of solar shadows during the solstices and equinoxes; latitude could also be determined by comparing shadow lengths at different locations (Needham et al., 1961). A table of tangent values was compiled from these measurements. The selected meridian lay between that used by Eratosthenes (ca. 200 BCE) and those later employed by astronomers of the Caliph al-Ma'mūn (ca. 827 CE) (Needham et al., 1961, p. 4). Beyond its scientific relevance, the survey had significant social implications for timekeeping and calendar regulation.

The gnomon appears in prehistoric rock art and across many ancient cultures worldwide, including Native American (Vogt, 2015), Inca (Malville, 2015), Egyptian, and Greek civilizations. A renowned application of the gnomon is attributed to the mathematician and astronomer Eratosthenes (ca. 276–195/194 BCE), who used shadow measurements at solar noon in Alexandria and Syene— two cities approximately aligned along the same meridian—to calculate the circumference of the Earth (Newton, 1980).

An important contribution to gnomonics in architecture was made by Vitruvius in his treatise *De Architectura*, written around 25 BCE (Vitruvius Pollio, 1914). Particularly detailed illustrations appear in Book IX of the Italian translation *I Dieci Libri dell'Architettura*, published in 1556 and edited by Daniele Barbaro (Vitruvius Pollio, 1557).

In this treatise, Vitruvius makes repeated references to Plato, Ptolemy, and Eratosthenes, explicitly acknowledging the foundational contributions of Greek philosophers, astronomers, and mathematicians. Both the text and the drawings demonstrate the depth of solar studies in architectural design, illustrating how sundials and solar measurements are integrated directly into buildings. These examples reveal a close interrelation between geometry, astronomy, and architecture, in which the movement of the Sun becomes a generative principle of spatial order (Figures 2–3).

Figure 2. Sun rays and geometric construction in I Dieci Libri dell'Architettura (Vitruvius Pollio, 1557, p. 401). Public domain.

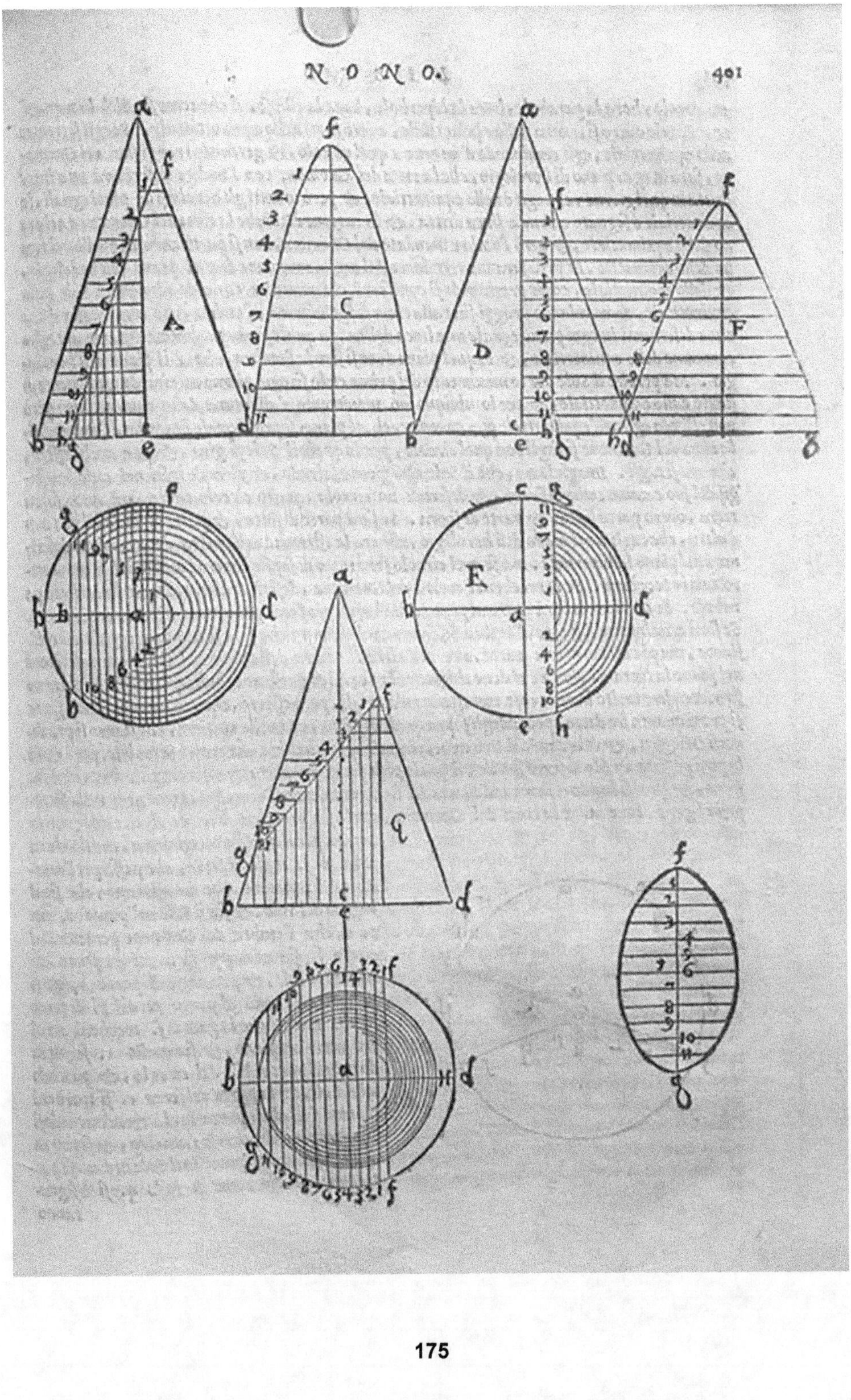

Figure 3. Solar measurement and buildings in Vitruvius' I Dieci Libri dell'Architettura (Vitruvius Pollio, 1556, p. 435). Public domain.

The gnomon represents a rudimentary form of cosmological architecture: a simple vertical element—of stone, wood, or metal—whose scale may range from the monumental, as in obelisks, to the intimate, as in garden sundials. Its defining feature is the creation of an architecture composed of light and continuously changing shadows. Complex spatial configurations emerge through the interaction between dynamic, ephemeral solar phenomena and the static, monolithic simplicity of the gnomon, which embodies an architecture of the Sun.

Measuring Shadows: Sundials

The term *sundial*—derived from the words "sun" and "dial"—clearly expresses the function of this instrument: the measurement of the Sun's position, as observed from a specific location on Earth, which corresponds to the measurement of time. A sundial therefore records the continuously changing position of the Sun through the shadows it casts.

The most basic sundial consists of a gnomon and a dial. Geometrically, the gnomon (or style) may be idealized as a straight line contained in a vertical plane and aligned parallel to the Earth's rotational axis. The dial is a surface—horizontal, vertical, equatorial, or inclined—onto which the shadow of the gnomon is projected. Hour lines are inscribed on the dial to measure the passage of time through the movement of the shadow (Figure 4) and are often labeled with Roman numerals.

Figure 4. Hour lines in the dial of an equatorial sundial.

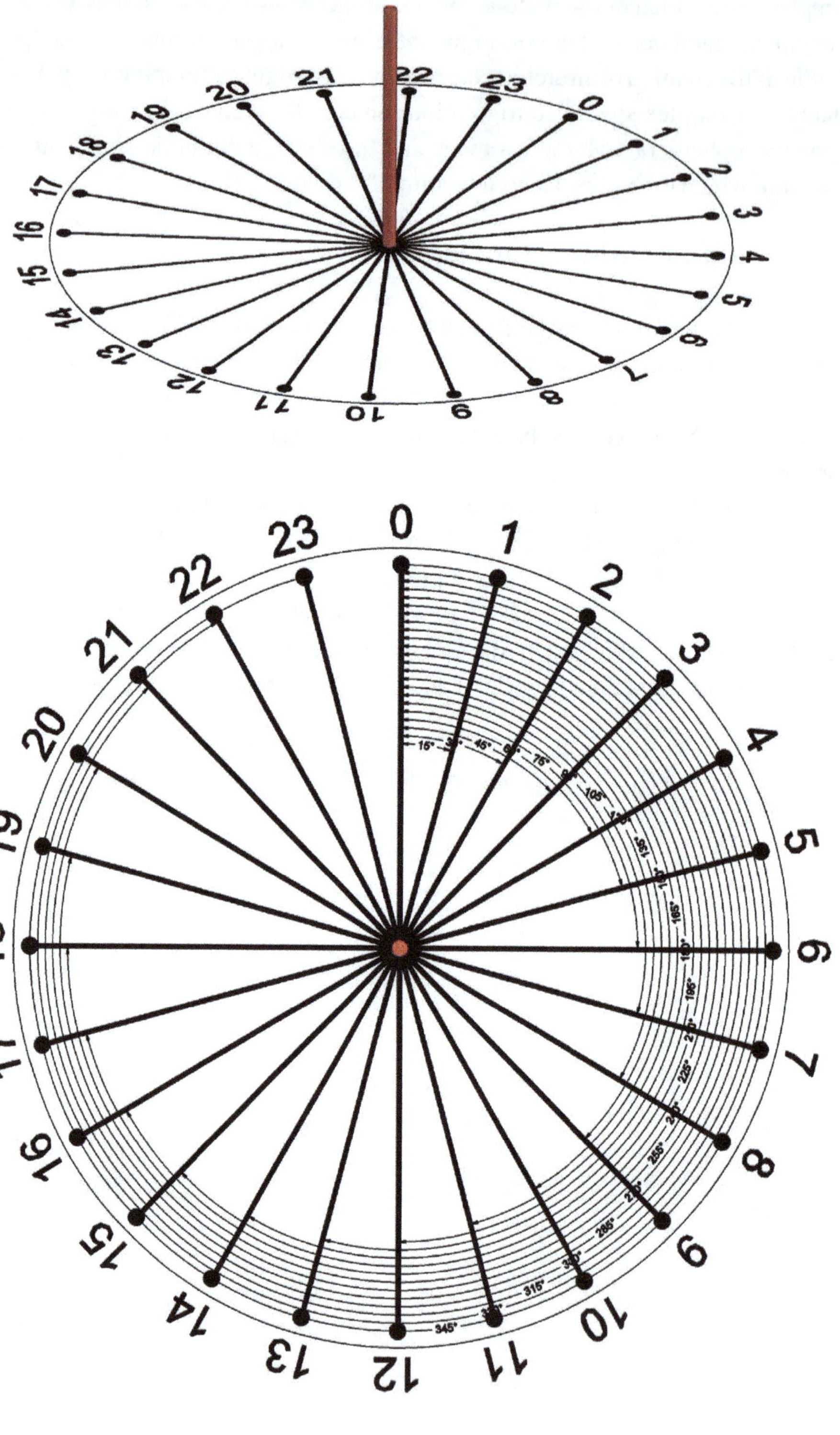

Solar noon—the moment when the Sun crosses the local meridian—is fundamental to sundial design. At solar noon, the shadow cast by the gnomon reaches its shortest length for the day and aligns with the meridian, defining a north–south line known as the meridian line. In the northern hemisphere, the Sun reaches its highest point toward the south; in the southern hemisphere, toward the north. The length of the noon shadow varies throughout the year: it is shortest at the summer solstice and longest at the winter solstice. Almost all sundials are aligned with the meridian line.

Sundials present hour lines at regular angular intervals corresponding to the Earth's rotation relative to the Sun. While the angular spacing of hour lines remains constant, the length of the gnomon's shadow varies throughout the year according to the Sun's changing altitude, which depends on the Earth's orbital motion and axial tilt (see Chapter 3). Sundials are classified according to the geometric relationship between the gnomon and the dial plane. The main types include (British Sundial Society, n.d.):

- *Horizontal sundials* present a gnomon aligned with the Earth's rotational axis and a horizontal dial plate tangent to the Earth's surface (Figure 5).

Figure 5. Diagram of horizontal sundials.

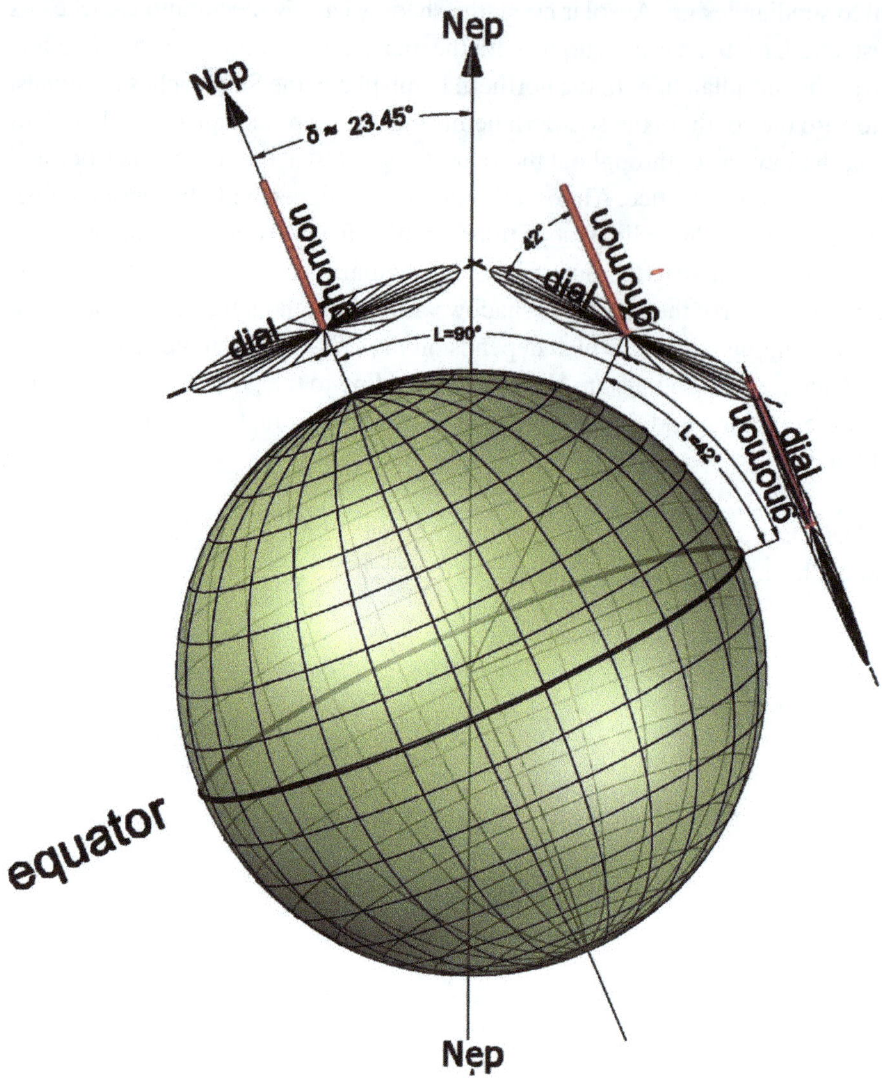

- *Equatorial sundials* have a dial plane parallel to the Earth's equatorial plane, with the gnomon perpendicular to the dial. Hour lines are evenly spaced at 15-degree intervals (Figure 6). Only at the North and South Poles does the gnomon align with the zenith.

Figure 6. Diagram of equatorial sundials.

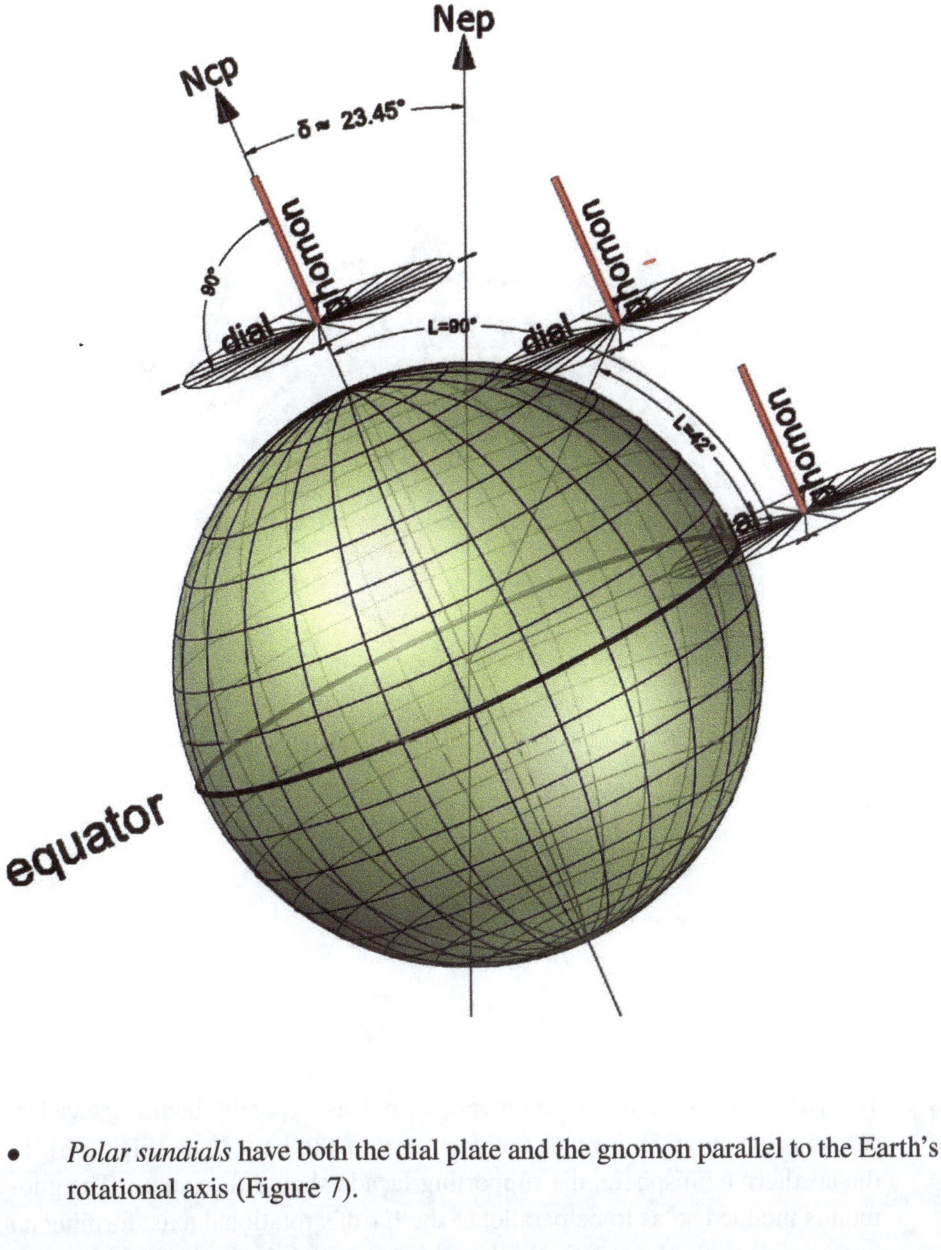

- *Polar sundials* have both the dial plate and the gnomon parallel to the Earth's rotational axis (Figure 7).

Figure 7. Diagram of polar sundials.

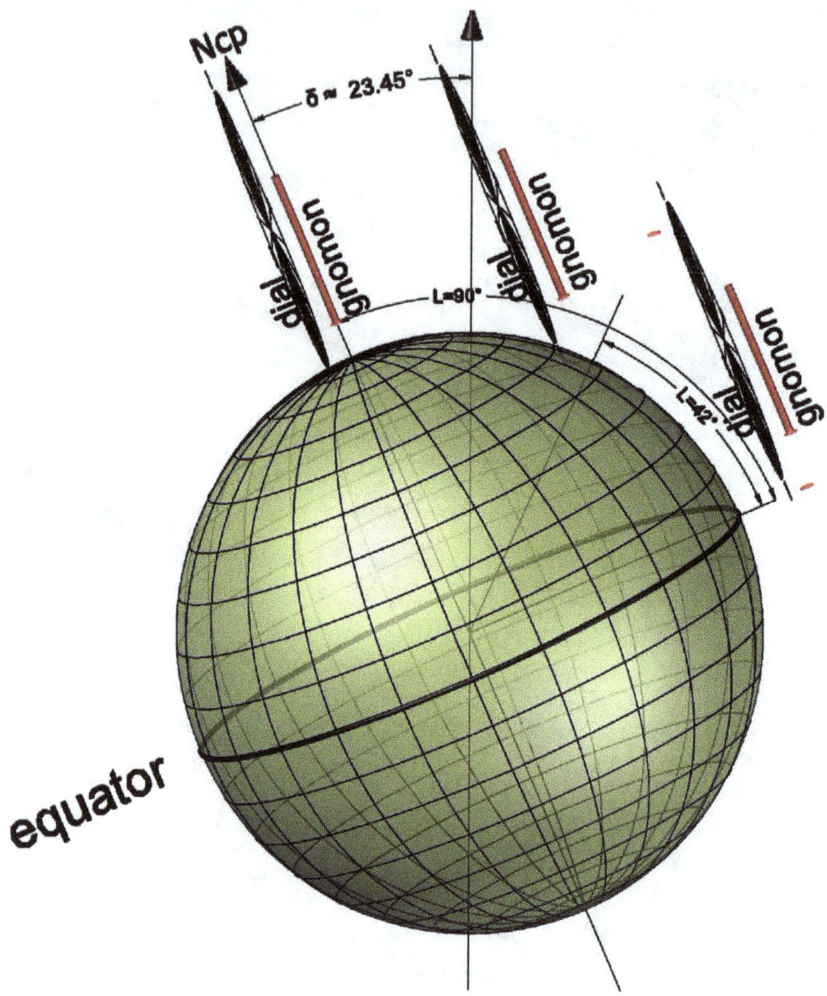

- *Vertical sundials* are mounted on vertical planes, typically building façades, whose orientation determines the geometry of the hour lines (Figure 8). In the northern hemisphere, the supporting façade should face south. The gnomon is inclined so as to be parallel to the Earth's rotational axis, forming an angle equal to the local latitude (L) with respect to the horizontal plane (or 90° – L with respect to the vertical plane), thereby aligning the gnomon parallel to the Earth's rotational axis.

Figure 8. Diagram of vertical sundials.

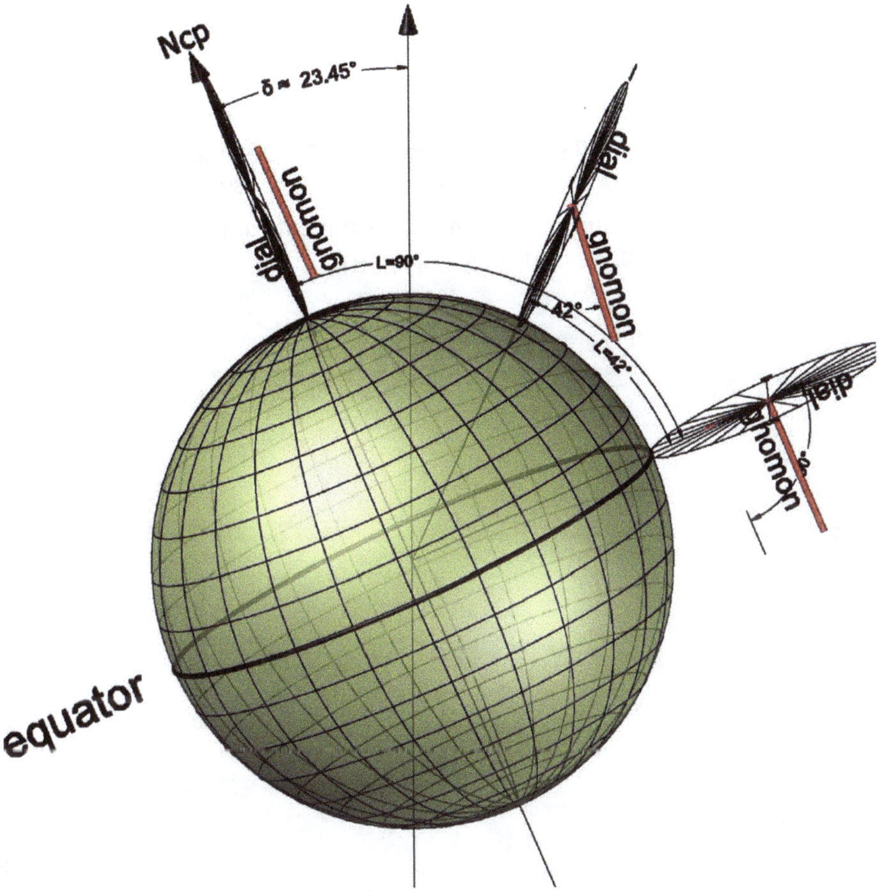

Figures 5–8 illustrate the geometric relationship between the gnomon and the dial at different latitudes (L) in the northern hemisphere: 0° (equator), 42°, and 90° (North Pole).

Other types of sundials include analemmatic, armillary, spherical, and cylindrical forms. In the latter three cases, the dial surface is not planar but curved, forming three-dimensional geometries.

The analemmatic sundial differs fundamentally from all other types. It employs a vertical gnomon that moves along a line according to the Sun's declination, which varies daily throughout the year. The gnomon may even be a person. This type of sundial is based on the figure-eight pattern of the analemma.[1]

Sundials constitute a clear example of cosmological architecture, integrating solar geometry and temporal measurement into the built environment. As architectural artifacts, sundials are constructed from durable materials suited to outdoor conditions, including bronze, cast iron, stainless steel, aluminum, and stone. Figures 9a–d present photos of horizontal sundials from different periods and geographic locations, all in the northern hemisphere. Figures 9a–c depict examples in the United States: (a) the Shakespeare Garden in Central Park, New York (40.77975° N, 73.97005° W); (b) the Blithewood Garden at Bard College, Annandale-on-Hudson, New York (42.02078° N, 73.91763° W); and (c) the Botanical Gardens of Cornell University, Ithaca, New York (42.449756° N, 76.471481° W). Figure 9d shows one of the fifteenth-century astronomical instruments at the Ancient Observatory in Beijing, China (39.90758° N, 116.43471° E).

Figure 9. Photos of horizontal sundials in the US (a, b. c) and China (d).

A notable example of a vertical sundial is located on the façade of the sixteenth-century church of Santissima Trinità dei Monti in Rome, Italy (41.90624° N, 12.48338° E). The church occupies a prominent urban position at the top of the Spanish Steps and serves as a focal point of Via Condotti. The façade is oriented west–southwest, at an azimuth of 163.8° measured eastward from north (Figure 10), and stands behind the Obelisco Sallustiano (Figure 11; see Chapter 7). The façade features two symmetrical bell towers; the sundial is located on the southern tower (Figure 12). Documentation of the dial is limited, and it includes an unusual horizontal line. Recent calculations suggest that this line marks hours counted from sunset (Taki, 2008).

Figure 10. Satellite view (from Google Maps) of Santissima Trinità dei Monti with superimposed axis of orientation.

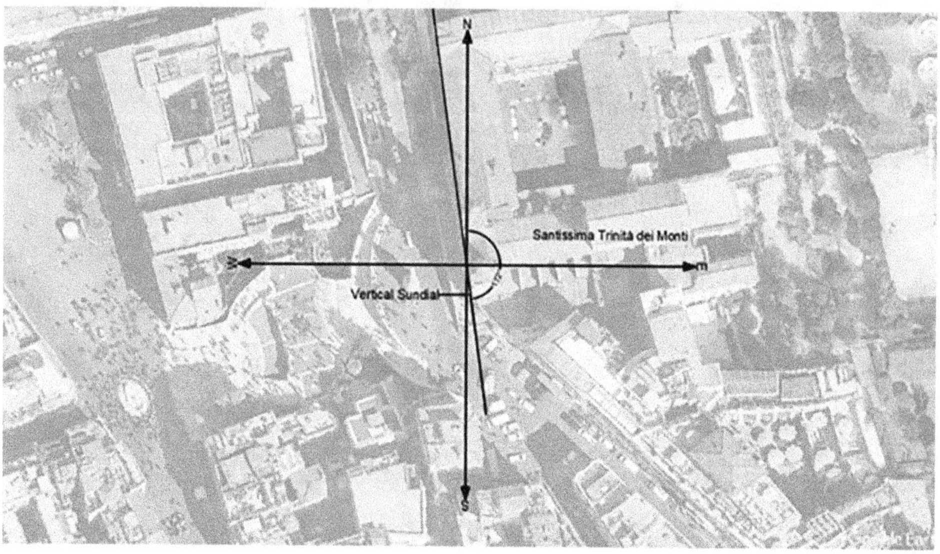

Figure 11. Obelisco Sallustiano in front of Santissima Trinità dei Monti.

Figure 12. Façade of Santissima Trinità dei Monti, with the vertical sundial below the southern bell tower.

Sundials Calculations

Sundial calculations derive from the geometry of the Sun's apparent position on the celestial sphere as observed from a given location on Earth. The mathematical framework of gnomonics relies on trigonometric relationships defined by the Sun's position and by the angles between the gnomon and the dial plane.

The principal angular parameters, measured in degrees, are the latitude (L), the solar declination (δ), the hour angle (h or HA), and the angle θ between the noon line and a given hour line. For horizontal sundials (North American Sundial Society, 2020), θ is given by:

$$\tan \theta = \tan (HA) \cdot \sin L$$

The solar altitude α is the angle used to calculate the length of the shadow cast by a sundial at a given hour angle h, and solar azimuth Φ. The solar altitude α is given by

$$\sin \alpha = \sin L \sin \delta + \cos L \cos \delta \cos h.$$

The solar azimuth Φ can be derived from α, δ, and L through standard trigonometric relations (Goswami, 2015):

$$\cos\Phi = \frac{\sin \delta - \sin L \sin\alpha}{\cos L \cos\alpha}$$

At local solar noon, h = 0. The solar declination δ varies between −23.45° and +23.45° over the course of the year, reaching its minimum at the winter solstice and its maximum at the summer solstice in the northern hemisphere. During the equinoxes, δ equals 0°.

Another important consideration in sundial design is the Equation of Time (EoT), defined as the difference between apparent solar time—measured by a sundial—and mean solar time, which is the averaged time kept by modern clocks. This difference arises from two factors: the ellipticity of the Earth's orbit and the inclination of the Earth's rotational axis relative to the ecliptic (U.S. Naval Observatory, Astronomical Applications Department, n.d.).

The Equation of Time may be expressed as:

$$\text{EoT} = \text{Apparent Solar Time} - \text{Mean Solar Time}$$

It can also be written as:

$$\text{EoT} = \frac{q}{15} - \text{RA},$$

where RA is the Sun's right ascension (in hours) and q is the mean longitude of the Sun (in degrees), based on the average position of the Sun considered as moving at constant angular speed along the ecliptic (U.S. Naval Observatory, Astronomical Applications Department, n.d.). Since EoT and RA use the hour unit, q is divided by 15 as a conversion factor to change degrees into hours.

JANTAR MANTAR: MEASURING THE COSMOS

Jantar Mantar is the name shared by complexes of naked-eye outdoor astronomical observatories located in five cities of northern India. The term derives from the Sanskrit *yantra* ("instrument") and *mantrana* ("calculation") (Rajawat, 1980). Conceived and developed by Maharaja Sawai Jai Singh II, the observatories were constructed between 1724 and 1738. The five complexes were built in Delhi (National Capital Region), Ujjain (Madhya Pradesh), Varanasi (Uttar Pradesh), Mathura (Uttar Pradesh), and Jaipur (Rajasthan). All have survived with the exception of the Mathura observatory, which was demolished prior to the revolt of 1857 (The Astronomical Observatories of Jai Singh, n.d.).

I visited the Jantar Mantar in Delhi and Jaipur in 1996. Walking through these structures was a defining experience and one of my earliest and most significant encounters with the architecture of cosmology, first encountered in Rome during childhood. These visits revealed how architecture can mediate a dialogue between science and art. The Jantar Mantar are not merely astronomical observatories; they are monuments to the cosmos. Their scale is monumental yet human, allowing visitors to walk through and inhabit the instruments themselves. The constructions function simultaneously as precise scientific devices and as inhabitable sculptures, publicly accessible and usable, enabling direct observation of celestial phenomena. Their composition—warm local stone set against the blue sky—creates an aesthetic in which the geometry of the universe is both measured and celebrated.

Earth–sky geometries are embodied in each structure of the Jantar Mantar. All instruments operate through naked-eye observation of the Sun and other celestial bodies on the celestial sphere. Terrestrial coordinates of latitude and longitude determine the orientation of the structures in both plan and elevation. Each instrument is aligned to the cardinal directions, enabling measurements of azimuth, altitude, declination, and time. While several instruments measure the positions of stars, the Sun remains the principal object of observation. The apparent solar path generates shadows that articulate an architecture of light and shadow, through which time of day and season are measured via solar altitude and azimuth.

Jantar Mantar in Jaipur

Jaipur is a walled city in Rajasthan, northern India. Often referred to as the "Pink City" for the color of its painted façades, Jaipur was founded in 1727 by Maharaja Sawai Jai Singh II. The city was planned according to the principles of *vāstu śāstra* (discussed in the section *Vāstu Śāstra and Śilpa Śāstra: Indian Cosmology, Art, and Urban Planning*), integrating nature, astronomy, and cosmology into urban and architectural design. Planning and design were undertaken by Jai Singh in collaboration with the architect Vidyadhar Bhattacharya, using a grid system subdividing the city into nine approximate squares, each symbolically associated with one of the planets recognized in Vedic astronomy.

Figure 13. City of Jaipur, from Google Earth, with superimposed outline of the chowkris and angle of deviation from the cardinal directions.

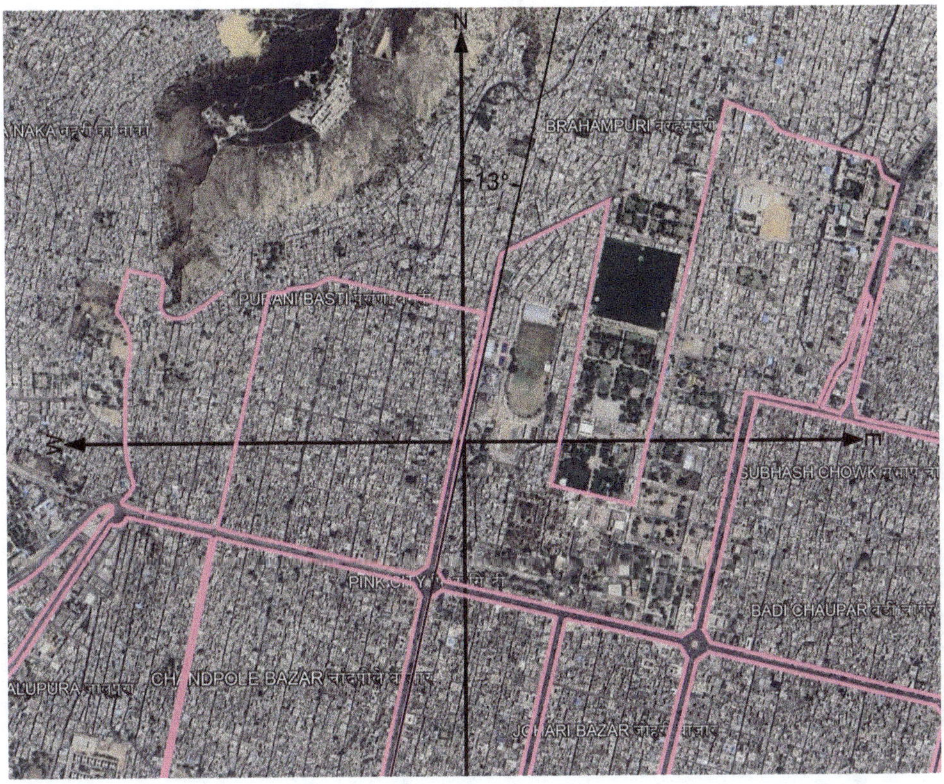

Defensive walls enclosed the city and were pierced by eight gates (Jaipur Municipal Corporation, n.d.), several of which were named after celestial bodies. The eastern gate, *Surajpole*, is associated with the Sun, while a western gate bears lunar associations. Prior to Jaipur, only one settlement from the Indus Valley Civilization (ca. 2600 BCE) is known to demonstrate comparable evidence of systematic grid planning (Patel, 2019). Astronomy and astrology were also employed to anticipate weather patterns and guide agricultural practices (Sarkar, 1984).

Although *vāstu śāstra* prescribes alignments to the cardinal directions, the Jaipur urban grid exhibits an angular deviation of approximately 13° eastward from true north (Figure 14). The city follows the *Vāstu Puruṣa Maṇḍala*, organized as a 9 × 9 grid (*chowkris*), each measuring approximately 0.5 miles (≈805 m). The street hierarchy comprises major east–west and north–south arteries (*Rajmarg*) measuring approximately 33 m in width, secondary streets of approximately 16.5 m, and tertiary residential lanes of approximately 8.25 × 4.00 m, arranged in a *prastāra* (chessboard) pattern (City Planning of Jaipur, n.d.). The City Palace occupies the central sector of the plan and combines Mughal and Rajput architectural traditions.

Figure 14. Satellite view of the Jaipur Jantar Mantar from Google My Maps.

The Jantar Mantar complex of Jaipur (26.92476°, 75.82455°) is located southeast of the City Palace and was completed in 1734. It is among the last observatories constructed under the patronage of Jai Singh II and the most prominent of the series, featuring the world's largest stone sundial, the *Bṛhat Samrāṭ Yantra*. In 2010, the complex was designated a UNESCO World Heritage Site for its innovations in architecture, astronomy, and cosmology, integrating knowledge from Western, Middle Eastern, Asian, and African traditions (UNESCO Portal to the Heritage of Astronomy, n.d.). A site plan of the complex observatory is shown in Figure 15.

Figure 15. Site plan of the Jantar Mantar complex, showing the location of the principal astronomical instruments.

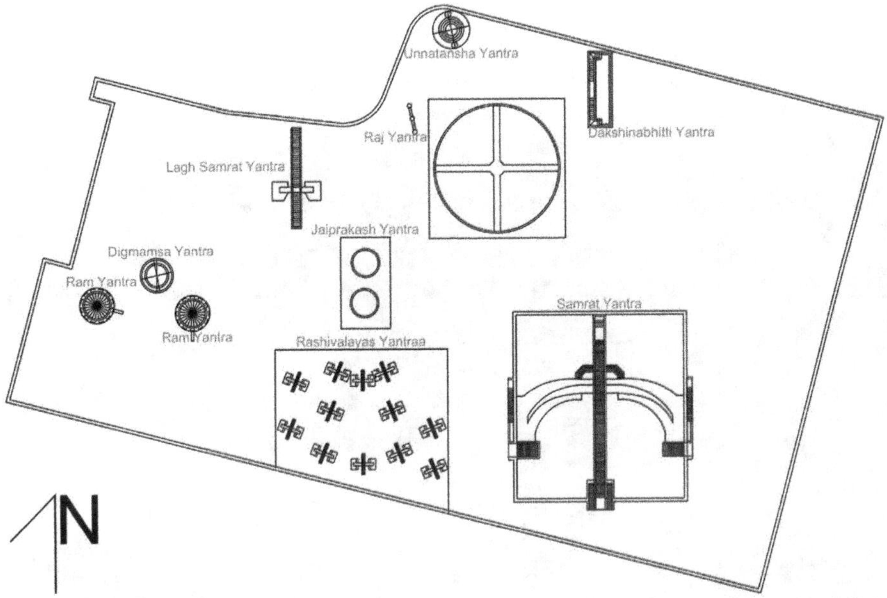

The Jaipur complex comprises 35 instruments grouped into 18 primary structures, all operating through naked-eye observation (Government of India, 2009). Each instrument employs one of the three coordinate systems used in spherical astronomy—horizontal, equatorial, or ecliptic—introduced in Chapter 2. Some instruments, such as the *Kapāla Yantraprakāra*, operate across multiple coordinate systems, transforming equatorial coordinates into horizontal ones (Sharma, 1991). Figure 16 shows a satellite view of the complex, with numbered references corresponding to the principal instruments described below.

1. *Bṛhat Samrāṭ Yantra* (26.92455° N, 75.82491° E), Sanskrit for "Great King of Instruments," is the largest stone sundial in the world, rising approximately 19 m above ground. It is shaped as a right triangle, with the hypotenuse parallel to the Earth's axis and its quadrants parallel to the equatorial plane (Figures 16–17). Based on the equatorial coordinate system, the instrument performs three primary functions with a timekeeping accuracy of ±3 seconds (Government of India, 2009): a. daytime timekeeping based on the Sun's position; b. nocturnal time measurement using the position of a celestial body, accord-

ing to right ascension and declination; and c. determination of the hour angle and declination of celestial objects. Its timekeeping accuracy is ±3 seconds (Government of India, 2009).

Figure 16. Samrat Yantra, 1996 (photo by David Foell).

Figure 17. Samrat Yantra, 1996 (photo by David Foell).

2. *Dakṣiṇottara Bhitti Yantra* (26.92520° N, 75.82491° E) consists of two instruments mounted on a vertical wall aligned along the local north–south axis. Measurements are based on the meridian coordinate system, allowing the determination of the altitude of a celestial body at its meridian transit (Government of India, 2009).
3. *Unnathāṃśa Yantr*a (26.92511° N, 75.82447° E) is based on the horizontal (altitude–azimuth) coordinate system. It consists of a suspended brass ring with an outer diameter of 5.35 m, hung from a beam supported by two columns measuring approximately 7 m in height and 1.5 m in width. The columns are aligned 12° west of the north–south axis (Government of India, 2009).
4. *Yantra Rāj* (26.92511° N, 75.82447° E) comprises two stone discs, each slightly over 2 m in diameter. One disc functions as an astrolabe based on the horizontal (azimuth–altitude) celestial coordinate system (Government of India, 2009).
5. *Nārivalaya Yantra* (26.92495° N, 75.82435° E) consists of two circular discs (Figure 18), one oriented northward and the other southward, each inclined parallel to the equatorial plane (Jantar Mantar, n.d.).

Figure 18. Nārivalaya Yantra, 1996 (photo by David Foell).

6. *Jai Prakāś Yantra* (26.92478° N, 75.82435° E) is based primarily on the horizontal (azimuth–altitude) coordinate system, with additional markings corresponding to the equatorial system. It consists of two hemispherical concave surfaces inscribed with inverted maps of the sky (Figure 19). Each hemisphere measures approximately 5.44 m in diameter and is aligned along the north–south axis. The instruments are constructed on a platform measuring approximately 20.5 × 11.22 m, partly above and partly below ground level. As noted by the Government of India (2009, 15), when viewed in plan the two instruments are exact complements; if their engraved surfaces were superimposed, they would form a single complete celestial surface.

Figure 19. Jai Prakāś Yantra, looking north on the same axis as the Nārivalaya Yantra, 1996 (photo by David Foell).

7. *Rāśivalaya Yantra* (26.92449° N, 75.82435° E) consists of a group of twelve instruments arranged on a rectangular platform (Figure 20), each corresponding to one of the twelve rāśis (zodiac signs) (Government of India, 2009, 51).

Figure 20. Rāśivalaya Yantra, 1996 (photo by David Foell).

8. *Laghu Samrāṭ Yantra* (26.92499° N, 75.82414° E), where *laghu* in Sanskrit means "small," is an equatorial sundial (Figure 21) that may be understood as a reduced-scale version of the *Bṛhat Samrāṭ Yantra.*

Figure 21. Laghu Samrāṭ Yantra with the author at the bottom of steps, 1996 (photo by David Foell).

9. *Digāṃśa Yantra* (26.92478° N, 75.82386° E) is based on the horizontal (azimuth–altitude) coordinate system. It consists of two concentric cylindrical walls with a central vertical pillar (Jantar Mantar, n.d.).
10. *Rām Yantra* (26.92469° N, 75.82384° E) is also based on the horizontal (azimuth–altitude) celestial coordinate system. It comprises two hollow cylindrical structures (Figure 22). The western structure contains openings inclined at 12°, and the eastern at 18°, beginning at a height of approximately 1.12 m above the base platform. Each cylinder is approximately 4.48 m tall with an internal diameter of 6.95 m (Government of India, 2009).

Figure 22. Rām Yantra, 1996 (photo by David Foell).

Geometrical and Urban Interpretation

From a geometrical standpoint, the Jantar Mantar observatory in Jaipur comprises a constellation of architectural forms unified by alignments derived from Earth–Sun geometry, both in plan and elevation. Right triangles—whose angles correspond to the local latitude—are among the most recurrent configurations, alongside cylindrical volumes and concave hemispherical surfaces.

Its integration within the planned urban fabric of Jaipur introduces an additional layer of astronomical meaning, establishing a rare and perhaps unprecedented relationship between urban space and architectural form as a reflection of remote celestial space. The observatory stands as an exceptional synthesis of architecture and astronomy within a public setting: an embodiment of the geometry of time, experienced not only by astronomers but by the broader population of the city.

Jantar Mantar in Delhi

The Jantar Mantar in Delhi is located in New Delhi, northern India. Its entrance lies at 28.62814° N, 77.21659° E, and the site extends approximately 550 feet north–south and 260 feet east–west. Construction began in 1723 as part of of the program by Maharaja Jai Singh II, ruler of the kingdom of Amber (later Jaipur), which involved major contributions to knowledge in mathematics, astronomy extending to city planning, architecture, education and social reform.

Smaller in scale than the Jaipur complex, the Delhi Jantar Mantar nevertheless includes the same principal typologies of astronomical instruments, categorized by geometric form and coordinate system (Government of India, 2009):

- *Samrāṭ Yantra*: equatorial sundials based on right-triangle geometry, with the hypotenuse parallel to Earth axis and equal to the latitude angle;
- *Rām Yantra*: concentric cylindrical instruments, with the center marked by a pillar using horizontal(azimuth-altitude) coordinates;
- *Jai Prakāś Yantra*: shaped as concave hemisphere and measuring celestial bodies according to both horizontal (azimuth-altitude) and equatorial coordinate systems (Jantar Mantar, n.d.).

The principal instruments are listed below following a north-to-south and west-to-east orientation. For their spatial organization within the site, refer to the labeled satellite view from Google Earth in Figure 23.

Figure 23. Satellite view of the Delhi Jantar Mantar, from Google Earth.

1. *Entrance*: located at 28.62817°, 77.21661°.

Figure 24. Miśra Yantra, 1996 (photo by David Foell).

2. *Miśra Yantra* (Figure 24): located in the north-west sector of the complex (28.62754° N, 77.21637° E), the Miśra Yantra—derived from the Sanskrit term meaning "composite instrument"—is an ensemble comprising five distinct astronomical instruments integrated into a single architectural configuration. The compound structure synthesizes multiple observational principles within a unified spatial and geometric framework, exemplifying the hybridization of celestial coordinate systems characteristic of the Jantar Mantar observatories.
3. *Samrāṭ Yantra* (Figures 25-27): situated approximately at the center of the complex (28.62705°N, 77.21658° E), the Samrāṭ Yantra functions as an architectural-scale gnomon for a horizontal sundial. It is shaped as a right triangle whose hypotenuse is aligned parallel to the Earth's rotational axis. The structure measures approximately 70 feet in height, 114 feet along the north–south axis, and 10 feet along the east–west direction. Quadrants positioned on both sides

of the triangular form are inscribed with calibrated markings that allow the measurement of hours, minutes, and seconds.

Figure 25. Samrāṭ Yantra, east façade, looking toward the Rām Yantra, 1996 (photo by David Foell).

Figure 26. Śaṣṭāṃśa at the Samrāṭ Yantra, east façade, 1996 (photo by David Foell).

Figure 27. East quadrant of the Samrāṭ Yantra, looking south, 1996 (photo by David Foell).

4. *Jaya Prakāś Yantra*: this instrument consists of two complementary concave hemispherical structures aligned along an east–west axis. The western hemisphere is located at 28.62675° N, 77.21651° E, and the eastern at 28.62675° N, 77.21670° E. Both hemispheres are engraved with coordinate markings and operate as naked-eye astronomical instruments, enabling the determination of celestial positions using both the horizontal coordinate system (altitude–azimuth) and an equatorial reference system. For definitions of celestial coordinate systems, see Chapter 2.
5. *Rām Yantra* (Figures 28–29): this instrument comprises two open-top cylindrical structures whose centers are aligned along a north–south axis, slightly west of and approximately 100 feet south of the Samrāṭ Yantra. The southern cylinder is located at 28.62627° N, 77.21653° E, and the northern at 28.62651° N, 77.21655° E. Together, these instruments are used to measure the altitude and azimuth of celestial bodies, allowing positional determination based on terrestrial latitude and longitude.

Figure 28. Rām Yantra, 1996 (photo by David Foell).

Figure 29. Rām Yantra: central pillar framed by an opening in the cylindrical wall, 1996 (photo by David Foell).

CONCLUDING THOUGHTS: FROM THEORY TO PRACTICE

The Jantar Mantar complexes—particularly the observatory in Jaipur—represent a paradigmatic instance of architecture embodying astronomy. Beyond providing built infrastructure for scientific observation, they materialize a Weltanschauung in which geometry mediates between cosmic order, timekeeping, architectural form, and urban structure. Geometry emerges as the connective framework through which the observation of the sky is translated into spatial, temporal, and cultural structures.

This chapter concludes the first part of the book, *Sky, Perception, and Geometry*, and initiates the transition to the second part, which examines how celestial perception has shaped civilizations through architectural and urban practices. The discussion has traced a progression from phenomenological experience of time to astronomical abstraction, culminating in the built expressions of cosmological knowledge exemplified by the Jantar Mantar observatories. The trajectory from perception to geometry to construction demonstrates that cosmology becomes operative only when translated into spatial practice. In this sense, these instruments and their architectural configurations may be understood as concrete manifestations of the Heideggerian fourfold—earth, sky, mortals, and divinities—gathered into relation through building (Heidegger, 1971).

The emphasis on architecture as an expression of knowledge finds continuity in Vitruvius's *De Architectura* (Vitruvius Pollio, 1557), which integrates gnomonics into architectural design as a foundational discipline. Together with the treatises discussed in Chapters 3 and 4, this lineage delineates a continuous trajectory from observation and measurement, through conceptualization and representation, to construction—marking the passage from theory to practice that underlies the architecture of cosmology. The following chapter opens Part II by situating these principles within archaeoastronomy, examining their expression in ancient Egypt and in the Neolithic temple complexes of Malta, where celestial alignment becomes foundational to monumentality and ritual space.

REFERENCES

Arya, R. (2000). *Vaastu: The Indian art of placement: Design and decorate homes to reflect eternal spiritual principles*. Destiny Books.

British Sundial Society. (n.d.). *How sundials work*. Retrieved September 12, 2025, from https://web.archive.org/web/20130802033857/http://www.sundialsoc.org.uk/HDSW.php

Brough, J. B., Dowden, B., & Fieser, J. (n.d.). Phenomenology and time-consciousness. In J. Fieser & B. Dowden (Eds.), *Internet encyclopedia of philosophy*. Retrieved September 6, 2025, from https://iep.utm.edu/phe-time/

Burgess, E. (Ed.). (1935). *Translation of the Sūrya-Siddhānta: A text-book of Hindu astronomy* (P. Gangooly, Ed.). Calcutta, India: University of Calcutta. (Original work published ca. 4th century CE)

City Planning of Jaipur. (n.d.). *City planning of Jaipur*. Retrieved September 17, 2025, from https://www.gopalancolleges.com/gsap/pdf/jaipur.pdf

EBSCO. (n.d.). Time measurement. In *Research starters: Science*. Retrieved September 9, 2025, from https://www.ebsco.com/research-starters/science/time-measurement

Encyclopædia Britannica. (n.d.). Mathematics in the Islamic world (8th–15th century). In *Britannica.com*. Retrieved September 9, 2025, from https://www.britannica.com/science/mathematics/Mathematics-in-the-Islamic-world-8th-15th-century

Goswami, D. Y. (2015). *Principles of solar engineering* (3rd ed.). CRC Press.

Government of India. (2009). *Nomination of the Jantar Mantar, Jaipur, for inclusion on the World Heritage List*. Paris, France: UNESCO World Heritage Centre. https://whc.unesco.org/uploads/nominations/1338.pdf

Heidegger, M. (1971). Building, dwelling, thinking. In *Poetry, language, thought* (Hofstadter, A., Trans.; pp. 141–160). Harper & Row.

Husserl, E. G. (1991). *On the phenomenology of the consciousness of internal time (1893–1917)* (Brough, J. B., Trans.). Kluwer Academic Publishers.

Jaipur Municipal Corporation. (n.d.). Our heritage. Retrieved September 18, 2025, from https://jaipurmc.org/Presentation/AboutJaipur/OurHeritage.aspx

Jantar Mantar. (n.d.). Digamsa Yantra. Retrieved September 19, 2025, from https://jantarmantar.org/learn/observatories/instruments/digamsa/index.html

Leybourne, W., & Morden, R. (1675). *An introduction to astronomy and geography: Being a plain and easie treatise of the globes*. Printed for R. Morden and W. Berry.

Li, G. (2015). Gnomons in ancient China. In Ruggles, C. L. N. (Ed.), *Handbook of archaeoastronomy and ethnoastronomy* (pp. 2095–2108). Springer.

Liddell, H. G., Scott, R., Jones, H. S., & McKenzie, R. (1940). *A Greek-English lexicon*. Clarendon Press.

Malville, J. M. (2015). Inca royal estates in the Sacred Valley. In Ruggles, C. L. N. (Ed.), *Handbook of archaeoastronomy and ethnoastronomy* (pp. 865–877). Springer.

Merleau-Ponty, M. (1962). Un inédit de Maurice Merleau-Ponty. *Revue de Metaphysique et de Morale*, *67*(4), 401–409.

Needham, J., Beer, A., & Pulleyblank, E. G. (1961). An 8th-century meridian line: I-Hsing's chain of gnomons and the pre-history of the metric system. *Vistas in Astronomy*, *4*, 3–26.

Newton, R. R. (1980). The sources of Eratosthenes' measurement of the Earth. *Quarterly Journal of the Royal Astronomical Society*, *21*(4), 379–387.

North American Sundial Society. (2020). *Sundial mathematics*. Retrieved September 11, 2025, from https://www.sundials.org/index.php/teachers-corner/sundial-mathematics

Oxford University Press. (n.d.). Time. In *Oxford English Dictionaries*. Retrieved from https://web.archive.org/web/20120704084938/http://oxforddictionaries.com/definition/time

Patel, K. (2019, August 23). The first planned city in India. *NASA Earth Observatory*. https://earthobservatory.nasa.gov/images/145509/the-first-planned-city-in-india

Puranavedas. (2021, February 26). Vedic paintings and sculpture. Retrieved September 10, 2025, from https://www.puranavedas.com/paintings-and-sculpture/

Rajawat, D. S. (1980). *Astronomical observatory of Jaipur*. Delta Publishers.

Sharma, V. N. (1991). The Kapāla Yantras of Sawai Jai Singh. *Indian Journal of History of Science*, *26*(2), 173–190.

Shi, Y. (2015). Ancient Chinese astronomy: An overview. In Ruggles, C. L. N. (Ed.), *Handbook of archaeoastronomy and ethnoastronomy* (pp. 2107–2118). Springer.

Shukla, D. N. (1993). *Vastu-sastra: Hindu science of architecture*. Munshiram Manoharlal Publishers.

Stanford Solar Center. (n.d.). Viewing and understanding the analemma. Stanford University. https://solar-center.stanford.edu/art/analemma.html

Taki, T. (2008, March 2). Sundial of Trinità dei Monti in Rome. http://ss977526.stars.ne.jp/sundial/sundial_trinita.htm

The Astronomical Observatories of Jai Singh. (n.d.). The observatory sites. Retrieved September 10, 2025, from https://www.jantarmantar.org/learn/observatories/sites/index.html

UNESCO Portal to the Heritage of Astronomy. (n.d.). Retrieved from https://web.astronomicalheritage.net/index.php/show-entity

U.S. Naval Observatory, Astronomical Applications Department. (n.d.). *Computing approximate solar coordinates*. Retrieved September 15, 2025, from https://aa.usno.navy.mil/faq/sun_approx

Vitruvius Pollio. (1914). *The ten books on architecture* (Morgan, M. H., Trans.). Harvard University Press.

Vitruvius Pollio, M. (1557). *I dieci libri dell'architettura* (D. Barbaro, Trans. & Ed.). Venice, Italy: Francesco de' Franceschi & Giovanni Chrieger. (Original work published ca. 15 BCE)

Vogt, D. E. (2015). Medicine wheels of Great Plains. In Ruggles, C. L. N. (Ed.), *Handbook of archaeoastronomy and ethnoastronomy* (pp. 541–550). Springer.

ADDITIONAL READING

Bowker, J. (Ed.). (1997). *The Oxford dictionary of world religions*. Oxford University Press.

Derrida, J., & Leavey, J. P. (1978). *Edmund Husserl's Origin of geometry: An introduction*. University of Nebraska Press.

Dirac, P. A. M. (1967). *The principles of quantum mechanics*. Oxford University Press.

Eliade, M. (1951). *Images and symbols: Studies in religious symbolism* (Mairet, P., Trans.). Harvill Press.

Eliade, M. (1961). *The sacred and the profane: The nature of religion*. Harper & Brothers.

Jenkins, A. (2013). The Sun's position in the sky. *European Journal of Physics, 34*(3), 633–652.

Mukherji, A. (2020). Time and space in the Jantar Mantars. In *Celestial mirror: The astronomical observatories of Jai Singh II* (pp. ix–xviii). Yale University Press., DOI: 10.12987/9780300252439-002

NOAA. (n.d.). *General solar position calculations.* Retrieved September 11, 2025, from https://gml.noaa.gov/grad/solcalc/solareqns.pdf

Sarkar, J. (1984). *A history of Jaipur: C. 1503–1938.* Orient Blackswan.

Sastri, B. D. (1861). *Translation of the Surya Siddhanta: A text-book of Hindu astronomy.* Baptist Mission Press.

Singh, A., & Sharma, S. (2019). *The spatial science of vāstushastra in traditional architecture of India. In 2019 Conference papers: Interpretation traditional built environment and practice. ANZAScA.*

ENDNOTE

1 The analemma is a pattern resembling a figure-eight shape and representing the apparent position of the Sun when observed from the same location on Earth at the same time each day over the course of a year. This pattern results from the tilt of the Earth's axis relative to the ecliptic and from the ellipticity of the Earth's orbit (Stanford Solar Center, n.d.).

Part II
Sky, Earth, and Culture

Chapter 6
Geometry Frames the Horizon:
Examples From Archaeoastronomy

ABSTRACT

Following the presentation of the concepts behind the architecture of cosmology, this chapter introduces archaeoastronomy—the interdisciplinary field concerned with the earliest human material expressions of observations of the sky. It examines how these first constructions establish a geometric relationship between the local terrestrial environment and the celestial sphere. The Neolithic megaliths in Malta and the Egyptian pyramids are presented as case studies. They are selected from the multitude of archaeoastronomical examples based on the phenomenological lived experience of the author's being-there at the sites. Both exemplify an architecture of cosmology characterized by a strong geometric intentionality—legible from satellite imagery and equally perceptible through embodied movement while walking through or around them.

DOI: 10.4018/979-8-3373-3399-1.ch006

Figure 1. Ħaġar Qim Archaeological Site, Malta, 2025.05.03.

A BRIEF INTRODUCTION TO ARCHAEOASTRONOMY

Cosmology and archaeoastronomy are deeply interconnected. Humankind has long looked toward the sky, and these observations led to interpretations of celestial phenomena in the universe. The apparent cycles of celestial bodies—most prominently the Sun—defined day and night and structured seasonal recurrence.

The quest to understand the cosmos beyond immediate reach generated ritual practices of spiritual and religious significance, as well as pragmatic applications such as timekeeping and agriculture. It also motivated the construction of markers within the local landscapes of early settlements. The alignments embodied in such markers guided the observation of remote celestial objects and events, generating the earliest examples of what this book defines as an architecture of cosmology—structures that function as spatial signposts to events in the sky. The systematic study of this ancient skywatching constitutes the field of archaeoastronomy.

Archaeoastronomy is an interdisciplinary field "drawing upon the written as much as the archaeological and iconographic record" (Aveni, 1980). Anthony Aveni, one of the leading researchers in the discipline, emphasizes that material culture constitutes the methodological foundation of archaeoastronomical inquiry.

The term itself indicates its scope. It derives from the Greek ἀρχαῖος (*archaios*, "ancient"), from ἀρχή (*archē*, "beginning"), and ἀστρονομία (*astronomia*), composed of ἄστρον (*astron*, "star") and νόμος (*nomos*, "law"). Sharing its prefix with archaeology, archaeoastronomy concerns the human past as interpreted through remains associated with astronomical practices.

Within the scope of this book, the focus narrows to three-dimensional constructions of monumental or architectural character, whether preserved intact or surviving as ruins. This distinction, however, cannot be absolute, since two-dimensional iconographic and symbolic elements—such as rock paintings or carvings—are often integrated with built forms.

Although archaeoastronomy investigates ancient and even prehistoric contexts, the discipline itself is relatively recent. According to the *Oxford English Dictionary*, "the earliest known use of the noun archaeoastronomy is in the 1960s," with documented usage in 1968 in *Current Anthropology* (Oxford University Press, n.d.).

Yet the intellectual roots of the field extend further back, to early modern antiquarian fascination with ruins and prehistoric monuments. The seventeenth-century investigations of Stonehenge, including excavations commissioned by the Duke of Buckingham in 1620 (English Heritage, n.d.), stimulated systematic inquiry into megalithic structures across Britain. Connections between these monuments and astronomical alignments began to be proposed.

As Johnson (1912) recounts in reference to Sir Norman Lockyer's research:

> Sir Norman Lockyer has made careful triangulations and measurements in connection with the prehistoric monuments of Cornwall, Wiltshire, and other counties, and his general conclusion is, that the axes of these monuments, and their mutual geographical relations, were carefully determined by their constructors. The assumption having been made that the monuments were originally aligned to sunrise, the problem has been to calculate when that arrangement could have existed, because now, owing to the alteration of the inclination of the earth's equator with regard of the ecliptic, the solstitial points have considerably shifted their positions. Stonehenge, which Sir Norman Lockyer considers to have been a solstitial temple, was accurately aligned by astronomical means when erected. (Johnson, pp. 252–253)

A major twentieth-century contribution to archaeoastronomy was made by the Scottish engineer and academic Alexander Thom (1894–1985), whose extensive field surveys of Neolithic and Bronze Age megalithic monuments proposed systematic astronomical alignments (Heath, 2003). His work encompassed stone circles and rows in Britain and France (Brittany), including Stonehenge and the Carnac Stones (Thom & Thom, 1971).

Archaeoastronomy is not limited to reconstructing ancient scientific observations of celestial phenomena. It also investigates astronomical practice within broader worldviews—religious, social, and cultural. In this sense, it extends beyond ancient astronomy strictly understood and parallels, in methodological scope, from expansion of astronomy to cosmology discussed in Chapter 4.

The methodology of archaeoastronomy intersects multiple disciplines. In addition to astronomy and archaeology, which form its foundation, it draws upon anthropology, architecture, urban planning, ethnography, geometry, mathematics, and surveying (Ruggles, 2005). Its methods range from archaeological excavation and field measurement to computational modeling, digital mapping, and simulations of past skies corresponding to the period of a site's construction.

Cultural astronomy encompasses archaeoastronomy and ethnoastronomy—the study of astronomical practices within contemporary societies. The term *ethnoastronomy* derives from the Greek ἔθνος (*ethnos*, "people" or "nation"), shared with ethnography, the study of cultural customs.

The principal distinction between archaeoastronomy and ethnoastronomy lies in temporal focus. Archaeoastronomy investigates past civilizations through material remains, whereas ethnoastronomy examines living traditions and contemporary cultural practices related to the sky. Methodologically, ethnoastronomy often incorporates interviews and participatory research in addition to observational data.

Despite these differences, both fields share a central concern: understanding how the perception and observation of the sky shape cultural expression, spatial organization, and cosmological imagination.

GEOMETRY OF ARCHAEOASTRONOMY

Since the earliest archaeoastronomical constructions, the primary concern was alignment with celestial bodies—most prominently the Sun. The horizon and the zenith, the cardinal directions, and the positions of sunrise and sunset during significant periods of the year—the equinoxes and solstices—provided the fundamental geometric references.

Such alignments offered stable points of orientation for the ever-changing apparent positions of celestial bodies.. These references were marked within the local landscape using stone—the material most readily available in many prehistoric contexts. Stones of varying size and morphology were positioned as individual elements or arranged in geometric configurations. These arrangements generated diverse typologies of construction: monuments, temples, enclosures, and other architectural compositions.

The geometric configurations of these early alignments are legible in both plan and elevation, defining what this book terms *Earth–Sky geometries*. A visual and spatial relationship is established between the observer and the azimuth and/or altitude on the horizon corresponding to the apparent path of the Sun—or other visually prominent celestial bodies observable with the unaided eye.

This relationship varies according to the Earth's daily rotation around its axis and its annual revolution around the Sun. As discussed in Chapters 3, 4, and 5, these motions are intrinsically linked to astronomical time and have historically enabled systems of timekeeping.

The principal geometric typologies found in archaeoastronomical constructions are summarized below and will be associated with built examples in Table 1.

Lines in Horizontal or Vertical Planes

1. **Linear Geometries**: lines may occur in horizontal or vertical planes.
   a. **Vertical linear elements** (e.g., gnomons, obelisks) (Figure 2a).
   These elements do not constitute alignments in themselves; rather, the shadow they cast establishes alignment with the Sun's altitude, varying according to time of day and season.
   b. **Horizontal alignments** (Figure 2b).
   Rows of stones define sightlines across the landscape. These alignments are typically associated with azimuthal orientations toward celestial events.

Figure 2. Linear geometries: diagram of a gnomon (a) and stone alignments (b).

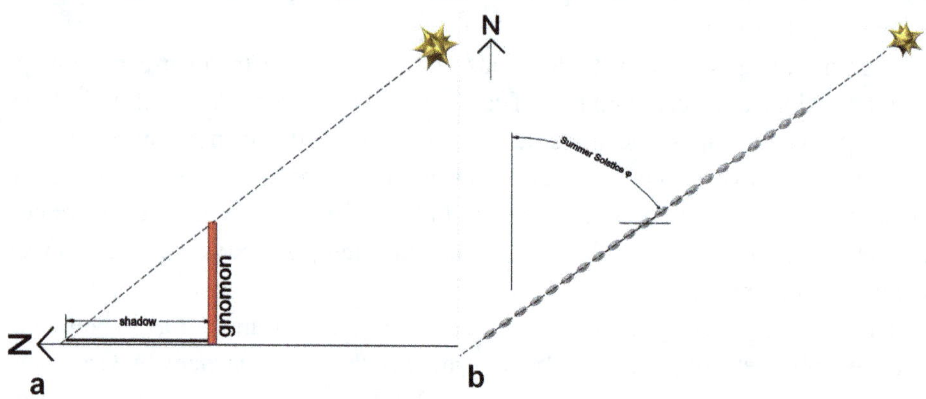

2. **Circular Geometries:** circles in horizontal planes are typically constructed from standing stones (Figure 3). Their geometry allows multiple sightlines radiating from a central observation point toward the horizon.

Figure 3. Diagram of a stone circle.

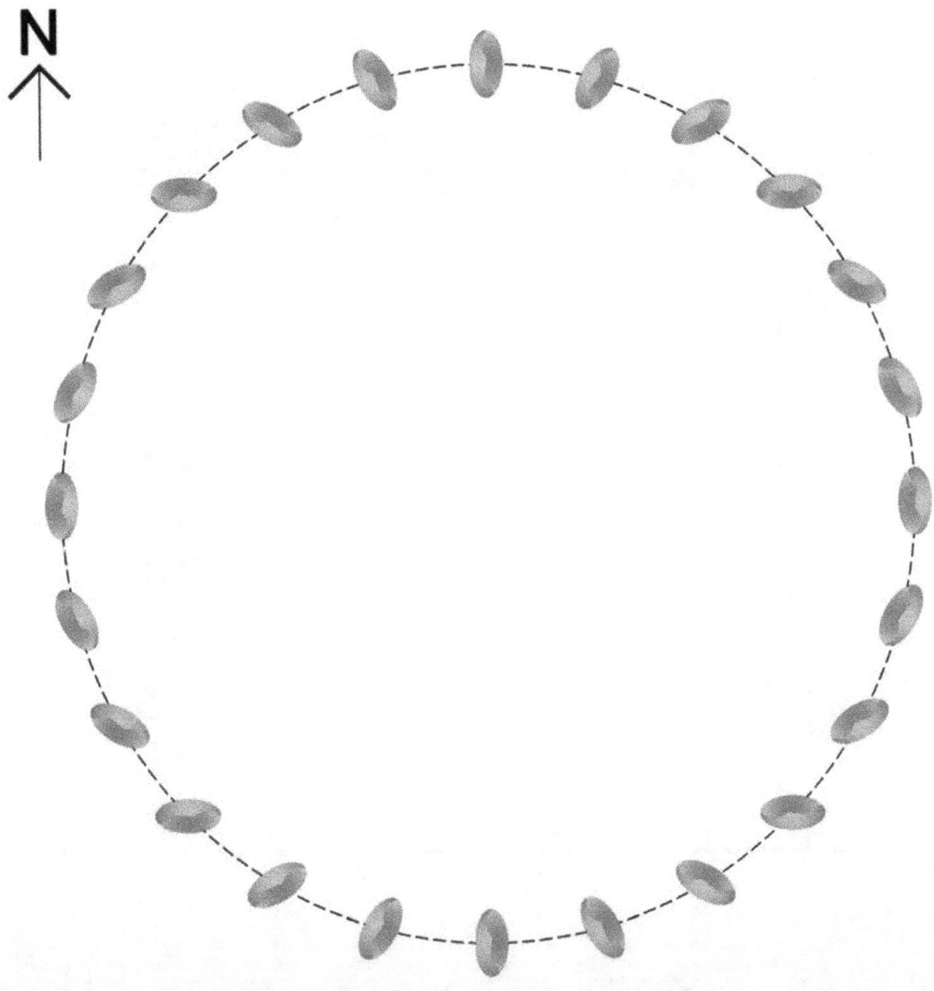

3. **Framed Geometries:** frames or arch-like constructions (Figure 4) are vertical planar compositions. They can be defined geometrically as rectilinear orthogonal elements forming an enclosure—often with the ground plane serving as the base. Curvilinear elements may also generate framed apertures. In such cases, the altitude of a celestial body as seen through the frame becomes the focal point of observation.

Figure 4. Diagram of a framed aperture.

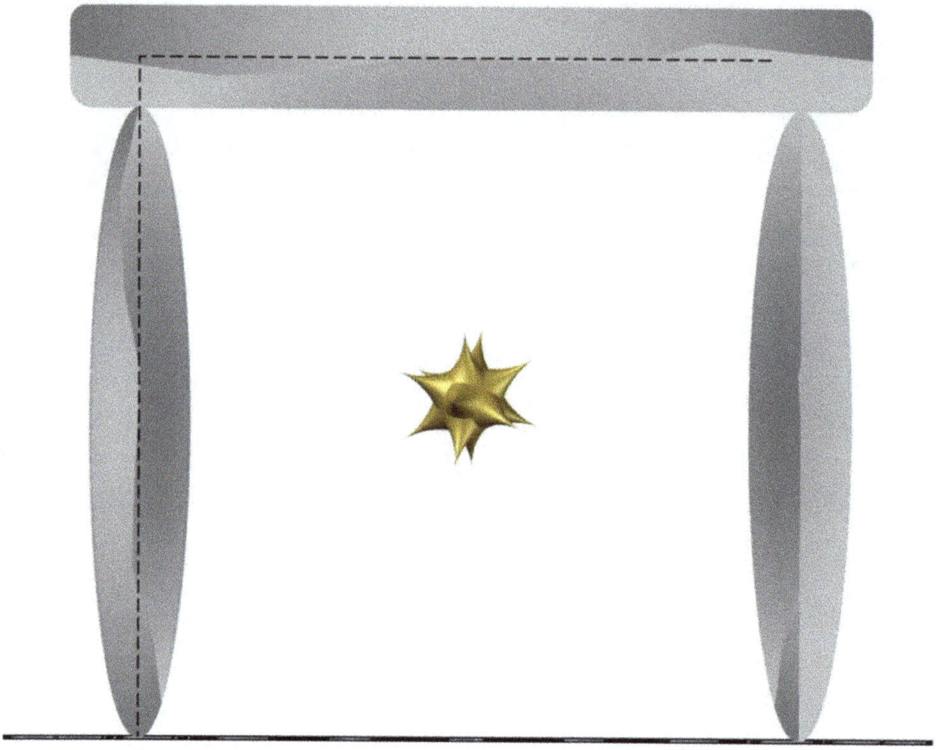

4. **Solid Geometries:** solid volumes—such as pyramids (Figure 5)—function as three-dimensional references for both azimuth and altitude. Their edges, vertices, or faces may encode directional or solar relationships.

Figure 5. Diagram of a solid volume.

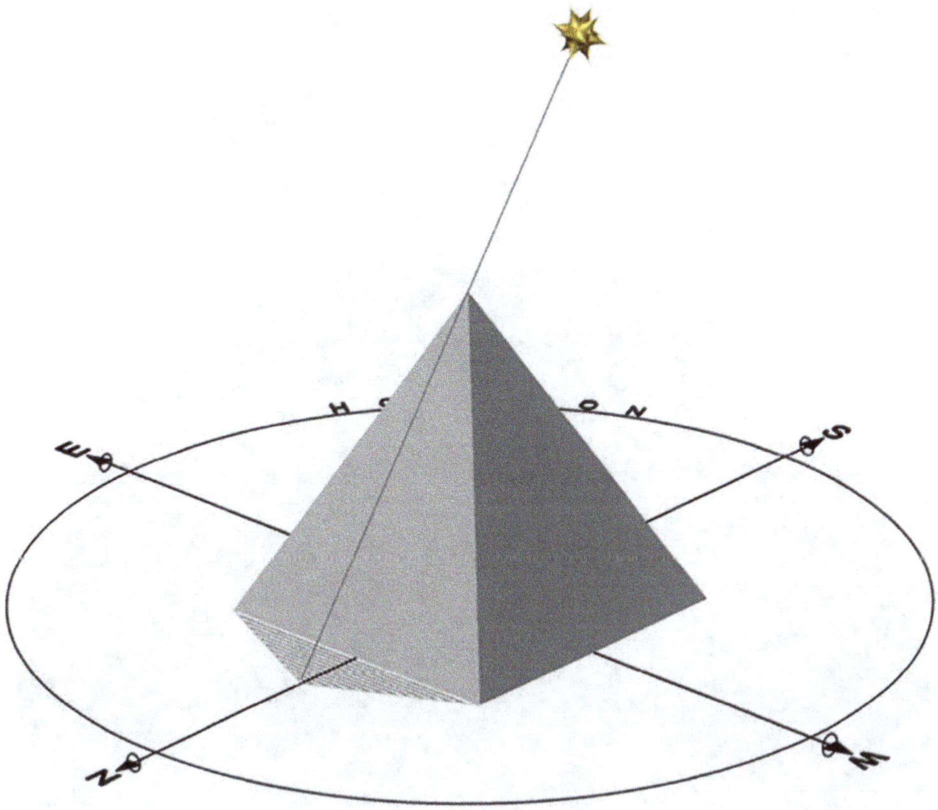

5. **Apertural Geometries:** openings within enclosing structures (Figure 6)—whether walls, passageways, or earthen mounds—create controlled sightlines. These apertures frame specific celestial events, often marking solstitial or equinoctial positions.

Figure 6. Diagram of an opening within an enclosure.

MEGALITHIC STRUCTURES IN ARCHAEOASTRONOMY

The term *megalithic* derives from the Greek μέγας (*megas*, "large") and λίθος (*lithos*, "stone"). The principal megalithic forms associated with archaeoastronomy include menhirs, dolmens, and cromlechs. Not all megaliths exhibit demonstrable astronomical alignments; however, many reveal spatial relationships with celestial events, particularly sunrise and sunset at equinoxes and solstices.

Menhir

A *menhir*—from the Breton *men* ("stone") and *hir* ("long")—is the most elementary megalithic form: a standing stone, typically monolithic. The earliest documented examples appear in Brittany around 5000 BCE (Tilley, 2004, p. 33).

Menhirs are among the earliest monumental markers and remained in use for more than 2,500 years, spanning the Neolithic through the Bronze Age (Tilley, 2004, p. 33). Their functions likely included territorial marking and the designation of settlement sites.

Geometrically, a single menhir can be assimilated to a vertical line. When arranged in rows, menhirs define horizontal alignments whose function has been interpreted—though not conclusively—as astronomical in character. The most extensive examples of such rows are the Carnac Stones in Brittany.

Dolmen and Trilithon

Dolmen derives from the Breton *tolmen*, meaning "stone table." It consists of two or more vertical stones supporting a horizontal capstone, forming a rudimentary framed structure. Dolmens are commonly associated with burial chambers.

Trollasten Dös (55.46546° N, 13.9499° E), \ near Ystad, Sweden, is considered one of the oldest dolmens associated with astronomical alignments. Dating to approximately 3300–3200 BCE, it faces the winter solstice sunset at an azimuth of approximately 241.2° (Stetson University, n.d.).

Trilithon (from the Greek τρίλιθος, "three stones") is geometrically analogous to the dolmen. It consists of two upright stones supporting a lintel. Notable trilithons are present at Stonehenge and within the prehistoric temples of Malta.

Cromlech

The term *cromlech* derives from the Welsh *crom* ("bent" or "arched") and *llech* ("flat stone"). It denotes a circular arrangement of standing stones and may be geometrically assimilated to a circle.

Cromlechs used for astronomical purposes operate through sightlines defined between an observer—often positioned near the center—and individual stones along the perimeter. Each sightline corresponds to a specific azimuth associated with a celestial event.

The Cromlech of Almendres (38.55751° N, 8.06123° W), Portugal is among the oldest known examples, dating back more than 7,000 years (BBC Reel, 2022).

Earthworks and Combined Typologies

Some megalithic structures are integrated with earthworks, occasionally excavated into the ground.

Henges are circular enclosures comprising an internal ditch and an external bank, sometimes containing timber or stone arrangements.

Cairns are artificial mounds of stone, frequently covering burial chambers or passage tombs.

Newgrange (53.69471° N, 6.47549° W), Ireland is a prominent example of a cairn with a passage grave exhibiting astronomical alignment. Dating to approximately 3200 BCE, its passage and inner chamber align with the sunrise at the winter solstice (Brú na Bóinne World Heritage Site, n.d.).

EARTH–SKY GEOMETRIC TYPOLOGIES IN MAIN ARCHAEOASTRONOMICAL SITES

The aim of this chapter is not a brief history of archaeoastronomy but to provide a chronological and geographic overview of the main sites which are relevant in a geometrical characterization.

Table 1 synthesizes principal archaeoastronomical sites across cultures through a refined Earth–Sky geometric taxonomy. Rather than classifying monuments solely by period or civilization, the table identifies their operative geometric logic: radial enclosures, azimuthal alignments, axial passage systems, framed apertures, solid cardinal geometries, and vertical gnomonic instruments. This typological refinement aligns with the theoretical framework developed in Chapter 3, where spatial orientation is understood as embodied mediation between terrestrial grounding and celestial reference. The comparison reveals recurring geometric strategies—circular, axial, orthogonal, pyramidal, and vertical—that articulate cosmological order through measurable astronomical events.

Table 1. Principal Archaeoastronomical Sites and Refined Earth–Sky Geometric Typologies

Site	Location	Coordinates (Lat°, Long°)	Date (Approx.)	Earth–Sky Geometric Typology	Scholarly Reference
Göbekli Tepe	Turkey	37.22329° N, 38.92182° E	9500–8000 BCE	Radial enclosure; orthostat pillars forming circular cosmogram	Schmidt (2010)
Nabta Playa	Egypt	22.76791° N, 30.91170° E	c. 7500 BCE	Stone circle; solar azimuthal alignment	Malville et al. (1998)
Cromlech of Almendres	Portugal	38.55751° N, 8.06123° W	c. 6000 BCE	Circular ring; horizon-based solar sightlines	Silva & Pimenta (2012)
Carnac Stones	France	47.59183° N, 3.08359° W	4600–4300 BCE	Linear megalithic azimuthal rows	Thom (1971)
Mnajdra Temples	Malta	35.82658° N, 14.43652° E	3600–3200 BCE	Framed trilithic aperture marking solstitial sunrise	Hoskin (2001)
Newgrange	Ireland	53.69471° N, 6.47549° W	c. 3200 BCE	Axial passage framing winter solstice sunrise	Ruggles (1999)
Giza Plateau (Pyramids)	Egypt	29.97693° N, 31.13265° E	c. 2600 BCE	Solid pyramidal geometry; cardinal and stellar orientation	Lehner (1997)
Stonehenge	England	51.17886° N, 1.82622° W	c. 2500 BCE	Circular trilithic structure; solstitial axial alignment	Ruggles (2005)
Early Chinese Gnomon	China	35.89292° N, 111.48962° E	c. 2300 BCE	Vertical linear instrument measuring solar altitude (shadow geometry)	Needham et al. (1961)
Uxmal (Governor's Palace)	Mexico	20.35879° N, 89.77113° W	c. 700 CE	Architectural aperture aligned to Venus cycles	Aveni (2001)
Chichén Itzá (El Castillo)	Mexico	20.68298° N, 88.56864° W	800–1200 CE	Pyramidal solar-shadow phenomenon (equinox serpent effect)	Aveni (2001)
Sun Dagger, Chaco Canyon	USA	36.06117° N, 107.96568° W	850–1250 CE	Vertical slab aperture marking solstices and lunar standstills	Sofaer (1997)
Angkor Wat	Cambodia	13.41262° N, 103.86696° E	c. 1150 CE	Orthogonal cardinal plan with equinoctial solar alignment	Magli (2013)
Machu Picchu	Peru	13.16418° S, 72.54464° W	1420–1530 CE	Trapezoidal apertures framing solstitial events	Malville & Ziegler (2007)

Thebes

The ancient city of Thebes is located approximately 650 km south of Cairo, on both the east and west banks of the Nile River. Originally known as Waset, Thebes was inhabited from at least the early third millennium BCE and became closely associated with the cult of Amun. It served as capital during significant periods of the Middle and New Kingdoms (Shaw, 2000).

Today, Thebes survives as extensive ruins integrated within the modern city of Luxor. In 1979, "Ancient Thebes with its Necropolis" was inscribed as a UNESCO World Heritage Site. The archaeological complex includes the temples of Luxor and Karnak on the east bank and major funerary complexes on the west bank.

The principal sites are listed from south to north and are shown in the satellite map in Figure 7.

Figure 7. Satellite map of principal ancient Theban sites, arranged from south to north. Map generated by the author using Googe My Maps. Retrieved 2026.02.14.

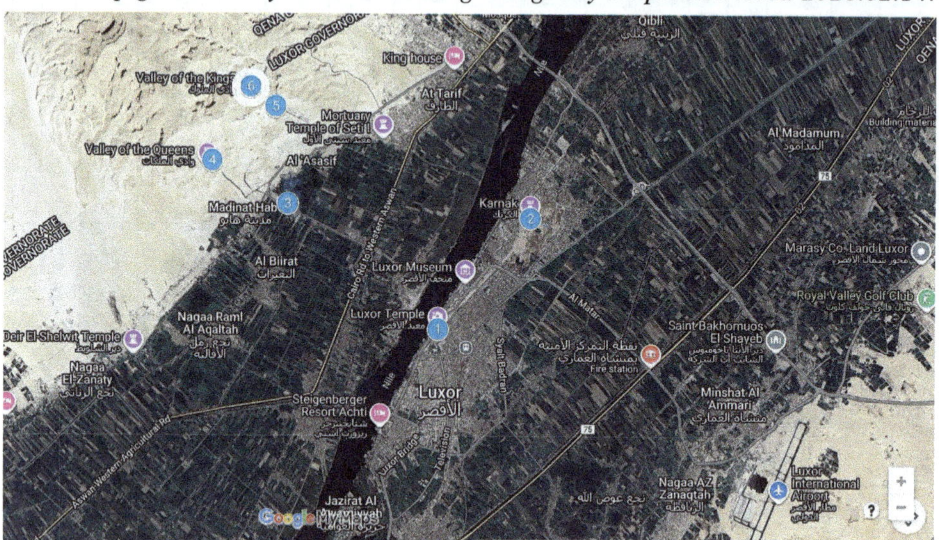

1. Luxor Temple (25.6995° N, 32.63905° E), Figure 8.
2. Karnak Temples (25.71883° N, 32.65727° E), see following section on Karnak
3. Amenhotep III Sun Temple (25.72139° N, 32.60888° E), Figure 9
4. Valley of the Queens (25.72765° N, 32.59308° E).
5. Mortuary Temple of Hatshepsut (Deir el-Bahari) (25.73827° N, 32.60649° E) – Figure 10
6. Valley of the Kings (25.74020° N, 32.60140° E)

Figure 8. Luxor Temple, east bank of the Nile. 2023.03.31.

Figure 9. Amenhotep III Sun Temple (Kom el-Hettan), west bank of the Nile. 2023.03.31.

Figure 10. Mortuary Temple of Hatshepsut, 2023.03.26.

The spatial organization of Thebes reflects an east–west cosmological polarity characteristic of Egyptian sacred geography, the east bank associated with solar renewal and temple ritual, and the west bank associated with funerary complexes aligned with the setting sun (Wilkinson, 2000).

ANCIENT EGYPT: GEOMETRY, COSMOLOGY, AND MONUMENTAL LANDSCAPE

Egyptian archaeoastronomical sites constitute some of the most significant historical expressions of cosmological architecture, both in geometric configuration and celestial alignment. Their spatial organization reflects deliberate relationships between built form and solar movement, particularly in monumental complexes such as the pyramids of Giza and the temples of Upper Egypt (Magli, 2013; Ruggles, 2005).

My visit to Egypt in 2023 confirmed that the experience of the pyramids cannot be reduced to monumental scale alone. The encounter with the Sahara Desert, despite contemporary tourism, foregrounded the perceptual relationship between the terrestrial plane and the celestial vault. The desert operates not merely as background but as spatial ground condition, capable of re-centering perception toward the Earth–sky continuum.

Unlike archaeoastronomical monuments situated in vegetated or topographically articulated landscapes, Egyptian sites emerge from expansive horizons with minimal visual obstruction. In satellite imagery, stone constructions blend into the chromatic continuum of sand. At eye level, however, the geometric mass of pyramids and temples defines a distinct figure against the sky. At latitudes between approximately 22° and 30° N, high solar altitude produces sharply articulated shadow geometries that reinforce axial orientation and volumetric clarity.

In Egyptian cosmological architecture, geometry can therefore be understood through a figure–ground relational logic, structured between desert plane, constructed mass, and celestial vault. The constructed form (pyramid, temple axis, stone alignment) emerges as a deliberate geometric figure against two grounds simultaneously: the terrestrial desert plane and the celestial vault. Moreover, this relationship is not static. It is continually redefined by the ephemeral architecture of light and shadow produced by the apparent solar trajectory. Diurnal and annual cycles inscribe temporal geometry upon built form, rendering cosmology both measurable and experiential.

Monumental complexes such as the Giza Plateau demonstrate cardinal orientation of high precision (Lehner, 1997), while temple complexes such as Karnak incorporate axial solar alignments embedded within successive building phases (Magli, 2013).

Figure 11 maps the principal Egyptian sites of archaeoastronomical significance discussed in this chapter. All are situated within desert landscapes and are listed from south to north:

1. Nabta Playa Ancient Astrological Site (22.76791° N, 30.91170° E) – prehistoric stone circle with probable solar alignments (Malville et al., 1998).
2. Temples of Karnak (25.71883°N, 32.65727°E).
3. Giza Plateau (29.97693°, 31.13265°).

Figure 11. Satellite map of principal Egyptian archaeoastronomical sites (Nabta Playa, Karnak, and Giza Plateau), arranged from south to north. Map generated by the author using Google My Maps. Retrieved 2026.02.14.

Karnak

The temple complex of Karnak extends over approximately 140 acres on the east bank of the Nile River near modern Luxor. It represents one of the largest religious complexes of the ancient world and was developed over more than a millennium, primarily during the Middle and New Kingdoms (Shaw, 2000; Wilkinson, 2000).

Among the archaeological sites of ancient Thebes, Karnak is of particular archaeoastronomical interest due to the axial orientation of its principal temple precinct, especially within the domain of Amun-Re. The longitudinal axis of the main temple has been associated with solar alignments, particularly at the winter solstice (Magli, 2013; Ruggles, 2005). This alignment will be examined in the following subsection.

The Karnak complex includes four principal precincts, each centered on a major cult temple (Wilkinson, 2000). These precincts are shown in Figure 12 and are listed below from south to north:

1. Precinct of Mut (25.71166° N, 32.65521° E)
   Located in the southernmost area of the complex, this precinct is dedicated to the goddess Mut, consort of Amun.
2. Precinct of Amun-Re (25.71939° N, 32.65616° E)
   The largest precinct and the core of the Karnak complex. It is the only precinct extensively accessible to visitors and contains the Great Hypostyle Hall and the primary east–west temple axis. See the following subsection, "Amun-Re."
3. Temple of Amenhotep IV (Akhenaten), Gem-pa-Aten (25.71755° N, 32.66256° E)
   Constructed during the 18th Dynasty under Pharaoh Akhenaten (formerly Amenhotep IV), this temple formed part of the king's Atenist religious reforms (Shaw, 2000). The structure was later dismantled following the restoration of the Amun cult, and only archaeological traces remain (Blyth, 2006). On modern site maps, the area is identified as *Gem-pa-Aten*.
4. Precinct of Montu (25.71990° N, 32.66080° E)
   Situated in the northernmost sector of the complex, this precinct was dedicated to Montu, a falcon-headed war deity associated with Theban theology (Wilkinson, 2000).

Figure 12. Satellite view of the four principal precincts of the Karnak temple complex (Mut, Amun-Re, Gem-pa-Aten, and Montu). Map generated by the author using Google My Maps. Retrieved 2026.02.14.

Amun-Re

The Precinct of Amun-Re is the best preserved and most visited of the four precincts comprising the Karnak temple complex. It is also the most extensively studied from both architectural and archaeoastronomical perspectives (Magli, 2013; Ruggles, 2005).

The name Amun-Re (also written Amun-Ra) combines Amun, the principal deity of Thebes, with Ra (Greek: Hēlios), the solar deity of ancient Egyptian religion. The fusion of these deities reflects the theological centrality of solar cosmology in Theban state religion during the New Kingdom (Shaw, 2000; Wilkinson, 2000). The importance of the precinct is expressed both in its monumental scale and in its axial organization.

Satellite imagery of the Karnak complex reveals two principal orthogonal axes. The dominant longitudinal axis runs approximately northeast–southwest, at an angle of about 23° from true north. Its perpendicular axis measures approximately 115° (and 295° in the opposite direction). These values are numerically close to the azimuth of sunrise at the winter solstice and sunset at the summer solstice at the latitude of Thebes (approximately 25.7° N). In the following discussion, this primary longitudinal axis is referred to as the "solstice axis."

The hypothesis that the main axis of the Temple of Amun-Re corresponds to solstitial solar events was first proposed by Sir Norman Lockyer in *The Dawn of Astronomy* (1894). Lockyer observed:

> At Karnak itself there are several temples so oriented, chief among them the magnificent Temple of Amen-Ra, one of the wonders of the world, to which a special chapter must be devoted. Suffice it to say here that the amplitude of the point to which the axis of the great temple of Amen-Ra points is 26° N. of W., which we learn from the table already given is the amplitude of the place of sunset at the summer solstice in the latitude of Thebes. The amplitude of the point to which the axis of an attached small temple is at angle 26° S. of E., exactly the position of sunrise at the winter solstice. (Lockyer, 1894, p. 78)

Subsequent studies have revisited and refined Lockyer's conclusions, confirming that the temple axis is closely aligned with solstitial solar positions (Figure 13), while also acknowledging topographical influences related to the Nile's course (Krupp, 1988; Belmonte & Shaltout, 2005; Magli, 2013). Although the local topography influenced urban development, solar symbolism and ritual orientation were central components of New Kingdom temple planning.

Figure 13. Satellite view of the Precinct of Amun-Re with superimposed solstice axis and approximate solar alignments. Map generated by the author using Google My Maps. Retrieved 2026.02.14.

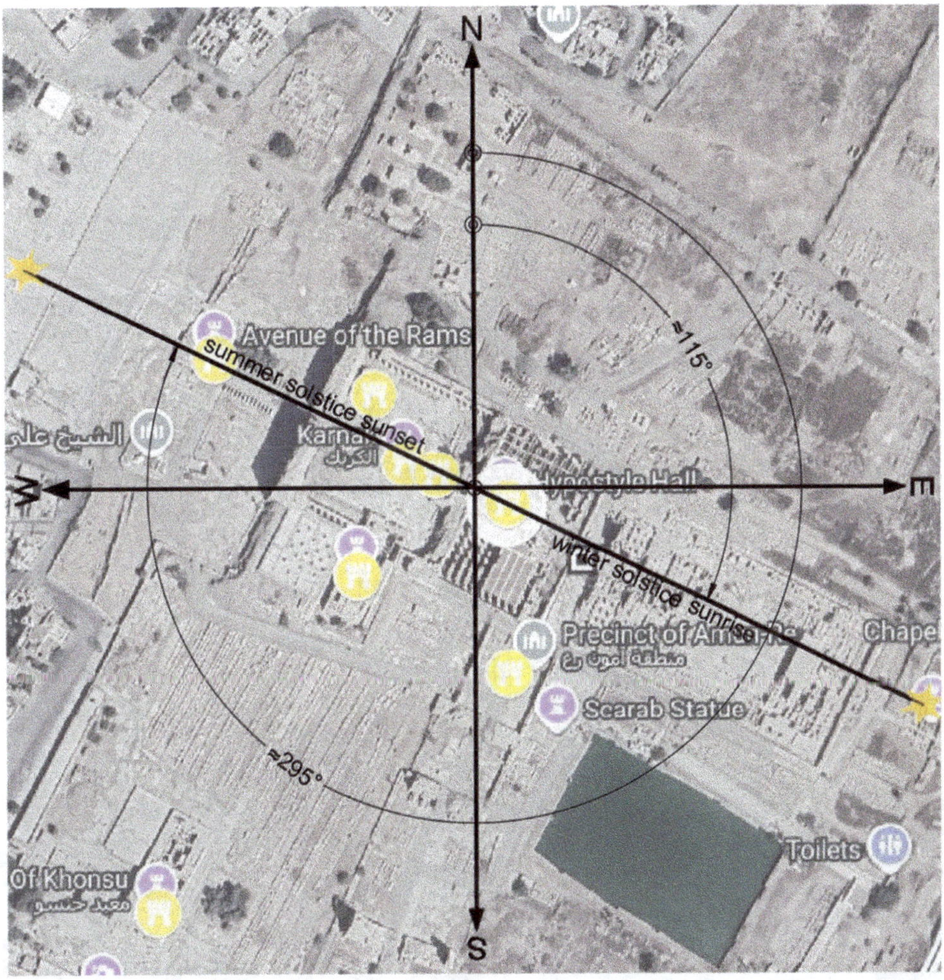

Chronology

The chronology of the Precinct of Amun-Re is complex and spans approximately 1500 years. Construction began during the Eleventh Dynasty under Intef II (c. 2112–2063 BCE). Major expansions occurred during the Eighteenth and Nineteenth Dynasties (1550–1070 BCE). The Great Hypostyle Hall was initiated under Seti I (1290–1279 BCE) and completed under Ramesses II (1279–1213 BCE) (Shaw, 2000; Wilkinson, 2000).

The precinct covers approximately 69 acres and is enclosed by massive mudbrick walls forming a roughly trapezoidal plan with a perimeter of approximately 1.4 miles.

Openings in the enclosure walls are framed by pylons (from Greek *pylōn*, πυλών), monumental gateways composed of two tapering trapezoidal towers flanking a central entrance. These pylons mark successive construction phases and articulate ceremonial progression through the complex (Wilkinson, 2000).

Principal Monuments Along the Solstice Axis

The sequential order in the list follows a west-east axis from the entrance.

- Avenue of the Sphinxes (Avenue of the Rams) (25.71939° N, 32.65616° E), leading to the Great Court (Figure 14).
- Temple of Seti II (25.71921° N, 32.65708° E), located slightly north of the main axis.
- Statue of Ramesses II with Nefertari (25.71878° N, 32.65744° E), dating to the reign of Ramesses II (1279–1213 BCE) (Figure 15).
- Temple of Ramesses III (25.71828° N, 32.65698° E). This structure measures approximately 61 m by 23 m (200 ft by 75 ft) and includes a hypostyle hall (Greek: *hypóstylos*, ὑπόστυλος, meaning "under columns"), supported by columns shaped as papyrus bundles (Figure 16). Its axis is approximately perpendicular to the solstice axis.
- Great Hypostyle Hall (25.71867° N, 32.65784° E) (Figures 17–19).

Figure 14. Avenue of the Sphinxes, or Rams. Photograph by the author, 2023.03.31.

Figure 15. Statue of Ramsses III with Nefretari, looking south. Photograph by the author, 2023.03.31.

Figure 16. Views of Temple of Ramesses III. Photographs by the author, 2023.03.31.

Figure 17. Great Hypostyle Hall, looking at the Sun on the south. Photograph by the author, 2023.03.31.

Figure 18. Great Hypostyle Hall, looking west. Photograph by the author, 2023.03.31.

Figure 19.Great Hypostyle Hall: column inscriptions. Photograph by the author, 2023.03.31.

Constructed during the Nineteenth Dynasty under Seti I and Ramesses II, the hall measures approximately 320 feet by 170 feet, enclosing about 54,000 square feet. The structure contains 134 sandstone columns arranged in 16 rows. Twelve central columns rise to approximately 80 feet and support architraves that once carried clerestory roofing. The central axis aligns with the solstice axis. The scale, structural organization, and preservation of the hall make it one of the most significant surviving examples of New Kingdom temple architecture (Shaw, 2000; Wilkinson, 2000).

- Obelisk of Thutmose I (25.71833° N, 32.65858° E), erected during the reign of Thutmose I (c. 1520–1492 BCE). It stands near the fourth pylon and is slightly south of the solstice axis (Figure 20).
- Obelisk of Hatshepsut (25.71831° N, 32.65857° E), erected during the reign of Hatshepsut (1479–1458 BCE), located between the fourth and fifth pylons (Figure 20).
- Festival Hall (Akh-men) of Thutmose III (25.71797° N, 32.65924° E), situated near the eastern end of the solstice axis.

Figure 20. Looking north: Obelisks of Thutmose (left) and Hatshepsut (right). Photograph by the author, 2023.03.31.

Southern Areas of the Precinct

The following areas are located south of the solstice axis; they are of lesser importance to astronomical practices.

- Cachette Courtyard (25.71777° N, 32.65787° E), where a large deposit of statues and ritual objects was discovered in 1903 by Georges Legrain (Shaw, 2000).
- Temple of Khonsu (25.71655° N, 32.65581° E), dedicated to Khonsu, son of Amun and Mut, constructed during the New Kingdom.
- Sacred Lake (25.71688° N, 32.65873° E), an artificial basin (Figure 21) excavated during the reign of Thutmose III. It functioned in ritual purification and ceremonial practice.

Figure 21. Sacred Lake, looking north. Photographs by the author, 2023.03.31.

Experiential Note

I visited the Precinct of Amun-Re on 2023-03-31 at approximately 11:00 EET. At that time, the Sun was near meridian transit, producing strong vertical shadows that accentuated the scale of columns and pylons. The contrast between sandstone surfaces and the blue sky intensified the perception of axial progression along the solstice axis. The monumental scale and solar orientation together reinforce the cosmological dimension embedded in the architecture.

Giza Necropolis

The Giza Plateau (Arabic: al-Jīzah, ةزيجلا; also historically spelled Gizeh) is a limestone plateau on the west bank of the Nile River forming the site of the Giza Necropolis. The entrance to the necropolis lies at approximately 29.97737° N, 31.13277° E, about 5.6 km southwest of central Cairo.

The Giza Necropolis includes three principal Fourth Dynasty pyramids, associated queen pyramids, mastabas, temples, and the Great Sphinx. The site forms part of the UNESCO World Heritage property "Memphis and its Necropolis – the Pyramid Fields from Giza to Dahshur," inscribed in 1979.

A satellite overview of the plateau is shown in Figure 22.

Figure 22. Satellite view of the Giza Necropolis. Map generated using Google My Maps. Retrieved 2026.02.14.

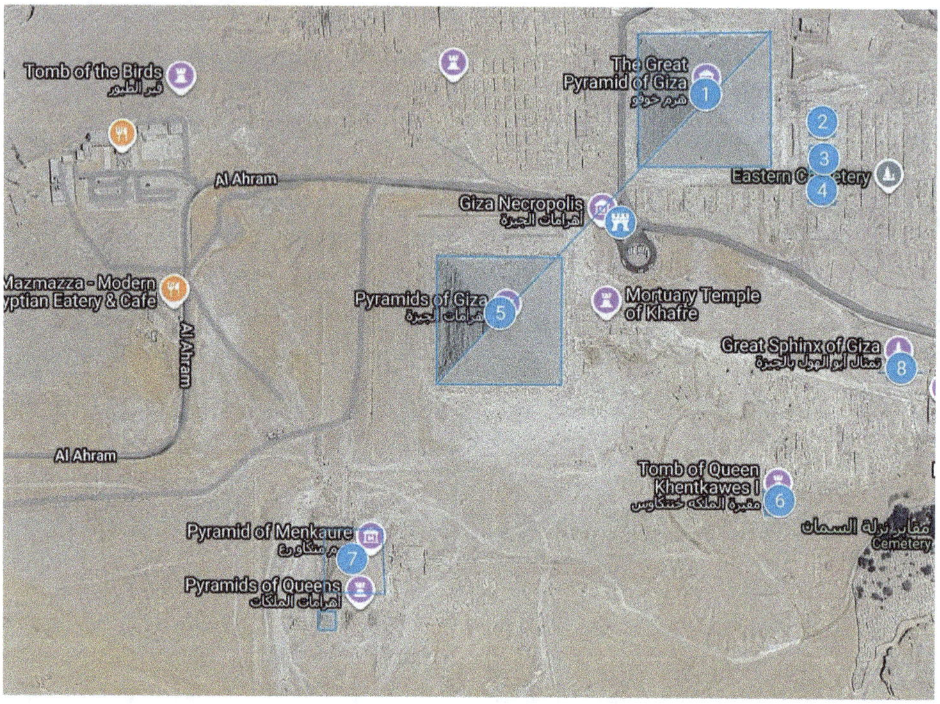

Principal Monuments of the Giza Necropolis

There are seven principal pyramid structures within the Giza Necropolis (Lehner, 1997; Verner, 2001). The three largest pyramids—Khufu, Khafre, and Menkaure—were constructed during the Fourth Dynasty (ca. 2575–2465 BCE).

Figure 23. The three Fourth Dynasty pyramids and the Great Sphinx viewed from the east, 2023.03.31.

The following monuments are listed from north to south (see Figure 22):

1. Great Pyramid of Khufu (Cheops in Greek)
   29.97922° N, 31.13422° E (Figure 24)
   Constructed for Pharaoh Khufu (2589–2566 BCE), this is the largest pyramid in Egypt. The square base measures approximately 230 m (756 ft) per side, and the original height was about 146.6 m (481 ft) (Lehner, 1997; Verner, 2001). The sides are aligned with high precision to the cardinal directions.
2. Pyramid G1-A (Queen Hetepheres I)
   29.97883° N, 31.13622° E
   Likely associated with Hetepheres I, mother of Khufu. Originally about 30 m in height; now significantly reduced (Lehner, 1997).
3. Pyramid G1-B
   29.97831° N, 31.13624° E
   Attributed to a queen of Khufu, possibly Meritites I, though attribution remains debated (Verner, 2001).
4. Pyramid G1-C
   29.97786° N, 31.13622° E

Associated with Queen Henutsen. Base side approximately 44 m (144 ft).

5. Pyramid of Khafre (Chephren in Greek)
   29.97605° N, 31.13074° E (Figure 25)
   Built for Khafre (2558–2532 BCE). Base side approximately 216 m (708 ft); original height approximately 143 m (471 ft). Remnants of limestone casing remain near the apex (Lehner, 1997).
6. Tomb of Queen Khentkawes I
   29.97334° N, 31.13550° E
   Geometrically, this tomb is a stepped monument rather than a true pyramid (Verner, 2001).
7. Pyramid of Menkaure (Mykerinos in Greek)
   29.97248° N, 31.12824° E (Figure 26)
   Constructed for Menkaure, grandson of Khufu. Base side approximately 102 m (335 ft); height approximately 65 m (218 ft) (Lehner, 1997).

Experiential Observation

I visited the Giza Necropolis on 2023-03-31, following my visit to the Karnak complex. Unlike the framed axial experience of the Great Hypostyle Hall at Karnak, the Giza Plateau presents an open horizon condition. The sky is not architecturally enclosed but serves as a vast celestial ground against which the pyramids emerge as geometric figures.

Despite the density of tourism, the clarity of the pyramidal forms against the desert and sky reinforced the perceptual experience of cosmological architecture. The geometry of the square bases and the sharply inclined faces produces a strong figure–ground relationship that is inseparable from the celestial backdrop.

Figures 23–27 document a chronological photographic sequence of the site, beginning from the east with the Great Sphinx in the foreground and proceeding counterclockwise around the plateau. These views illustrate the compositional relationships among the monuments, a spatial condition less common in single-axis temple complexes.

Figure 27. The three pyramids viewed toward the southeast: Khufu at northeast (left), Khafre center, Menkaure right, 2023.03.31.

The Great Sphinx of Giza

Unlikely the pyramids, the Great Sphinx of Giza is a monolithic sculptural monument (Figures 28–29). Carved from limestone bedrock, it represents a hybrid creature combining a human head with a lion's body. The monument is generally attributed to the reign of Khafre (Fourth Dynasty), though alternative hypotheses exist (Lehner, 1997).

Figure 24. Pyramid of Khufu (Cheops), viewed from the west, 2023.03.31.

Figure 25. Pyramid of Khafre (Chephren) viewed from the west; the skyline of Cairo visible on the horizon, 2023.03.31.

Figure 26. Pyramid of Menkaure (Mykerinos), camels resting in the foreground, 2023.03.31.

Figure 28. The Great Sphinx with the Pyramid of Khafre in the background (left) and the Great Pyramid of Khufu (right), 2023.03.31.

Figure 29. The Great Sphinx viewed from the southeast, 2023.03.31.

The term sphinx derives from Greek *sphinx* (σφίγξ), though the Egyptian form represents a distinct iconographic tradition associated with royal guardianship.

The Great Sphinx of Giza, like the pyramids, presents an alignment to the four cardinal points. The center of the Great Sphinx is located at approximately 29.97529° N, 31.13758° E. It faces due east and measures about 73.5 m (242 ft) in length and 20 m (66 ft) in height. Its eastward orientation aligns it with sunrise, reinforcing the solar symbolism associated with the Fourth Dynasty necropolis (Lehner, 1997).

Geometrical and Astronomical Significance

In satellite view (Figure 22), the Giza pyramids display highly regular geometric organization. Each of the principal pyramids has a square base with sides aligned very closely to the cardinal directions (Figure 30). The precision of this alignment has been studied extensively.

One hypothesis suggests that stellar observations—possibly involving circumpolar stars such as Phecda (γ Ursae Majoris) and Megrez (δ Ursae Majoris)—were used to determine true north (Belmonte, 2001). Other methods have also been proposed. The detailed technical debate exceeds the scope of this chapter; however, the astronomical intentionality of cardinal orientation is widely acknowledge (Magli, 2013).

Cardinal alignment reinforces solar orientation. At the equinoxes, sunrise and sunset occur approximately along the east–west axis of each pyramid. At local solar noon, the Sun aligns with the meridian, emphasizing the north–south faces of the pyramids.

Geometric relationships also exist among the pyramids themselves. The diagonal extending from the northeast to southwest corners of the Pyramid of Khufu aligns approximately with that of the Pyramid of Khafre (see Figure 30). The three queen pyramids (G1-A, G1-B, G1-C) are also approximately aligned and are of comparable scale, reinforcing the geometric coherence of the overall layout.

Figure 30. Diagram of the Giza pyramids superimposed on satellite imagery from Google Earth, emphasizing the geometric alignment to the four cardinal directions.

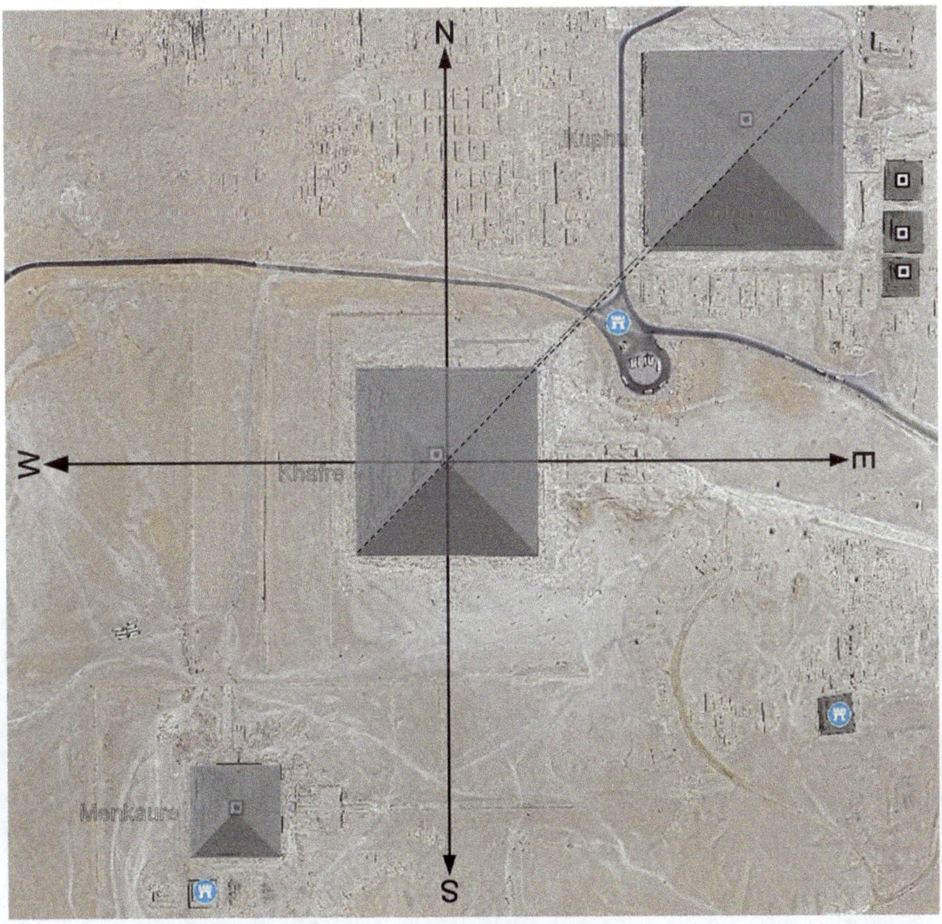

The so-called Orion Correlation Theory, proposing correspondence between the three pyramids and the three stars of Orion's Belt, remains controversial. While some statistical analyses have explored geometric parallels (Orofino & Bernardini, 2016), the hypothesis is not universally accepted within Egyptology.

THE NEOLITHIC TEMPLES IN THE MALTESE ISLANDS

The Mediterranean islands of Malta and Gozo—collectively referred to as the Maltese Islands—host some of the world's earliest freestanding megalithic temples and are widely regarded among the oldest known archaeoastronomical sites (Figure 31). Distributed across different locations throughout the archipelago, these prehistoric structures share common architectural typologies, construction techniques, and chronological phases rooted in the Neolithic period.

Figure 31. Geographical distribution of the principal Neolithic temple sites on Malta and Gozo. Satellite base map generated with Google My Maps.

The principal temple-building phases span the fourth to the third millennium BCE. These are conventionally classified as the Ġgantija phase (ca. 3600–3200 BCE), the Saflieni phase (ca. 3300–3000 BCE), and the Tarxien phase (ca. 3000–2500 BCE), named after representative sites (Malta Tourism Authority, n.d.; Trump, 1999). The "Megalithic Temples of Malta" were inscribed on the UNESCO World Heritage List in 1980, initially with Ġgantija (Gozo), and subsequently extended to include

Ħaġar Qim, Mnajdra, Skorba, Ta' Ħaġrat, and Tarxien. UNESCO describes them as "amongst the earliest free-standing stone buildings in the world" and highlights their architectural diversity and decorative richness (United Nations Educational, Scientific and Cultural Organization [UNESCO], n.d.).

Geographical and Cultural Context

As islands situated in the central Mediterranean, approximately 80 km south of Sicily and about 240 km from the North African coast, Malta and Gozo developed distinctive cultural trajectories while remaining positioned within broader maritime exchange networks. Archaeological evidence indicates human presence as early as ca. 5000 BCE, including cave habitation at Għar Dalam and early village settlements such as Skorba (Encyclopædia Britannica, n.d.; Trump, 1999).

Material culture—particularly pottery—constitutes a key diagnostic element of Maltese Neolithic phases. Funerary practices are exemplified by the Ħal Saflieni Hypogeum (ca. 3300–3000 BCE), an underground complex carved into globigerina limestone and also inscribed as part of the UNESCO World Heritage property "Megalithic Temples of Malta" (UNESCO, n.d.). The Hypogeum demonstrates advanced spatial articulation and acoustic properties, reinforcing the ritual dimension of Maltese prehistoric architecture.

Although early temple builders operated in relative insular conditions, Malta's strategic location later attracted successive Mediterranean powers, including Phoenicians, Romans, Arabs, Normans, and others. The Maltese language reflects this layered history: primarily Semitic in origin, with strong Sicilian and Romance influences, and written in the Latin alphabet (Encyclopædia Britannica, 2025).

Architectural and Geometrical Typology

As diagrammatically represented in Figure 32, the Maltese Neolithic temples exhibit recurring geometric and architectural characteristics.

Figure 32. Schematic diagram of the typical Maltese Neolithic temple plan, showing axial symmetry and apsidal chambers (after Trump, 1999).

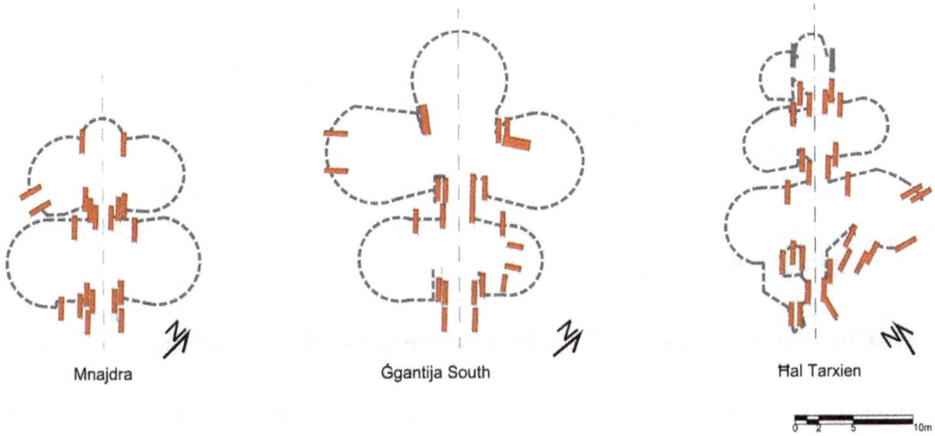

Most temples are organized along a principal axis of bilateral symmetry terminating in a focal apse. Curvilinear chambers (apses) flank the main axis, often themselves symmetrically arranged (Trump, 1999). The façade typically presents a concave forecourt articulated by a trilithon—two vertical orthostats supporting a horizontal lintel—which frames the axial entrance. The internal spatial sequence proceeds through an ellipsoidal forecourt into progressively articulated apsidal chambers, often culminating in an altar or ritual focus at the terminal apse.

Construction materials consist primarily of limestone. Exterior walls were frequently constructed of harder coralline limestone, while interior architectural elements often employed softer globigerina limestone. Roofing was achieved through corbelling, a progressive cantilevering technique forming domical or vault-like enclosures.

Despite strong typological coherence, orientations vary considerably among temples—even among structures belonging to the same archaeological phase.

Orientation and Archaeoastronomical Interpretations

Unlike the strict cardinal alignments observed in Old Kingdom Egyptian pyramids, Maltese temple orientations do not follow a single uniform astronomical scheme.

Ventura and Hoskin (2015) observed that many temples exhibit orientations between southeast and southwest, suggesting possible alignments with sunrise positions between solstices or with the heliacal rising of the Pleiades around 3000 BCE. An exception frequently cited is the Mnajdra South Temple, whose orientation corresponds closely to solar positions at equinox and solstice events.

Interpretative complexity is compounded by axial precession. The slow precessional movement of Earth's rotational axis alters stellar rising points over millennia, meaning that any stellar alignment hypothesis must account for astronomical retrocalculation (Foderà Serio et al., 1992). Additional hypotheses propose lunar alignments or orientations toward southern constellations such as Centaurus. However, no single astronomical model accounts comprehensively for all temple orientations.

Given the maritime context of island societies, stellar observation likely played a role in navigation and seasonal reckoning. Nevertheless, the architectural evidence suggests a combination of ritual, symbolic, and landscape-oriented factors rather than a singular astronomical determinism.

Experiential and Contemporary Condition of the Sites

During my visit in May 2025, nearly all major temple complexes were covered by large-scale protective shelters installed to mitigate environmental degradation. These tensile membrane structures, supported by steel arch frameworks and spanning significant lengths (Figures 33–35), currently shelter Tarxien, Mnajdra, and Ħaġar Qim.

Figure 33. Protective tensile canopy structure installed over the Tarxien temple complex. Satellite view generated with Google My Maps.

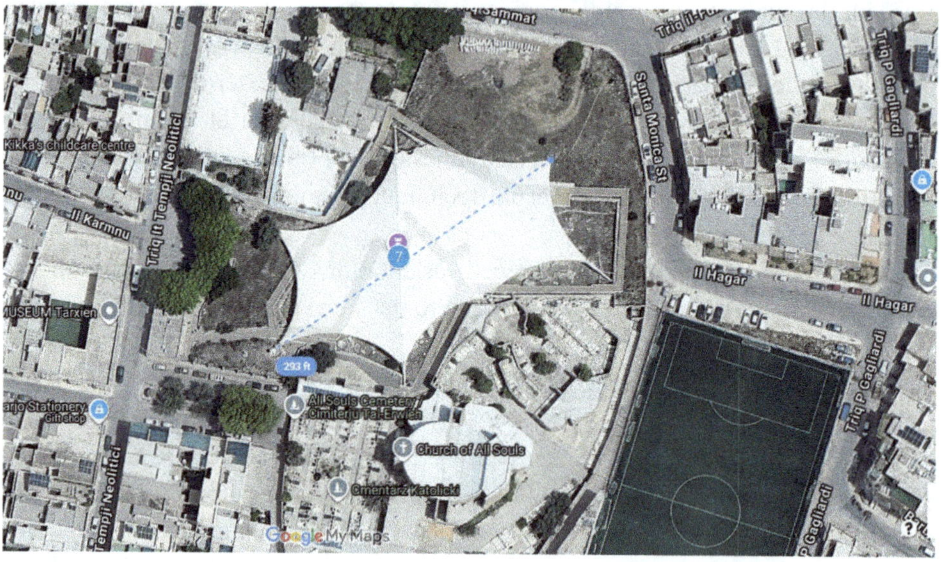

Figure 34. Protective tensile canopy structure over the Mnajdra temple complex. Satellite view generated with Google My Maps.

Figure 35. Protective tensile canopy structure over the Ħaġar Qim temple complex. Satellite view generated with Google My Maps.

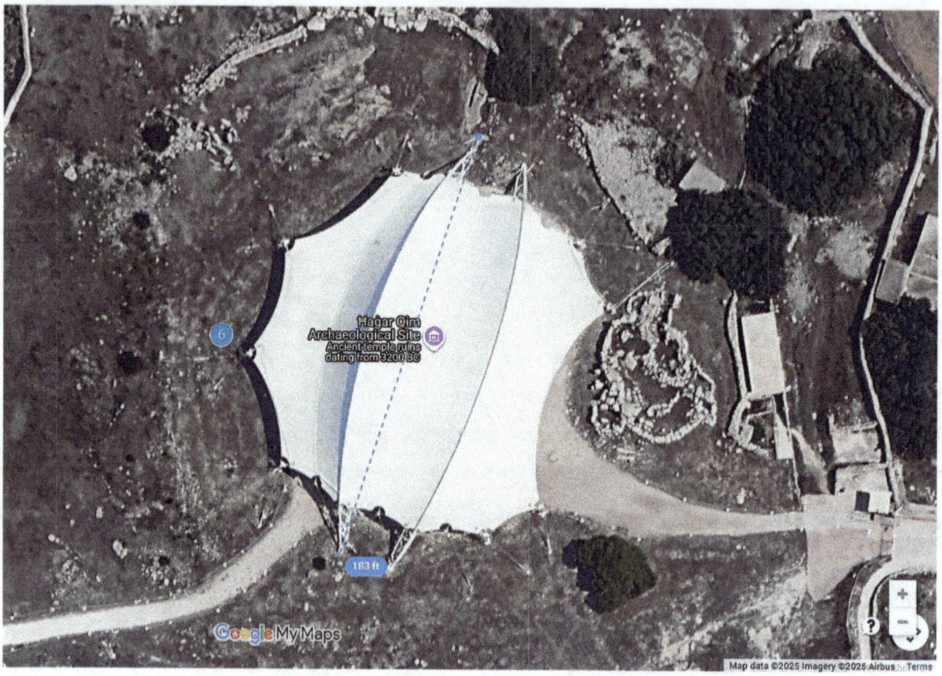

While structurally effective, these protective installations alter perceptual relationships between temple, sky, and landscape. Direct observation of celestial alignments is partially obstructed, and the architectural experience becomes mediated by contemporary infrastructural interventions.

Despite this mediation, the temples maintain a distinctly human scale. The spacing of orthostats at entrances approximates the span of extended arms, reinforcing bodily engagement. With the exception of Tarxien and Ġgantija—situated within more urbanized contexts—most sites remain embedded within Mediterranean landscapes, approached through gradual immersion. The chromatic interplay between limestone, vegetation, sea, and sky contributes to a layered experiential sequence, often intensified by successive megalithic doorways that frame and reframe the central axis in a progressive “zooming-in” perception.

Seascape and Ħaġar Qim (3600–3200 BCE)

The Neolithic temple complexes of Ħaġar Qim and Mnajdra form part of the same archaeological park, located near the village of Qrendi along the southern coast of Malta. The two sites are connected by a pedestrian pathway approximately 0.5 km in length. I visited both complexes on May 3, 2025.

The approach from Qrendi to Ħaġar Qim involves a gradual transition from the village fabric into an open Mediterranean landscape overlooking the sea. The 18th-century Parish Church of the Assumption in Qrendi—constructed in local limestone—provides an architectural prelude to the Neolithic structures. Although separated by several millennia, both employ globigerina limestone and share a warm chromatic palette of beige, honey, and ochre tones set against the saturated blue of the Mediterranean sky. This continuity of material produces a perceptual dialogue across time between sacred architecture of different epochs.

Beyond the village, immersion in vegetation and topography leads toward the temple plateau, where expansive views of the Mediterranean Sea become integral to the spatial experience. The interplay of sea, sky, and limestone contributes to a perceptual environment in which architecture, landscape, and horizon are inseparable.

Site and Chronology

Ħaġar Qim, whose name in Maltese is commonly translated as "standing stones" or "worshipping stones," lies at approximately 35.8277° N latitude and 14.44176° E longitude, east of Mnajdra. Chronologically, the complex belongs to the Ġgantija phase (ca. 3600–3200 BCE) (Trump, 1999; UNESCO, n.d.).

The temple plan presents a clover-leaf configuration composed of multiple apsi dal chambers—four clearly defined, with evidence suggesting a possible fifth. The architectural composition develops along a principal axis, articulated by successive trilithon doorways.

Astronomical Features

From an archaeoastronomical perspective, Ħaġar Qim is frequently associated with solar phenomena. During the June solstice, sunlight at sunrise enters through an aperture located in an apse on the northeastern side of the structure. The projected beam creates a luminous form on an interior stone slab; as the Sun rises, the projected light gradually transforms in shape—from circular to crescent, then elongated—before descending out of view (Heritage Malta, 2023). This dynamic projection suggests deliberate engagement with solar cycles, although the full interpretative framework remains debated.

Architectural Sequence and Spatial Experience

The temple is experienced as a succession of spatial thresholds. Entry occurs through trilithon portals composed of orthostats supporting horizontal lintels. In some instances, openings are carved directly into single megaliths, forming so-called "porthole" doorways (Heritage Malta, n.d.). Evidence indicates that portions of the structure were originally roofed using corbelling techniques, whereby successive stone courses were progressively cantilevered inward to create enclosed volumes.

Decorative elements in low relief—now replaced onsite with replicas—were originally carved into the limestone surfaces. The originals are conserved at the National Museum of Archaeology in Valletta for preservation purposes (Heritage Malta, n.d.).

Figures 36–40 document my walkthrough experience of Ħaġar Qim on May 3, 2025. Movement along the principal axis from the southeastern entrance reveals a dynamic sequence of framed views. Trilithon openings operate as spatial filters, guiding perception while simultaneously directing attention toward lateral apses. The presence of multiple perforations within certain megaliths encourages speculative interpretation regarding controlled light entry and the projection of solar beams onto interior surfaces.

Figure 36. Ħaġar Qim temple complex, exterior view looking southwest, 2025.05.03.

Figure 37. Trilithon doorway at Ħaġar Qim with human figure for scale, 2025.05.03.

Figure 38. Perforated megalith at Ħaġar Qim, 2025.05.03.

Figure 39. Interior spatial sequence at Ħaġar Qim showing axial alignment through successive openings and with human figure for scale, 2025.05.03.

Figure 40. Megalithic framing elements at Ħaġar Qim illustrating human scale and material texture, 2025.05.03.

The contemporary protective shelter—an open-sided tensile membrane supported by steel arches—ensures preservation but mediates celestial observation. While structurally necessary, it limits direct perception of the Sun's apparent path and modifies the relationship between temple, seascape, and sky.

Mnajdra (3600 –2500 BCE)

The Neolithic temple complex of Mnajdra lies west of Ħaġar Qim and is connected to it by a pedestrian pathway of approximately 0.5 km. The toponym "Mnajdra" is generally considered to derive from Arabic and has been interpreted as referring either to cultivated land or to a site commanding extensive views (Aquilina, 1987–1990).

Mnajdra consists of three adjacent temples constructed during different phases of the Maltese Neolithic (Figure 41). The earliest structure, known as the East Temple (or East Trefoil Temple), dates to the Ġgantija phase (ca. 3600–3200 BCE) (Heritage Malta, 2007). Located at approximately 35.82679° N, 14.43650° E, it presents a relatively simple trefoil plan.

Figure 41. Mnajdra temple complex. Left: satellite imagery from Google Earth (2007). Right: aerial imagery from Google My Maps (2025).

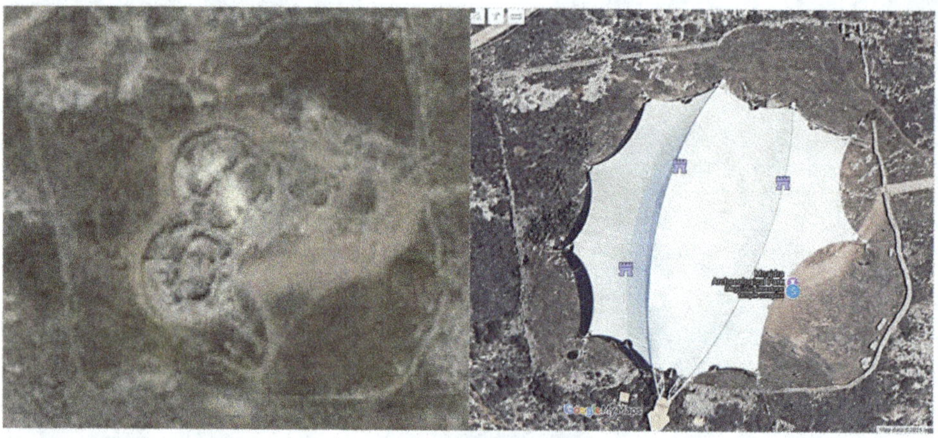

The South Temple, dated between ca. 3150 and 2500 BCE (Tarxien phase), lies at approximately 35.82663° N, 14.43613° E. It is the most architecturally complex of the three, composed of four apses arranged around a principal axis. The northwestern apse contains an additional internal chamber, creating a nested spatial configuration of increasing geometric articulation.

The Middle Temple, located at approximately 35.82682° N, 14.43625° E and also dated to the Tarxien phase (ca. 3150–2500 BCE), presents four symmetrically arranged apses organized along a central axis defined by the southeastern entrance and the northwestern terminal chamber (Heritage Malta, 2007).

Figure 41 compares earlier satellite imagery with more recent aerial imagery, illustrating both the temple layout and the contemporary protective canopy structures. Figure 42 shows a scale model of the three temples displayed at the Visitor Centre of the Ħaġar Qim and Mnajdra Archaeological Park; a diagram is superimposed to indicate the principal sunrise axes associated with solstitial and equinoctial events.

Figure 42. Scale model of the Mnajdra temple complex at the Visitor Centre of the Ħaġar Qim and Mnajdra Archaeological Park, 2025.05.03. Superimposed diagram indicates principal sunrise axes at solstices and equinoxes.

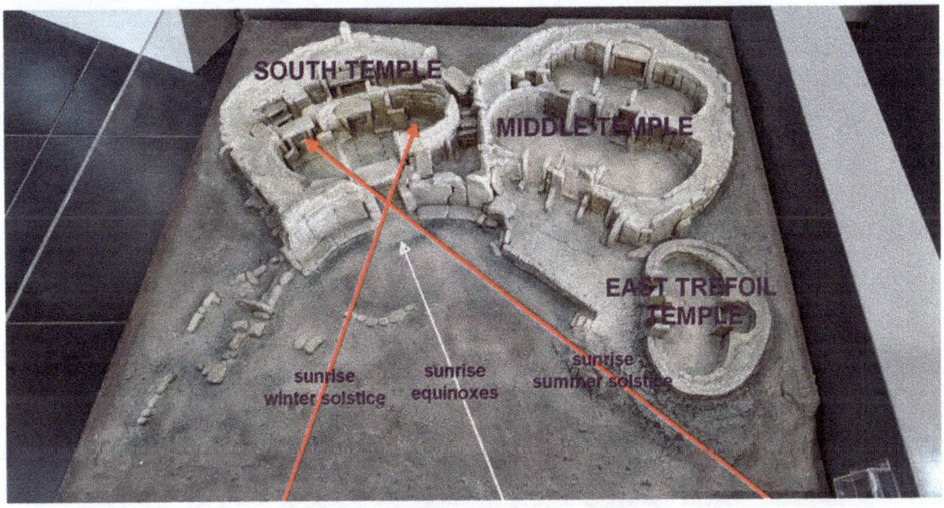

Astronomical Orientation

Mnajdra—particularly the South Temple—is frequently cited in archaeoastronomical literature for its solar alignments (Figure 42). The South Temple follows an east-west alignment. During the equinoxes, sunlight at sunrise penetrates the entrance and illuminates specific interior stones; at the solstices, the light shifts to mark lateral architectural elements. These solar interactions reinforce the hypothesis that controlled light penetration was integrated into the architectural conception, although interpretative models vary and must account for long-term astronomical changes, including axial precession.

Spatial Experience and Landscape Context

Figures 43 through 49 document my visit to Mnajdra on 2025.05.03. The approach from Ħaġar Qim follows an east–west trajectory across the Mediterranean landscape, gradually revealing the temple complex against the sea horizon.

Figure 43. Approach toward Mnajdra from Ħaġar Qim across the Mediterranean landscape, 2025.05.03.

Upon arrival at the South Temple, the architectural experience echoes that of Ħaġar Qim: a sequence of trilithon openings framing interior and exterior views. Movement along the central axis culminates visually at the altar-like focal stone within the South Temple (Figures 44–46). The trilithons are particularly expressive of scale and tectonic clarity.

Figure 44. Human figure framed by a trilithon in the South Temple, Mnajdra, 2025.05.03.

Figure 45. South Temple, Mnajdra, interior view with focal stone elements, 2025.05.03.

Figure 46. South Temple, Mnajdra, at the end of the principal east-west axis, 2025.05.03.

The contemporary tensile shelter filters direct sunlight, softening contrasts of illumination and shadow. While essential for conservation, the canopy mediates the direct experience of solar alignment phenomena, leaving their full effect partially to reconstruction and interpretation.

Departing the temple complex, the immersion in landscape reasserts itself. The Mediterranean Sea lies less than 300 meters from the site, separated by steep, rugged limestone formations (Figure 48). Along this stretch of coastline, the 17th-century Tal-Ħamrija Coastal Tower stands as the only later architectural intervention, with views toward the uninhabited islet of Filfla on the horizon (Figure 49). The juxtaposition of Neolithic megalithic architecture, early modern protective tensile structures, and open seascape underscores the enduring strategic and symbolic significance of this coastal promontory.

Figure 47. Mnajdra complex looking southwest, 2025.05.03.

Figure 48. Mediterranean Sea between Mnajdra and Ħaġar Qim, 2025.05.03.

Figure 49. Tal-Ħamrija Coastal Tower south of Mnajdra, with the islet of Filfla on the horizon, 2025.05.03.

Ġgantija: Solstices and Goddesses

The Neolithic temple complex of Ġgantija is located on the island of Gozo, near the town of Xagħra. Gozo, the second-largest island of the Maltese archipelago after Malta, lies approximately 5 km northwest of Malta and is connected by ferry. The Ġgantija site is situated inland, with the northern coastline approximately 2 km to the north-northeast.

The name "Ġgantija" derives from the Maltese word *ġgant* ("giant"), referring both to the monumental scale of the megaliths and to local folklore attributing construction of the temples to a giantess. According to legend, the giant goddess Sansuna built the temples while carrying a child on her shoulders (De Soldanis, 1746/1995). Archaeologically, the term "Ġgantija phase" also designates the earliest major temple-building period of Maltese prehistory (ca. 3600–3200 BCE). In 1980, Ġgantija was the first site inscribed under the UNESCO designation "Megalithic Temples of Malta" (UNESCO, n.d.).

Figure 50 presents a satellite view of the site, illustrating the layout and orientation of the two adjacent temples.

Figure 50. Ġgantija North and South; satellite imagery from Google Earth shows clearly the temples' layout and alignment.

Architectural Configuration

The Ġgantija complex consists of two freestanding temples: Ġgantija South (approximately 36.04722° N, 14.26893° E) and Ġgantija North (approximately 36.04735° N, 14.26907° E). The temples share a common boundary wall but have separate entrances.

Constructed between ca. 3600 and 3200 BCE (Heritage Malta, n.d.), they are among the best-preserved Neolithic structures in Malta. As with other Maltese temples, harder coralline limestone forms the exterior walls, while softer globigerina limestone is used for interior spaces and doorways.

The South Temple, the larger of the two, consists of five apses: four arranged symmetrically about a principal axis and a fifth terminal apse centered on the axis opposite the entrance. The axial distance from entrance to terminal apse is approximately 25 meters (ca. 81 feet). The North Temple presents a slightly more compact configuration but follows a similar axial organization.

Astronomical Orientation

The principal axis of both temples has an azimuth of approximately 130° measured clockwise from true north, producing a northeast–southwest alignment. This orientation closely corresponds to the position of sunrise near the December (winter) solstice, when the solar azimuth at this latitude is approximately 133°. The proximity between architectural axis and solstitial sunrise suggests deliberate orientation.

In addition to solar alignment, some studies propose lunar associations, particularly with major lunar standstills; such hypotheses have been explored through digital simulation models (Barratt, 2022). As with other Maltese temples, however, interpretative caution is required, given the effects of long-term astronomical variation and the absence of explicit textual evidence.

Ritual Interpretation and Material Culture

Archaeological findings indicate that Ġgantija functioned as a ritual and ceremonial center. Figurines and anthropomorphic statues—often referred to colloquially as "Fat Ladies"—have been recovered and are commonly interpreted as representations of female deities associated with fertility (Rountree, 2002). The association with a "Goddess Movement" reflects later interpretative frameworks rather than direct archaeological attribution.

Evidence of ritual activity includes animal bone remains suggesting sacrifice, as well as stone hearths indicating the use of fire. The forecourt to the southeast of the temples likely accommodated gatherings of participants (Heritage Malta, n.d.).

Perforations visible in both horizontal and vertical megaliths have been interpreted functionally rather than astronomically. Holes in horizontal slabs may have been used for libations—pouring liquid offerings—whereas vertical perforations likely supported barriers or closure systems (Heritage Malta, n.d.).

Experiential Observation

I visited Ġgantija on 2025.05.04. Unlike Ħaġar Qim and Mnajdra, the temples are not currently covered by large protective canopy structures. However, overcast weather conditions prevented direct observation of solar alignment phenomena.

Figures 51 through 56 document the walkthrough experience. Movement along the axial walkway emphasizes the temple's longitudinal geometry (Figures 52 and 54), while lateral apses introduce spatial complexity (Figure 55). The perforations in vertical slabs are visible in Figure 53.

Figure 51. Ġgantija North, looking southwest, 2025.05.04.

Figure 52. On the central walkway of Ġgantija North Temple, illustrating axial alignment and human scale, 2025.05.04.

Figure 53. Perforations in vertical limestone slabs at Ġgantija, 2025.05.04.

Figure 54. Ġgantija South Temple, axial walkway looking northwest; structural scaffolding visible in the background, 2025.05.04.

Figure 55. Apsidal chambers of Ġgantija South adjacent to the main axial walkway, 2025.05.04.

Although the Mediterranean Sea lies relatively nearby, it is not visually dominant from within the temple enclosure. This raises questions regarding the perceptual role of the seascape in the cosmological framework of an island culture whose navigation would have depended upon stellar observation. The subdued sunlight visible in Figure 56 reinforces the atmospheric conditions of the visit and the partial mediation of solar perception.

Figure 56. Ġgantija South Temple under overcast sky, looking southwest, 2025.05.04.

The Spirals of Ħal Tarxien

The Neolithic site of Ħal Tarxien is located at geolocation 35.86918° N, 14.51193° E, in the town of Tarxien, southeast of Malta's capital, Valletta. The name derives from the Maltese language: the prefix Ħal means "place of," while Tarxien is interpreted as "large stone."

The archaeological remains were discovered in 1913 by the local farmer Lorenzo Despott. Excavations followed between 1915 and 1919 under the direction of Sir Themistocles Zammit, then director of the National Museum of Archaeology (Heritage Malta, n.d.). Restoration campaigns were completed by 1956 (Wells, 2021). The site was inscribed in 1992 within the UNESCO "Megalithic Temples of Malta."

The Tarxien cultural phase (ca. 3150–2500 BCE), which follows the Saflieni phase (ca. 3300–3000 BCE), represents the apogee of Maltese temple architecture in artistic and architectural refinement. Spiral decoration, although present in earlier phases, becomes particularly characteristic during this period (Wells, 2021).

Ħal Tarxien was a place of cult and ritualistic celebrations; it consists of four separate temple structures (Figure 57), listed below in chronological order. The temples are conventionally named according to their relative orientation within the complex.

Figure 57. Ħal Tarxien complex. Historical satellite imagery (Google Earth, 2013), prior to installation of the tensile protective structure.

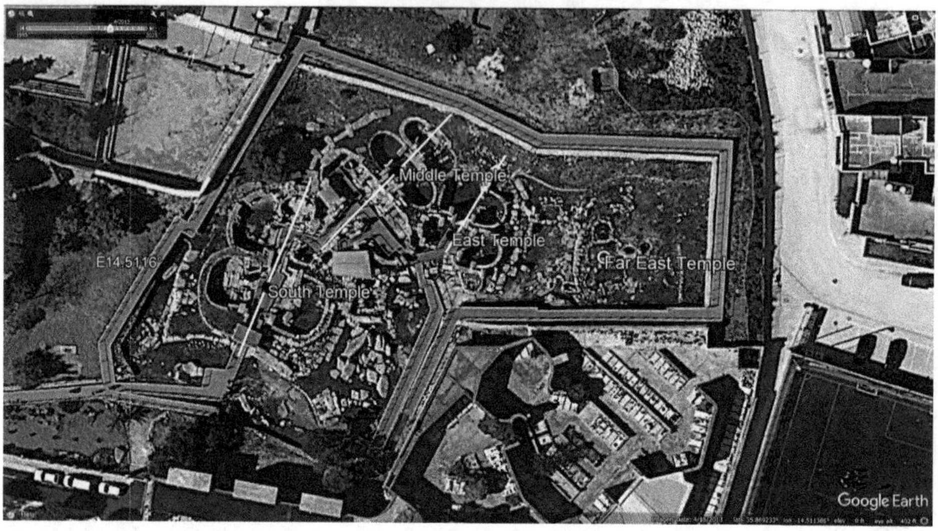

- Far East Temple (35.86919° N, 14.51246° E)
  Constructed during the Ġgantija phase (ca. 3600–3000 BCE), this earliest structure originally followed a five-apse plan (Odyssey: Adventures in Archaeology, n.d.). Its present condition makes the original layout difficult to reconstruct. The azimuth of its principal axis measures approximately 351.8° from true north (Figure 58).

Figure 58. Site plan of the Ħal Tarxien temples. 2013 Google Earth satellite image with superimposed diagram indicating measured azimuths of principal temple axes and calculated solstitial solar azimuths.

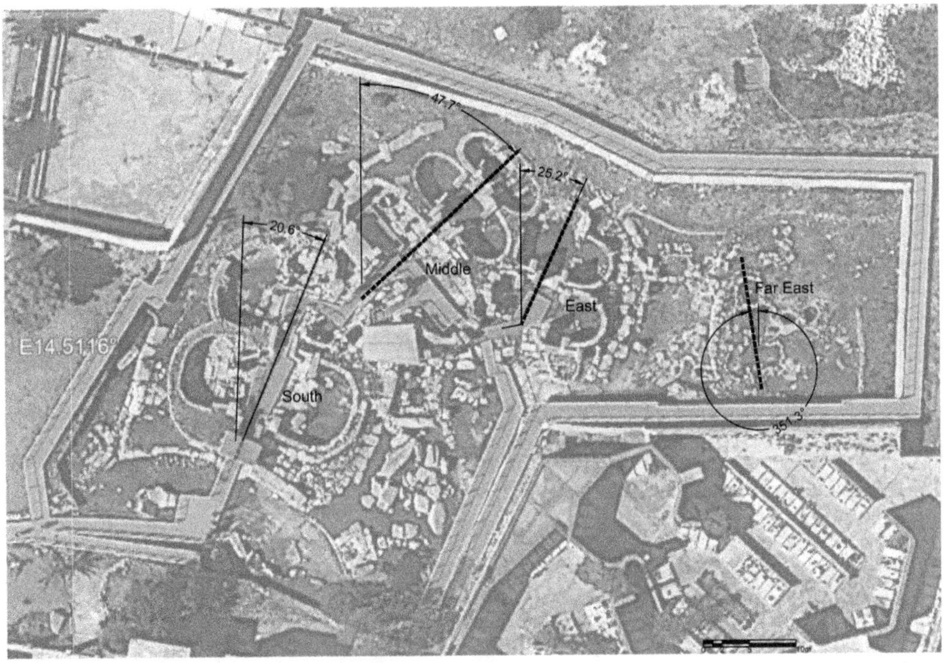

- East Temple (35.86926° N, 14.51216° E)
  Built during the Tarxien phase (ca. 3150 BCE) (Trump & Cilia, 2002), this temple presents a simpler plan consisting of two semicircular apses symmetrically arranged about a central corridor. The main axis is oriented approximately 25.2° from true north.
- South Temple (35.86922° N, 14.51193° E)
  Constructed ca. 3000–2500 BCE (Superintendence of Cultural Heritage Malta, 2012), the South Temple is the most architecturally elaborate. Originally a five-apse structure, it was later modified to integrate with the Central Temple. The northeast apse became a connecting passage between the two. The axis measures approximately 20.6° from true north. This temple contains the iconic colossal statue fragment, often interpreted as a fertility figure (Heritage Malta, n.d.), and is distinguished by extensive bas-relief spiral and animal motifs.
- Middle Temple (35.86921° N, 14.51199° E)
  The last structure constructed in the complex (after 3000 BCE), it presents an unusual six-apse configuration arranged symmetrically along a central axis

oriented approximately 47.7° from true north. Archaeological evidence suggests the possible presence of a roof covering (Heritage Malta, n.d.).

Astronomical Considerations

Unlike Mnajdra or Ġgantija, Ħal Tarxien does not present clear correspondence between temple axes and solstitial solar azimuths (60.6°, 119.4°, 240.6°, 299.4° at latitude 35.87° N). The principal axes span multiple orientations (20.6°, 25.2°, 47.7°, 351.8°), none coinciding with calculated solstitial sunrise or sunset bearings.

Entrances likewise do not align with cardinal directions. This dispersion of orientations suggests that solar alignment was not a primary geometric determinant in the planning of the complex.

Digital simulations (Barratt, 2022) have proposed potential stellar correspondences. Some authors (Silva & Lomsdalen, 2025) suggest that decorative motifs in the South Temple may correspond to stellar patterns. However, claims concerning visibility of the Southern Cross constellation from Malta during the Neolithic should be treated cautiously, as precessional effects would have placed it low on the southern horizon and only marginally observable.

Spiral Motif

Spiral ornamentation is the defining artistic feature of the Tarxien phase. Spirals appear carved or sculpted in relief in limestone and globigerina elements: altars, thresholds, orthostats, and screens.

Comparative examples of prehistoric spiral iconography occur at:

- Newgrange (Ireland)
- Knowth (Ireland)
- Dowth (Ireland)
- Chaco Canyon (New Mexico, USA)

Spirals have long been interpreted as symbols of continuity, cyclical time, fertility, and cosmological expansion. At Ħal Tarxien, only replicas remain in situ; the original decorated slabs are conserved at the National Museum of Archaeology in Valletta.

Site Experience (Visit: 2025.05.06)

My visit on 6 May 2025 revealed an experience shaped by two contemporary conditions:

1. The tensile protective membrane structure installed in 2015.
2. The embedding of the site within a dense urban fabric.

This contrasts with the more isolated landscape settings of other Maltese Neolithic temples.

Movement through the interconnected South, Middle, and East Temples produces a complex spatial sequence in which axes of differing orientations intersect and flow into one another. The result is a nearly labyrinthine spatial perception, loosely structured by central corridors but dynamically reoriented at each transitional threshold.

Framed doorway alignments create perspective effects reminiscent of one-point spatial compositions, particularly within the South Temple. The photographic sequence (Figures 59–65) documents a walkthrough from the southwest entrance through the South Temple corridor, continuing into the Middle and East structures.

Figure 59. The walkway intersecting the tensile cover, looking east, 2025.05.06.

Figure 60. The South Temple, framed views sequence. 2025.05.06.

Figure 61. South Temple, on the right copy of the statue "Fat Lady" or "Mother Goddess": a fertility goddess, 2025.05.06. The original is in the National Museum of Archaeology in Valletta.

Figure 62. South Temple spiral decorations, 2025.05.06.

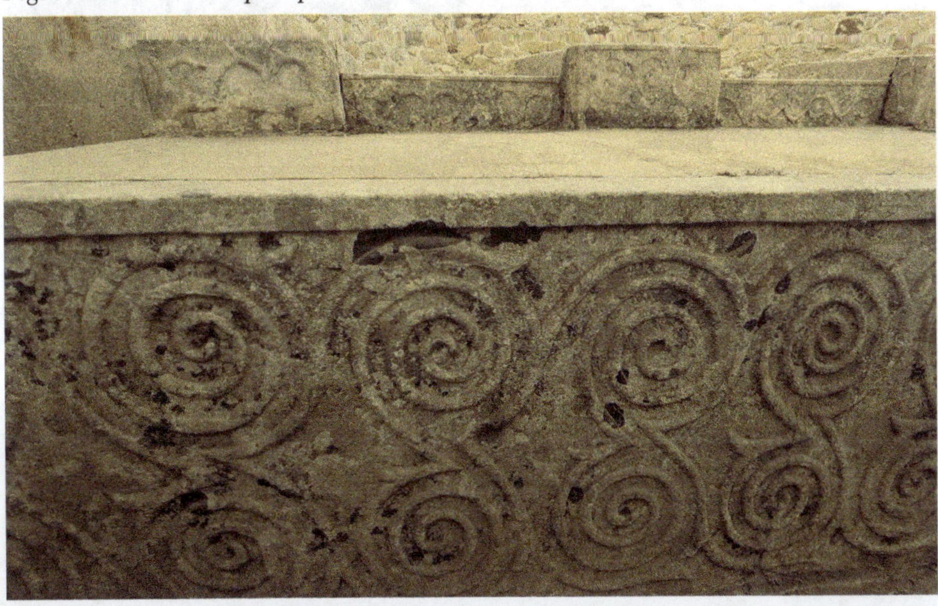

Figure 63. East Temple, viewed from above the walkway, 2025.05.06.

Figure 64. Far East Temple, 2025.05.06.

Figure 65. Metal gate casting sharp spiral shadow, 2025.05.06.

Spiral motifs recur throughout the spatial sequence. Particularly striking was the shadow cast by a contemporary metal gate, whose projected spiral form evoked resonances with my laser-cut spiral installations at S.U.N.F.A.R.M. (see Chapter 8).

Tal-Qadi

The visit to the ruins of the *Tal-Qadi* temple was my most intimate experience with an archaeological site in Malta. Unlikely from the other temples I visited in the Maltese islands, Tal-Qadi is out of the tourist itinerary; probably, because the ruins are in very poor condition, and no restoration or preservation has been made so far.

Tal-Qadi is located at geolocation 35.93665° N, 14.42055° E in the Burmarrad plain, town of Salina, and overlooking Salina Bay; the Mediterranean Sea is about one mile away (Figure 66). The ruins were discovered in 1916 by Henry Sant, a Malta government civil engineer (M.A.R. 1916-1917: 8) (Micallef, 2001). Later excavations were performed in 1927 and a survey was done in 1952. According to the National Inventory of the Cultural Property of the Maltese Islands [NICPMI], the survey of the ruins shows an almost precise east-west central axis—which I verified from a Google Earth satellite view (Figure 67)—and "the temple belongs to the four-apse type, typical of the late Temple Period" (National Inventory of the Cultural Property of the Maltese Islands [NICPMI], 2012). There is uncertainty about the date of construction: "the site seems to have been frequented during the Tarxien phase, although it is possible that use of the site was as early as the Ġgantija phase. The presence of some Tarxien Cemetery pottery sheds hint to a later use of the site" (National Inventory of the Cultural Property of the Maltese Islands [NICPMI], 2012).

Figure 66. Tal-Qadi site, current satellite imagery from Google Earth.

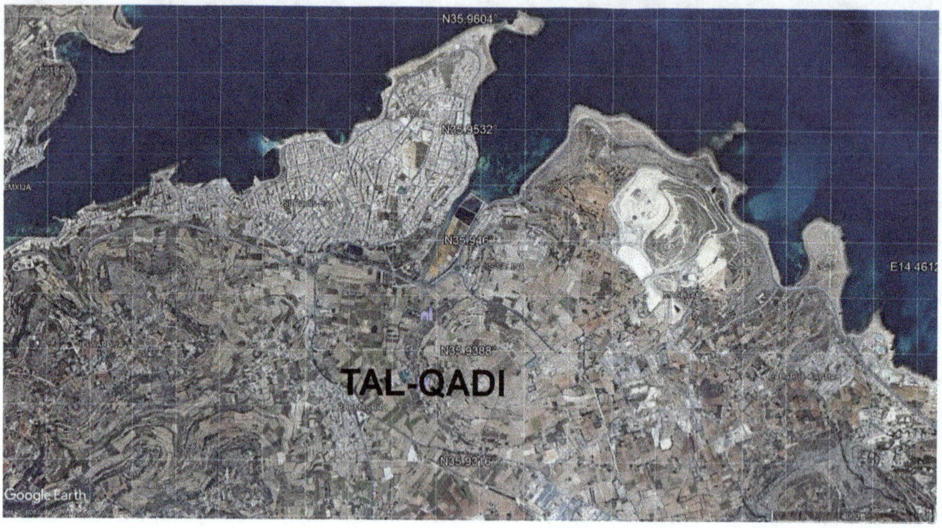

Figure 67. Tal-Qadi site, current satellite imagery from Google Earth showing east-west direction.

The archaeological remains of Tal-Qadi have been considered relevant from an archaeoastronomical perspective mainly for a globigerina stone slab which was found on the site and is known as the *Tal-Qadi Stone*, now on display at the National Museum of Archaeology in Valletta (Figure 68). The stone, which measures 29 cm × 24 cm × 5 cm (Micallef, 2001), is incised with symbols of stars and crescent moon symbols. There is no certain conclusion about the use of the stone "to read the number of days for the rising or setting of the first quarter, full, last quarter and new moon phases" but "it is interesting to note that the interval of days between these moon phases fits in the Tal-Qadi stone sequence" (Micallef, 2001).

Figure 68. Tal-Qadi stone: left, National Museum of Archaeology, Valletta; right diagram showing stars symbols, presumably of different constellations.

(source https://commons.wikimedia.org/wiki/File:Himmelstafel.Tal-Qadi.Zuordnung.png, available under the Creative Commons CC0 1.0 Universal Public Domain Dedication)

My visit to Tal-Qadi was on May 6, 2025. It was a clear, sunny day. The presence of the Sun and its casting shadows enhanced greatly the experience of the ruins. Figures 69–72 are excerpt of the photographic sequence from my walkthrough. On my walk to the site, I encountered the Chapel of Saint Michael (35.94117° N, 14.42075° E), a 17th century building presenting a limestone façade with a belfry in the middle, which has been recently restored (Times of Malta, 2006 October 3). Uncannily, the Sun appeared centered on the façade (Figure 69).

Figure 69. Chapel of Saint Michael (1652).

Figure 70. Approaching Tal-Qadi from west, facing east, 2025.05.06; the stone steps were installed in the 1920s to facilitate access.

Figure 71. Among Tal-Qadi temple remains, looking north. 2025.05.06.

Figure 72. Flattened photosphere of Tal-Qadi temple remains. 2025.05.06.

THE SKY IN THE PAST AND THE ARCHITECTURAL SUBLIME

Archaeoastronomical structures represent some of the earliest built attempts to articulate a relationship between terrestrial environment and celestial order. The Maltese Neolithic temples and the Egyptian pyramids and temple complexes, examined in this chapter, constitute two distinct yet comparable expressions of what may be termed an architecture of cosmology.

Both case studies reveal geometrical intentionality perceptible at multiple scales. From satellite imagery, principal axes and orientation patterns become legible as abstract geometric constructs inscribed upon the landscape. At ground level, these same geometries are experienced phenomenologically through movement, framing, enclosure, and light. The architecture mediates between the local and the remote: between stone and sky.

The Maltese temples operate at a human scale. Their apsidal interiors, trilithic doorways, and corbelled chambers suggest enclosure and bodily proximity. While walking through these spaces, I repeatedly extended my arms to measure distance against orthostats and lintels; the architecture remained within reach. Orientation in Malta is complex and varied, yet the embodied experience of axis, aperture, and framed horizon is immediate.

In contrast, the Egyptian pyramids and temple complexes exceed bodily proportion. Their geometry is apprehended primarily from without, where monumental mass defines a figure against the expansive desert and sky. The alignment of pyramids and temple axes toward cardinal directions or solar events situates architecture within cosmic order at a scale beyond direct corporeal engagement.

Despite these differences, both contexts evoke what Immanuel Kant described as the sublime—"that which is absolutely great" (Kant, 1914, p. 109). The Maltese spirals carved into limestone and the vast pylons of Karnak operate differently, yet each generates awe that transcends aesthetic beauty. The experience is not merely visual but existential: an encounter with magnitude, temporality, and celestial immensity.

Especially for the Neolithic temples of Malta, the visit to the archaeological sites is only part of the experience, which is also provided by immersion in the surrounding landscape. The phenomenological understanding of the site as figure-ground leads to interpreting why and how these examples of architecture of cosmology were important in the past beyond their built features, as pointers to the remote sky—intangible yet present in myths and legends inhabited by the deities to which the temples were dedicated. The blue Mediterranean Sea surrounding the Maltese islands and the yellow desert landscape of the Egyptian necropolis, provide a palette of contrasting colors, forming a background for the architecture, which in turn becomes a figure against the ever-changing background of the sky.

The experience of both examples of architecture of cosmology presented in this chapter inspires a sense of awe, not only in the encounter with the archaeological remains but also for what these constructions embody and meant to express. From the mysterious spiral symbols in the carving of the Maltese temples, to the beyond-human scale of the Egyptian necropolis, the power of these structures goes beyond the experience of the architecture and converse with our imaginary in the search for meaning in symbols. The 18th century German philosopher Immanuel Kant wrote "We call that sublime which is absolutely great" (Kant, 1914, p. 109), and associated the two different types of sublime—mathematical and dynamical—with a sense of awe, beyond beauty. The experience of the Maltese spirals, of the Egyptian *Great Hypostyle Hall*, or the pyramids at Giza, reminded me of the Kantian sublime, inspiring a sense of reverence, wonder and awe. The infinite sky was reflected in the greatness of the architecture, whose monumental scale conversed with the Sun—called not by chance *Amun-Ra*—the *Supreme Sun*.

These architectures of cosmology demonstrate that the sky was never a neutral backdrop to human construction. It functioned as measure, orientation system, symbolic order, and metaphysical reference. Whether approached through intimate enclosure or overwhelming monumentality, architecture became a medium through which terrestrial communities inscribed themselves into celestial cycles. What remains today is not only stone, but a geometric dialogue between earth and sky that continues to structure perception, imagination, and inquiry.

METHODOLOGICAL NOTE

Throughout this chapter, orientation was examined using publicly available georeferenced satellite imagery (Google Earth historical satellite imagery) combined with digital drafting tools. Principal axes were traced directly from high-resolution orthophotos; azimuths were measured relative to geographic north; and solar azimuth values at solstices were computed using the standard astronomical relation:

$$cosA = \frac{\sin\delta}{\cos\varphi}$$

where A is the sunrise or sunset azimuth relative to due east or west, δ is solar declination (±23.44° at solstices), and φ is site latitude.

This accessible methodology allows preliminary verification of alignment hypotheses without invasive field instrumentation. It does not replace in-situ survey, but it demonstrates how digital tools can support archaeoastronomical interpretation and make geometric analysis replicable.

The sky observed by Neolithic builders and by Egyptian architects was not fundamentally different from ours, yet the cultural frameworks through which it was interpreted were profoundly distinct. The temples and pyramids examined here stand as material witnesses to past cosmologies—stone mediators between earthbound communities and the perceived order of the heavens.

Their geometry remains; their myths persist in fragments; and the sky continues its cycles above them.

REFERENCES

Aquilina, J. (1987–1990). *Maltese-English dictionary* (Vols. 1–2). Midsea Books Ltd.

Aveni, A. F. (1980). *Skywatchers of ancient Mexico*. University of Texas Press.

Barratt, R. P. (2022). The crux of astronomical alignment in Neolithic Malta: Using 3D simulation to produce new data. *Digital Applications in Archaeology and Cultural Heritage*, *26*, e00229.

Belmonte, J. A. (2001). On the orientation of Old Kingdom Egyptian pyramids. *Journal for the History of Astronomy*, *32*(1), 1–20.

Belmonte, J. A., & Shaltout, M. (2005). On the orientation of ancient Egyptian temples. *Journal for the History of Astronomy*, *36*(3), 273–298.

Blyth, E. (2006). *Karnak: Evolution of a temple*. Routledge.

Brú na Bóinne World Heritage Site. (n.d.). *Newgrange*. Retrieved September 25, 2025, from https://www.newgrange.com/

De Soldanis, G. P. F. (1995). *Gozo, ancient and modern, religious and profane* (Aquilina, J., Trans.). Malta University Press. (Original work published 1746)

Encyclopædia Britannica. (n.d.). Malta. In *Britannica academic*. Retrieved October 10, 2025, from https://www.britannica.com/place/Malta

Foderà Serio, G., Hoskin, M., & Ventura, F. (1992). The orientations of the temples of Malta. *Journal for the History of Astronomy*, *23*(2), 107–119.

Heath, R. (2003). *Alexander Thom: Cracking the Stone Age Code*. Bluestone Press.

Heritage, E. (n.d.). *Research on Stonehenge*. English Heritage. Retrieved on September 21, 2025 from https://www.english-heritage.org.uk/visit/places/stonehenge/history-and-stories/history/research/

Johnson, W. (1912). *Byways of British Archaeology*. Cambridge University Press.

Kant, I. (1914). *Kant's critique of judgement* (Bernard, J. H., Trans.). Macmillan. (Original work published 1790)

Krupp, E. C. (1988). Light in the Temples. In Ruggles, C. L. N. (Ed.), *Records in Stone: Papers in Memory of Alexander Thom* (pp. 473–499). CUP.

Krupp, E. C. (1988). Light in the temples. In *Echoes of the ancient skies*. Harper & Row.

Lehner, M. (1997). *The complete pyramids*. Thames & Hudson.

Lockyer, N. (1894). *The dawn of astronomy*. Macmillan.

Magli, G. (2013). *Architecture, astronomy and sacred landscape in ancient Egypt*. Cambridge University Press.

Malta, H. (2007, May). *Megalithic Temples of Malta: Description*. https://web.archive.org/web/20071020170410/http://www.heritagemalta.org/Consultation/Documents/Temples%20-%20Description.pdf

Malta, H. (2023). *The Equinox & Solstice at Ħaġar Qim and Mnajdra* [Brochure]. https://heritagemalta.mt/app/uploads/2023/06/HQMnajdra-leaflet.pdf

Malta, H. (n.d.). *Ħaġar Qim and Mnajdra Archaeological Park*. Retrieved October 9, 2025, from https://heritagemalta.mt/explore/hagar-qim-and-mnajdra-archaeological-park/

Malta, H. (n.d.). *Ġgantija Archaeological Park*. Retrieved October 10 2025 from https://heritagemalta.mt/explore/ggantija-archaeological-park/

Malta, H. (n.d.). *Ħal Tarxien Prehistoric Complex*. Retrieved October 12, 2025, from https://heritagemalta.mt/explore/hal-tarxien-prehistoric-complex/

Malville, J. M., Wendorf, F., Mazar, A. A., & Schild, R. (1998). Megaliths and Neolithic astronomy in southern Egypt. *Nature*, *392*(6674), 488–491.

Micallef, C. (2001). The Tal-Qadi Stone: A Moon Calendar or Star Map. *The Oracle. Journal of the Grupp Arkeologiku Malti*, 2, 1–9.

National Museums Scotland. (n.d.). *Casing stone from the Great Pyramid of Giza* [Stone]. Museum reference A.1955.176. Edinburgh, Scotland. https://www.nms.ac.uk/discover-catalogue/casing-stone-from-the-great-pyramid-of-giza

Neolithic temples of Malta. (n.d.). In *Encyclopædia.com*. Retrieved October 3, 2025, from https://www.encyclopedia.com/humanities/encyclopedias-almanacs-transcripts-and-maps/neolithic-temples-malta

Odyssey: Adventures in Archaeology. (n.d.). *Tarxien Temples*. Retrieved October 12, 2025, from https://www.odysseyadventures.ca/articles/malta_temples/maltaTemples03h_tarxien.htm

Orofino, V., & Bernardini, N. (2016). New evidence on the Orion Correlation Theory. *Mediterranean Archaeology & Archaeometry. International Journal*, *16*(4), 1–12.

Oxford University Press. (n.d.). Archaeoastronomy. In *Oxford English dictionary*. Retrieved September 14, 2025, from https://www.oed.com/dictionary/archaeoastronomy_n

Reel, B. B. C. (2022, February 23).*The mysteries of Portugal's 7,000-year-old 'Stonehenge'*. BBC. Retrieved on August 26 20025 from https://www.bbc.com/reel/video/p0bbrfm1/the-mysteries-of-portugal-s-7-000-year-old-stonehenge-

Rountree, K. (2002). Re-inventing Malta's neolithic temples: Contemporary interpretations and agendas. *History and Anthropology, 13*, 31–51.

Ruggles, C., Prendergast, F., & Ray, T. (Eds.). (2001). *Astronomy, cosmology and landscape*. Ocarina Books.

Ruggles, C. L. N. (2005). *Ancient astronomy: An encyclopedia of cosmologies and myth.* ABC-CLIO.

Serio, G. F., Hoskin, M., & Ventura, F. (1992). The orientations of the temples of Malta. *Journal for the History of Astronomy*, *23*(2), 107–119.

Shaw, I. (Ed.). (2000). *The Oxford history of ancient Egypt.* Oxford University Press.

Silva, F., & Lomsdalen, T. (2025). 'No easy way from the earth to the stars': A new statistical approach to the orientation of the Maltese temples. *Archaeological and Anthropological Sciences*, *17*(96). Advance online publication. DOI: 10.1007/s12520-025-02208-4

Silva, I. (2015). *Latin American Perspectives on Science and Religion*. Routledge.

Stetson University. (n.d.). *Trollasten Dös (Dolmen), Stora Köpinge, near Ystad, Skåne, Sweden.* Neolithic Studies. Retrieved September 25, 2025 fromhttps://www2.stetson.edu/neolithic-studies/simple-dolmens/trollasten-dos-dolmen-stora-kopinge-near-ystad-skane-sweden/

Superintendence of Cultural Heritage. (2012, March 30). *NICPMI Inv. No. 00059* [Data set]. https://ps/neolithic-temples-malta

Superintendence of Cultural Heritage. Malta. (2012). *Tarxien Temple.* http://www.culturalheritage.gov.mt/filebank/inventory/00046.pdf

Taylor, K. (2012). *Celestial geometry: Understanding the astronomical meanings of ancient sites.* Watkins Publishing.

The Editors of Encyclopædia Britannica. (2025, March 7). *Karnak*. In *Encyclopædia Britannica*. Retrieved August 31, 2025 from https://www.britannica.com/place/Karnak

The Editors of Encyclopædia Britannica. (2025, July 31). Pyramids of Giza. In *Encyclopædia Britannica*. Retrieved September 21, 2025, from https://www.britannica.com/topic/Pyramids-of-Giza

The Editors of Encyclopædia Britannica. (2025, September 3). *Maltese language*. In *Encyclopædia Britannica*. Retrieved October 10, 2025, from https://www.britannica.com/topic/Maltese-language

Thom, A., & Thom, A. S. (1971). The Astronomical Significance of the Large Carnac Menhirs. *Journal for the History of Astronomy, 2*(3), 147-160. https://doi.org/ (Original work published 1971).DOI: 10.1177/002182867100200301

Tilley, C. (2004). *The materiality of stone: Explorations in landscape phenomenology*. Routledge.

Times of Malta. (2006, October 3). "17th century chapel restored".*Times of Malta*. Retrieved October 14 2025. https://timesofmalta.com/article/17th-century-chapel-restored.39531

Trump, D. H. (1999). The architecture of the Maltese temples. In Mifsud, A., & Savona Ventura, C. (Eds.), *Facets of Maltese prehistory* (pp. 91–99). Prehistoric Society of Malta.

Trump, D. H., & Cilia, D. (2002). *Malta, Prehistory and Temples*. Midsea Books.

United Nations Educational, Scientific and Cultural Organization. (n.d.). *Megalithic Temples of Malta*. UNESCO World Heritage Centre. Retrieved October 3, 2025, from https://whc.unesco.org/en/list/132/

Ventura, F., & Hoskin, M. (2015). Temples of Malta. In Ruggles, C. L. N. (Ed.), *Handbook of archaeoastronomy and ethnoastronomy* (pp. 1421–1430). Springer.

Verner, M. (2001). *The pyramids: The mystery, culture, and science of Egypt's great monuments*. Grove Press.

Wells, O. (2021, February 17). *The megalithic temples of Malta*. World History Encyclopedia. Retrieved October 6 2025 from https://www.worldhistory.org/article/1678/the-megalithic-temples-of-malta/

Wilkinson, R. H. (2000). *The complete temples of ancient Egypt*. Thames & Hudson.

Zammit, T. (1927). *The Neolithic Temples of Hal-Tarxien, Malta: A short description of the monuments*. Empire Press.

ADDITIONAL READING

Akkerman, A. (2016). *Phenomenology of the Winter-City.* Springer International Publishing., DOI: 10.1007/978-3-319-26701-2

American Research Center in Egypt. (n.d.). *Khonsu Temple.* Retrieved September 29, 2025, from https://arce.org/khonsu-temple/

Ancient Wisdom Foundation. (n.d.). *Karnak (Thebes)*, Egypt. http://www.ancient-wisdom.com/egyptkarnak.htm

Bard, K. A. (2015). *An introduction to the archaeology of ancient Egypt* (2nd ed.). Wiley-Blackwell.

Burl, H. A. W. (1993). *From Carnac to Callanish: the prehistoric stone rows and avenues of Britain, Ireland and Brittany.* Yale University Press.

Cox, J., & Lomsdalen, T. (2010). Prehistoric Cosmology: Observations of Moonrise and Sunrise from Ancient Temples in Malta and Gozo. *Journal of Cosmology, 9*, 2217–2231.

Eliade, M. (1961). *The sacred and the profane: the nature of religion.* Harper & Brothers.

Fagan, B., & Durrani, N. (2016). *A brief history of archaeology* (2nd ed.). Routledge.

Jäger, T. (2004). The art of orthogonal planning: Laparelli's trigonometric design of Valletta. *Journal of the Society of Architectural Historians*, *63*(1), 4–31.

Jarvis, D. (2016, November 6). *Egyptian obelisks, Karnak* [Photograph]. World History Encyclopædia. Retrieved August 31, 2025 from https://www.worldhistory.org/image/6008/egyptian-obelisks-karnak/

Jones, S., & Jones, E. (n.d.). *Hal Tarxien, Malta.* Ancient-Wisdom. Retrieved October 12 2025 from http://www.ancient-wisdom.com/maltahaltarxien.htm

Magli, G. (2008). Akhet Khufu: Archaeoastronomical hints at a common project of the two main pyramids of Giza. *Nexus Network Journal, 11*, 35–50.

Magli, G. (2009). Power and Replica. In *Mysteries and Discoveries of Archaeoastronomy: From Giza to Easter Island* (pp. 289–303). Springer New York.

Magli, G. (2017). Archaeoastronomy in the Khmer Heartland. *Studies in Digital Heritage*, *1*(1), 1–17. DOI: 10.14434/sdh.v1i1.22846

Meaden, G. T. (2017). Advances in understanding megaliths and related prehistoric lithic monuments. *Journal of Lithic Studies*, *4*(4), 1–4.

Michell, J. (2001). *A Little History of Astro-Archaeology*. Thames & Hudson.

Nagase, H., & Okamoto, S. (n.d.). Obelisks at the Karnak temple. In *Obelisks of the World*. Retrieved August 31, 2025, from http://www.obelisks.org/en/karnak.htm

National Aeronautics and Space Administration. (n.d.) "Problem 6: Rebuilding a high-altitude community". Space Math @ NASA. https://spacemath.gsfc.nasa.gov/SED11/P6MachuPichu.pdf

Neugebauer, O. (1942). The origin of the Egyptian calendar. *JNES*, *1*(4), 396–403.

Neugebauer, O. (1943). Demotic horoscopes. *JAOS*, *63*, 115–127.

Neugebauer, O. (1947). Studies in ancient astronomy. VIII. The water clock in Babylonian astronomy. *Isis*, *37*, 37–43. PMID: 20247883

Neugebauer, O. (1980). On the orientation of the pyramids. *Centaurus*, *24*, 1–3.

Neugebauer, O., & Parker, R. A. (1960–1969). Egyptian astronomical texts, vol. I–III. Brown University.

Neugebauer, O., Parker, R. A., & Zauzich, K.-T. (1981). A demotic lunar eclipse text of the first century, B.C. *Proceedings of the American Philosophical Society*, *125*, 312–327.

Neugebauer, O., & Van Hoesen, H. B. (1987). Greek Horoskopes: American Philosophical Society [1959].

Papadopoulos, C., & Moyes, H. (Eds.). (2021). *The Oxford Handbook of Light in Archaeology*. Oxford University Press.

Penprase, B. E. (2011). *The power of stars: how celestial observations have shaped civilization*. Springer Verlag.

Ruggles, C., & Urton, G. (Eds.). (2007). *Skywatching in the ancient world: New perspectives in cultural astronomy: Studies in honor of Anthony F. Aveni*. University Press of Colorado.

Silva, F., & Campion, N. (2012). Role and Importance of the Sky in Archaeology. In *34th Annual Conference of the Theoretical Archeology Group* (pp: 16-20). University of Liverpool.

Silva, F., & Campion, N. (Eds.). (2015). *Skyscapes*. Oxbow Books.

Team, E. T. P. (2021, April 20). *The Karnak Cachette discovery by Georges Legrain in 1903*. Egypt Tours Portal. https://www.egypttoursportal.com/en-us/blog/luxor-attractions/karnak-cachette

United Nations Educational, Scientific and Cultural Organization. (n.d.). *Memphis and its Necropolis – the Pyramid Fields from Giza to Dahshur*. World Heritage Centre. Retrieved September 21, 2025, from https://whc.unesco.org/en/list/86/

United Nations Educational, Scientific and Cultural Organization. (n.d.). *Ancient Thebes with its Necropolis*. UNESCO World Heritage Centre. Retrieved September 26, 2025, from https://whc.unesco.org/en/list/87

University of Memphis. (2024, October 26). *Great columns*. Hypostyle. https://www.memphis.edu/hypostyle/tour_hall/great_columns.php

University of Memphis. (n.d.). *Hypostyle: The Karnak Great Hypostyle Hall Project*. Retrieved September 29, 2025, from https://www.memphis.edu/hypostyle/welcome/

Chapter 7
Rome, the Sky, and Time

ABSTRACT

This chapter examines selected examples of architecture of cosmology in Rome, including Castel Sant'Angelo, the Egyptian obelisks, and St. Peter's Square. It analyzes how cardinal geometry and solar alignments operate across multiple scales—from individual monuments to urban fabric—and interprets these spatial configurations as expressions of imperial and religious ideology. Particular attention is given to meridian lines associated with specific obelisks, examining their transformation from symbolic solar monuments into calibrated astronomical instruments. The architectural and urban composition of St. Peter's Square is further interpreted as a large-scale solar device, in which axial geometry, light, and shadow register celestial phenomena. While extensive scholarship has addressed the urban history and architectural evolution of St. Peter's, the systematic relationship between its built geometry and solar orientation remains comparatively underexplored. This chapter proposes that discrepancies in its geometrical alignments warrant further investigation.

DOI: 10.4018/979-8-3373-3399-1.ch007

Figure 1. Obelisco Vaticano, Piazza San Pietro at solar noon, 2025.05.12.

TIME AND THE "ETERNAL CITY"

Rome is often defined as the "Eternal City" for its many layers of history: time is manifested in the complexity of its architecture and art, omnipresent throughout the urban fabric. Timelessness and timekeeping seem to be both a major component of the city, in many of its buildings, monuments, and urban features.

The nearly three millennia of history can be observed not only by visiting museums or archeological sites, but by taking walks through its alleys, streets and squares. Walking is the best way to know the eternal city, which offers unlimited sights and discoveries. The phrase "It's not the destination, it's the journey", attributed to the American philosopher and essayist Ralph Waldo Emerson, is particularly true for Rome. I was born there and after several decades of walking I can still encounter new narratives in the "vicoli" (alleys) and "piazze" (squares) where the boundary between private and public space appears to blur in an urban-architectural continuum. It seems that every stone, sometimes thousands of years old, has stories to tell. Definitely knowing by walking can be a methodology for learning about the genius loci of the city, a living statement about culture generated by the timeless interaction between built places and humanity.

In Rome, time and timekeeping are embodied by its ancient architecture and urban fabric. Looking at the Sun in the sky "above", can provide orientation and wayfinding "below", in this city, where human interventions have transformed the original landscape into an urbanscape that mirrors the skyscape. Rome, despite its troubled history, has been called *Caput Mundi* (from the Latin, meaning "head of the world") and is a religious center for one of the main religions, the Roman Catholic Church. Time in Rome is not only history, but a presence in its architectural and urban development. Here—perhaps unprecedented in other cities—history and storytelling, time and space are intertwined in a quest for power, beauty, and meaning, which defines the shaping of the built environment according to Earth–sky geometries.

Rome may be described as a scenographic construction in the theater of life. In this metaphor, the buildings of the eternal city stand in the foreground in the backdrop of the urban fabric, which is shaped by topography, religion and ideology, defining ideal, sometimes sacred, geometries. In this chapter I will outline how time is the definer of architectural and urban spaces.

TIMELINE OF BUILDING ROME

It is not the aim of this chapter to present a summary of the history of Rome. However, outlining a historical timeline connected to geographical places provides a framework for understanding the examples of architecture of cosmology discussed in the following sections.

- **April 21, 753 BCE** | Rome, according to mythology, was founded by the demigod twins Romolo e Remo[1] (Romulus and Remus). This birthday, even if derived from a legend, is even now celebrated with several public events. Legend says that the first settlement was on the Palatino (Palatine Hill) near the river Tevere (Tiber) and that this initial hamlet merged with other villages on the surrounding hills, which still comprise the center of Rome. The Sette Colli (seven hills), which played an essential role in Rome urban development, are located on the east bank of the Tiber and are, listed here from north to south: Quirinale (Quirinal Hill), Viminale (Viminal Hill), Esquilino (Esquiline Hill), Campidoglio (Capitoline Hill), Palatino (Palatine), Celio (Caelian Hill), Aventino (Aventine Hill). Besides the Sette Colli, there are other elevations which are most notable in the history of Rome: Pincio (Pincian Hill), Colle Vaticano (Vatican Hill) and Gianicolo (Janiculan Hill); the last two hills are located on the west bank of the Tevere (Tiber), (Figures 7.2–7.3).
- **Early 4th century BCE (traditionally 378–352 BCE)** | Construction of the Mura Serviane (Servian Walls) (Fields & Dennis, 2008).
- **27 BCE–476 CE** | Roman Empire. Under Imperatore Traiano (Emperor Trajan; 98–117 CE), the Empire reached its maximum territorial extent, estimated at approximately five million square kilometers (Southern, 2001; Turchin et al., 2006).
- **c. 30–67 CE** | According to Christian tradition, San Pietro Apostolo (Saint Peter the Apostle) was martyred and buried at the Circo di Nerone (Circus of Nero) in the *Ager Vaticanus*.
- **271–275 CE** | Construction of the Mura Aureliane (Aurelian Walls) under Imperatore Aureliano (Emperor Aurelian) (Dyson, 2010).
- **326–333 CE** | Emperor Costantino (Constantine I) commissioned the first basilica over the presumed burial site of San Pietro on Colle Vaticano (Krautheimer, 1980).
- **848–852 CE** | Construction of the Mura Leonine (Leonine Walls) under Papa Leone IV (Pope Leo IV) (Krautheimer, 1980).
- **1506–1626 CE** | Construction and consecration of the nuova Basilica di San Pietro (St. Peter's Basilica) (Hibbard, 1986).
- **1585–1590 CE** | Urban reforms under Papa Sisto V (Pope Sixtus V), including the erection of obelisks and axial reorganization of the city (Petrucci, 1993).
- **1656–1667 CE** | Planning and construction of Piazza San Pietro (St. Peter's Square), designed by Gian Lorenzo Bernini and commissioned by Papa Alessandro VII (Pope Alexander VII) (Wittkower, 1999).

- **1929 CE** | Patti Lateranensi (Lateran Treaty) recognize the sovereignty of the Stato della Città del Vaticano, establishing Vatican City as an independent state.

The timeline of the major dates of Rome's history concludes with the Patti Lateranensi of 1929, which formally recognized the Stato della Città del Vaticano. This outline is relevant to the inquiry into the architecture of cosmology in Rome. Many Roman monuments and architectural constructions associated with astronomical practices emerged within a culture shaped by religion and territorial expansion. Architecture becomes an expression of power, mediating between the sacred and the profane, and between celestial and terrestrial space.

This historical sequence can be learned by walking and is narrated by maps, which perhaps more effectively than words reveal how the configuration of the city is shaped by geography while simultaneously intertwined with political and religious authority. Figures 2–4 derive from Robert Burn's nineteenth-century historical guide to the archaeological sites of Rome, which emphasizes the relationship between topography and architectural remains.

Figure 2. Sette Colli (Seven Hills) and surrounding topography of Rome. Source: Burn, Rome and the Campagna (1880), public domain.

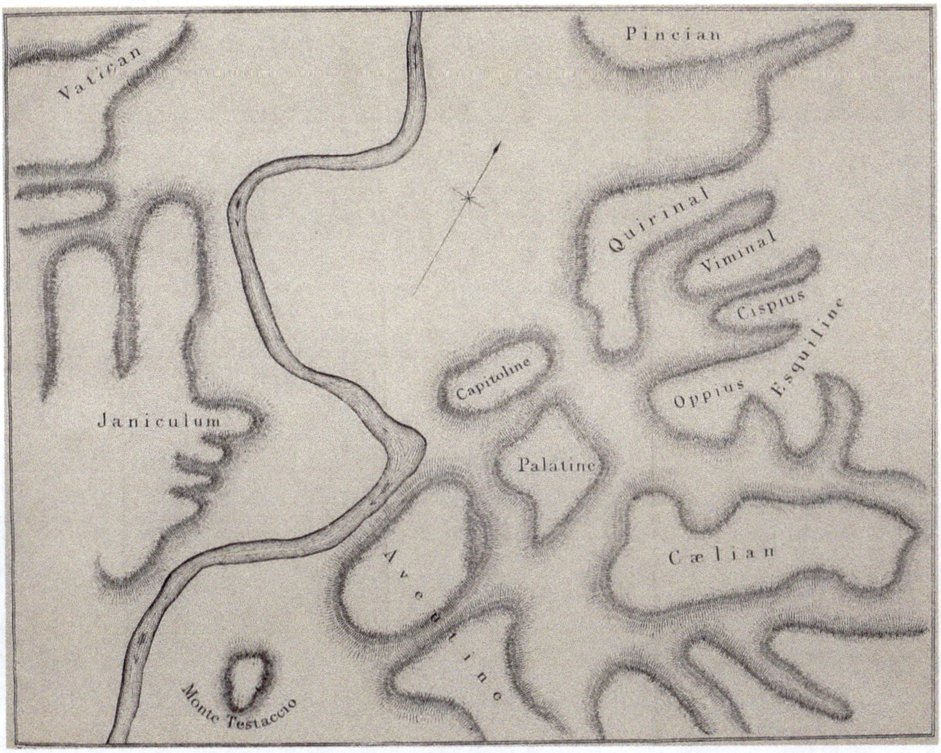

Figure 3. Mura Aureliane (Aurelian Walls). Source: Burn, Rome and the Campagna (1880), public domain.

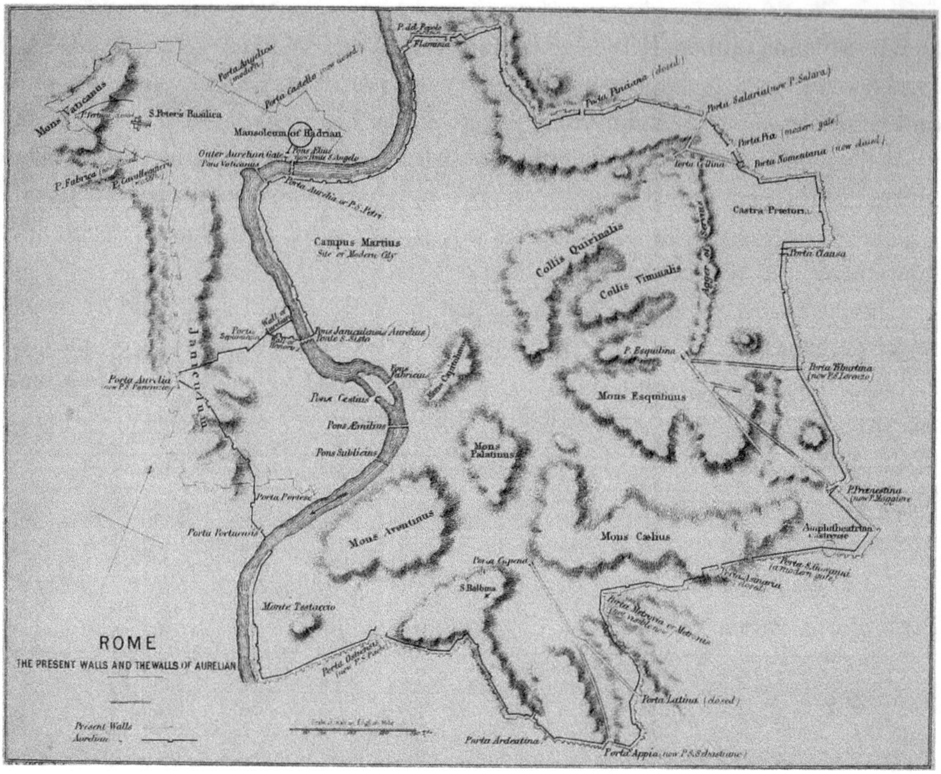

Figure 4. Foro Romano (Roman Forum). Source: Burn, Rome and the Campagna (1880), public domain.

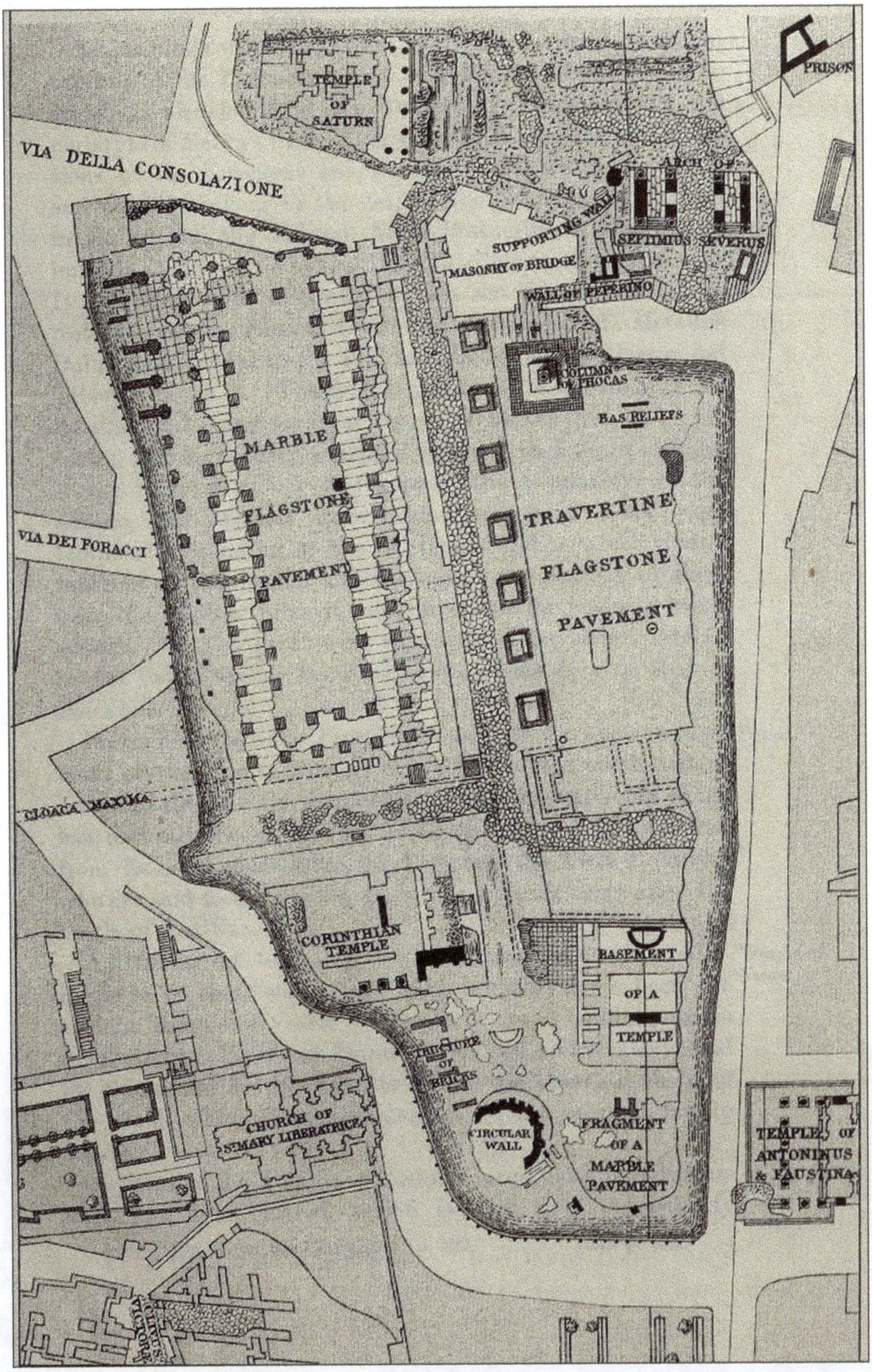

Figure 5 reproduces the 1748 map of Rome by the architect and surveyor Giovan Battista Nolli, commissioned by Papa Benedetto XIV (Pope Benedict XIV). Comprising twelve engraved plates and representing fourteen *rioni*, the map is oriented toward magnetic north. In Rome, magnetic north currently deviates approximately 4° east of true north (value varying slightly by historical period). Buildings are rendered as black poché against the white field of public space, establishing a figure–ground representation that continues to inform urban analysis.

Figure 5. Map of Rome (Nuova Pianta di Roma). Engraving by Giovan Battista Nolli, 1748. Public domain.

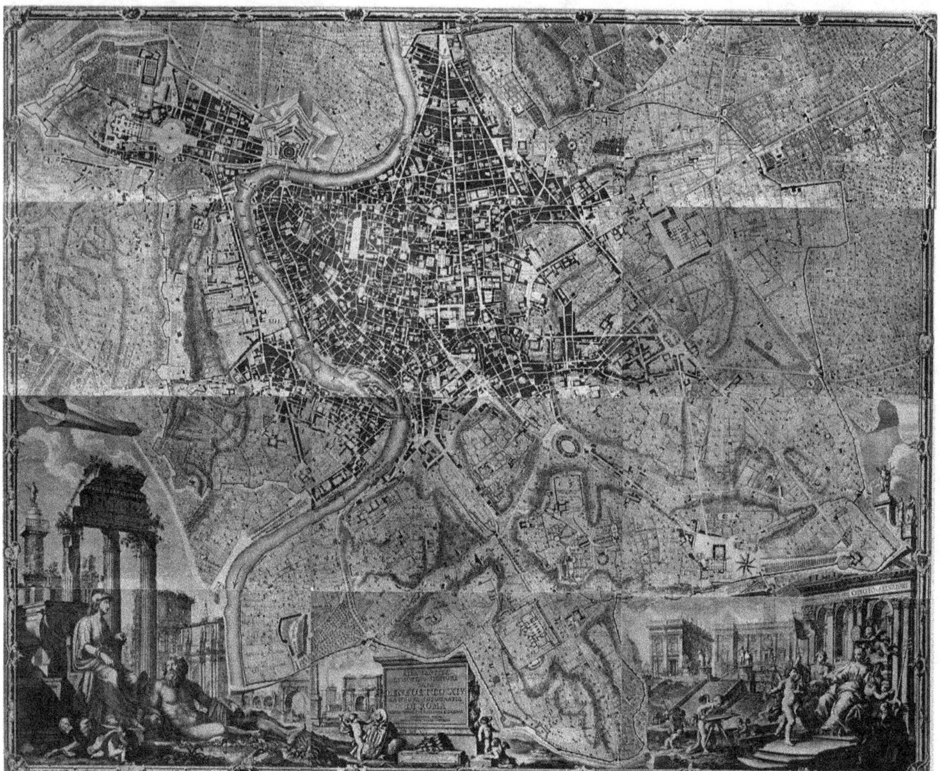

ALIGNMENTS: CARDO, DECUMANUS & SOLSTICES

The *cardo* and *decumanus* constituted the primary axes of Roman urban planning (Ward-Perkins, 1994). The *cardo*—from the Latin *cardo*, meaning "hinge" (Wayte & Marindin, 1890)—designated the north–south axis, while the *decumanus*—from *decumanus*, "tenth"—defined the east–west alignment. In planned Roman settlements, from colonial foundations to military camps (*castra*), these axes structured an orthogonal grid system organized around the *cardo maximus* and *decumanus maximus*. The forum, as civic and symbolic center, was typically positioned at or near their intersection.

Beyond their practical surveying function, the *cardo* and *decumanus* carried cosmological meaning. Their orthogonal configuration established a terrestrial coordinate system corresponding to the cardinal directions. For Christian Norberg-Schulz (1971), such orientation reflects existential space grounded in natural phenomena: east, as the origin of light, is associated with life; west, as the setting of the sun, evokes transition and mortality. In certain cosmological traditions, the cardinal axes intersect with the *axis mundi*, generating a comprehensive spatial schema.

Similarly, Cassirer (1955) describes the east–west solar line and its perpendicular north–south bisector as the earliest symbolic articulation of a cosmic schema:

> The east-west line, established by the course of the sun, was bisected by a vertical from north to south. With this intersection of the two lines—the decumanus and the cardo, as they were called in the language of the priests—religious thinking created its first basic schema of coordinates (Cassirer, 1955, p. 100)

This coordinate system may be understood, within the Earth–sky geometries framework developed in previous chapter, as a translation of celestial order into spatial practice. The cardinal cross becomes both surveying instrument and cosmological diagram.

Rome itself, however, was not founded as a planned orthogonal colony. Its morphology evolved incrementally, shaped by the topography of the Tiber and the Seven Hills. No singular *cardo maximus* or *decumanus maximus* governs the city at large scale. Yet local implementations of cardinal geometry appear in specific complexes.

The *Castra Praetoria* (21–23 CE), constructed under Tiberius, offers a clear example. The rectangular camp measured approximately 440 meters along its north–south axis and about 380 meters east–west (Platner & Ashby, 1929), closely approximating canonical orthogonal alignment to the cardinal directions. Here, the Roman military grid materializes as an imposed geometry within the otherwise organic urban field.

Astronomical considerations were also integral to Etruscan ritual practices associated with town foundation. The designation of a *templum* aligned to the four cardinal directions formalized the relation between celestial observation and terrestrial demarcation (Magli, 2008). The *Corpus Agrimensorum Romanorum* describes *centuriation*, a geometric subdivision of land into square plots organized through orthogonal axes. This procedure established a coordinate matrix in which spatial identity derived from relational positioning within a cardinal grid (Magli, 2008).

Although Rome lacks a unified orthogonal structure, several alignments correspond to solar azimuths near the winter solstice. Notably, the Via Sacra within the Roman Forum aligns toward the rising sun over the Colli Albani in the southeast during the December solstice (Quintela, 2022). Such alignments indicate that solar geometry informed specific spatial sequences even within an organically evolved urban fabric. The absence of a master grid does not imply the absence of cosmological intention.

CASTEL SANT'ANGELO: OF ANGELS, EMPERORS & ALIGNMENTS

An extremely significant astronomical reference to the cardinal points can be found in the *Mausoleo di Adriano* (Mausoleum of Hadrian), later known as Castel Sant'Angelo (Castle of the Holy Angel), located at 41°54′11″ N, 12°27′59″ E. The Mausoleum was built between 134 CE and 139 CE under Imperatore Adriano, conceived as his monumental above-ground tomb and that of his successors. It stands on the left bank of the Tevere (Tiber) (Figure 6).

Figure 6. Satellite imagery of Castel Sant'Angelo and Mole Adriano, from Google My Maps.

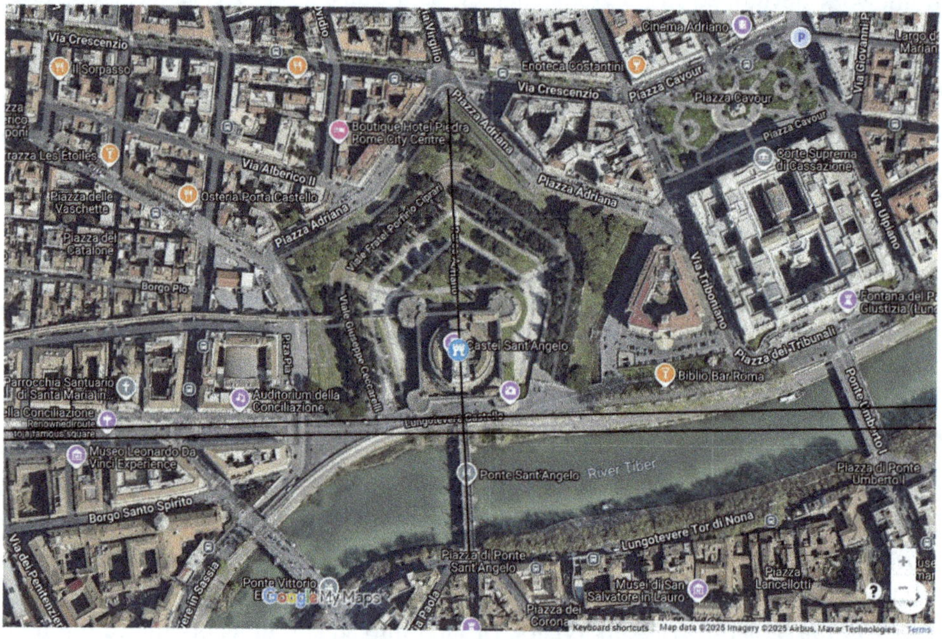

The Mausoleum was constructed in conjunction with *Pons Aelius* (Aelian Bridge), today Ponte Sant'Angelo (Bridge of the Holy Angel), which connected it to the Campus Martius (Platner & Ashby, 1929, p. 336). In 271 CE, the construction of the *Mura Aureliane* (Aurelian Walls) began under Imperatore Aureliano. The walls were extended to incorporate the Mausoleum as a fortified outpost controlling the river crossing. The building thus changed function and, from mausoleum, became a *castellum*. It assumed the current name of Castel Sant'Angelo in the late sixth century. According to tradition, in 590 CE Papa Gregorio Magno saw the Archangel Michael sheathing his sword above the monument during a plague epidemic, signifying its end; from this episode derive both the structure's modern name and the statue of the archangel that surmounts it (Encyclopædia Britannica, n.d.).

The Mausoleum consists of a cylindrical concrete structure measuring approximately 21 meters in height and 64 meters in diameter. It rests upon a square travertine podium measuring about 84 meters per side and 10 meters in height (Platner & Ashby, 1929, p. 337). A helicoidal ramp connects the internal levels, culminating at the rooftop terrace, *Terrazzo dell'Angelo* (Angel's Terrace), from which panoramic views extend toward San Pietro and across the city. The present bronze statue of *San Michele Arcangelo* (St. Michael the Archangel), executed in 1752 by Peter Anton von Verschaffelt, stands slightly off the central axis of the cylindrical drum.

The most significant interior space of the *Mausoleo di Adriano* (Mausoleum of Hadrian) is the *Sala delle Urne* (Hall of Urns), which once housed the cinerary remains of Imperatore Adriano and members of his family (Platner & Ashby, 1929, p. 336). Located on the third level, the chamber is shaped in plan as a *crux quadrata* (Greek cross), with its four arms oriented toward the cardinal directions (Figure 7).

Figure 7. Diagram of geometry and alignments of Castel and Ponte Sant'Angelo.

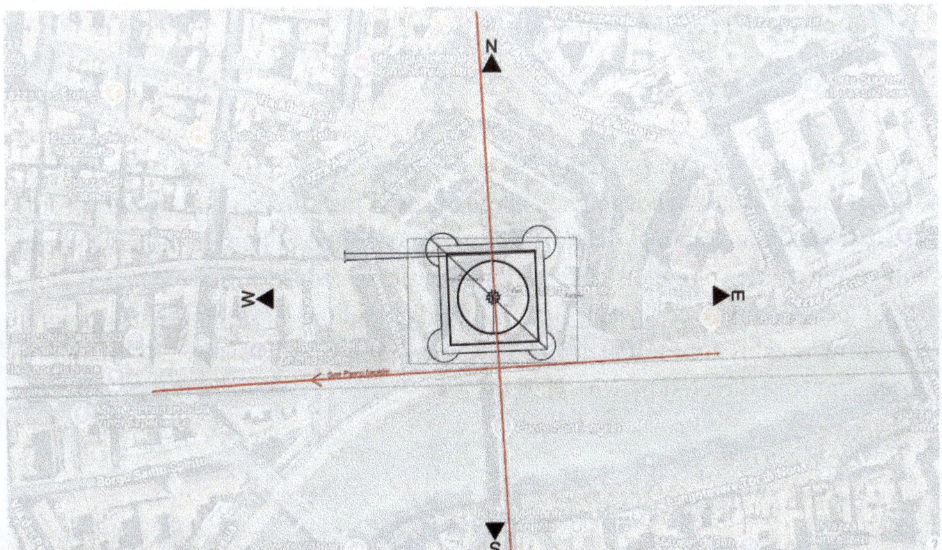

Openings above the east and west niches admit natural light into the interior. Given the precise cardinal alignment of the monument, seasonal solar variations produce shifting patterns of illumination within the space. The relationship between architecture and light in Hadrian's monuments has been examined by De Franceschini and Veneziano (2018), who analyze the symbolic use of solar phenomena in the *Mausoleo di Adriano* and other buildings commissioned by the emperor. Their study situates the monument within a broader cosmological framework in which solar light reinforces imperial ideology and notions of apotheosis, while noting the complexity of verifying specific equinoctial penetration within the present structural configuration.[2]

The western niche contains a marble plaque inscribed with a poem attributed to Hadrian. The text gained renewed visibility through Marguerite Yourcenar's novel *Memoirs of Hadrian* (1951). The original Latin verses read:

> Animula vagula blandula,
> Hospes comesque corporis,
> Quae nunc abibis in loca,
> Pallidula, rigida, nudula,
> Nec, ut soles, dabis iocos.

The poem is translated in English as:

> Little soul, gentle and drifting,
> guest and companion of my body,
> now you will dwell below in pallid places,
> stark and bare;
> there you will abandon your play of yore.

Pons Aelius, later Ponte Sant'Angelo, has undergone several restorations but remains one of the best-preserved bridges of ancient Rome (Encyclopædia Britannica, n.d.). In the sixteenth century, Papa Clemente VII installed statues of the apostles San Pietro and San Paolo at the northern entrance to the bridge. In the seventeenth century, ten statues of angels designed by Gian Lorenzo Bernini were added along the parapets (Wittkower, 1999). Both the bridge and the Castel has entered popular culture, notably through Dan Brown's novel *Angels & Demons* (2000) and the subsequent film directed by Ron Howard (2009).

The geometrical and astronomical alignments of Castel Sant'Angelo and Ponte Sant'Angelo are remarkable. The diagram of the Mausoleum is that of a circle inscribed within a square defined by the outer fortifications; the sides of the square align with the four cardinal directions (Figure 7). The plan of the Sala delle Urne, positioned at the geometric center of the cylinder, reinforces this orientation through the crux quadrata configuration.

Ponte Sant'Angelo (Bridge of the Holy Angel) follows an almost precise north–south alignment. As will be discussed in the following section, "Obeliscus Vaticanus & the Meridian Line," the axis connecting the midpoint of the façade of Basilica di San Pietro (St. Peter's Basilica) to the center of the Obelisco Vaticano (Vatican Obelisk) deviates by approximately 1° from the true east–west direction. When this axis is extended eastward across the Tevere, it intersects Ponte Sant'Angelo near its northern end at approximately 41.90244° N, 12.46641° E, slightly south of the statue *Angelo con la Spugna* (Angel with the Sponge), located at 41.90247° N, 12.46633° E.

Measured through satellite imagery and angular calculation using Google Earth, this orthogonality between the basilica–obelisk axis and the longitudinal axis of the bridge reveals a geometric correspondence may suggest intentionality rather than coincidence. The diagram in Figure 8 illustrates these relationships, derived from connecting the referenced coordinates and analyzing their angular intersections.

Figure 8. Diagram of orthogonality between the axis of Basilica di San Pietro and the longitudinal axis of Ponte Sant'Angelo, derived from Google Earth measurements.

My experience of Castel Sant'Angelo dates back to childhood memories. I still cross Ponte Sant'Angelo several times a week when I am in Rome. The Mausoleo di Adriano remains the focal point when traversing the bridge from Rione Ponte toward Rione Borgo. The paired statues of angels frame the perspective (Figure 9), marking the encounter between sacred narrative, urban fabric, art, and architecture (Figure 10). The experience is sequential: the bridge prepares the gaze; the cylindrical mass of the Mausoleum anchors the horizon; the Vatican plateau emerges beyond.

Figure 9. Castel Sant'Angelo as focal point from Ponte Sant'Angelo, framed by a pair of statues of angels. 2025.01.16.

Figure 10. Ponte Sant'Angelo photographed from Castel Sant'Angelo. 2026.02.01.

I frequently visit the interior of Castel Sant'Angelo, now a national museum hosting art, historical, and military collections. Its architectural stratification narrates nearly 1,900 years of history: from imperial mausoleum to fortified *castellum*, to papal residence. My most recent visit took place on 2026.02.01; the walkthrough was documented in a photographic series, excerpts of which are presented in Figures 11–14. The ascent through the helicoidal ramp from the entrance to the rooftop terrace offers a dynamic, embodied reading of the monument's geometry. The curvature of the ramp, the thickness of the masonry, and the controlled apertures of light reinforce the sense of centripetal movement toward elevation.

Figure 11. Ascending the helicoidal ramp inside Castel Sant'Angelo. 2026.02.01.

Figure 12. Sala delle Urne, west niche; marble plaque with Hadrian's poem at center. 2026.02.01.

Figure 13. Via della Conciliazione and the dome of Basilica di San Pietro at sunset, viewed from the terrace of Castel Sant'Angelo. 2026.02.01.

Figure 14. Looking north, at San Michele Arcangelo on the rooftop terrace, 2026.02.01.

From the terrace, the visitor encounters a 360° panorama of the city. Figure 13 shows Via della Conciliazione and the dome of Basilica di San Pietro at sunset, the sun descending toward the south-west. The spatial continuity between river, bridge, Mausoleum, and Vatican becomes legible as a single articulated system.

The roof of Castel Sant'Angelo is crowned by the bronze statue of San Michele Arcangelo, sculpted in 1753 by the Flemish artist Peter Anton von Verschaffelt. The Archangel faces south toward the Tevere and the city and is represented sheathing his sword, symbolizing the end of the plague of 590 CE. Once again, art is integrated with architecture and the urban landscape, affirming the encounter between sky and earth that defines the cosmological reading of this monument.

THE OBELISKS & THE GREATEST SOLAR CLOCKS IN THE WORLD

Rome presents thirteen ancient obelisks distributed across the urban territory, often separated by considerable distances. Most were transported to Rome from Egypt during the Imperial period, where they functioned as solar monuments and symbols of divine kingship (Barrett, 2017). Their relocation to Rome represented not merely architectural reuse but the transposition of solar symbolism into the imperial capital.

Fascination with obelisks persisted from antiquity through the Renaissance. By the late sixteenth century, they were regarded as among the marvels of the ancient world (Mercati, 1589). Although Egyptian in origin, the term *obelisk* derives from the Greek ὀβελίσκος (*obelískos*), diminutive of ὀβελός (*obelós*), meaning "spit" or "pointed pillar," a reference to their slender vertical form.

Architecturally, an obelisk consists of:

- a monolithic shaft (*fustum*),
- a tapering quadrangular body,
- and a crowning pyramidion.

Most Roman obelisks are carved from red granite and inscribed with hieroglyphs. Their geometric form is that of a tall, attenuated truncated pyramid with square plan, terminating in a pyramidion. When mounted on a podium, the base remains centered on the same vertical axis, reinforcing geometric stability.

The diagram in Figure 15 represents the thirteen obelisks currently standing in Rome as nodes within a spatial graph. The connecting lines do not imply direct historical relationships but rather visualize their geometric and geographic correspondences within the urban fabric. This network representation allows the city to be read as a field of axial and radial relationships structured by vertical markers.

Figure 15. Spatial graph network of the thirteen obelisks of Rome as nodes and their geographical connections. Diagram derived from Google My Maps.

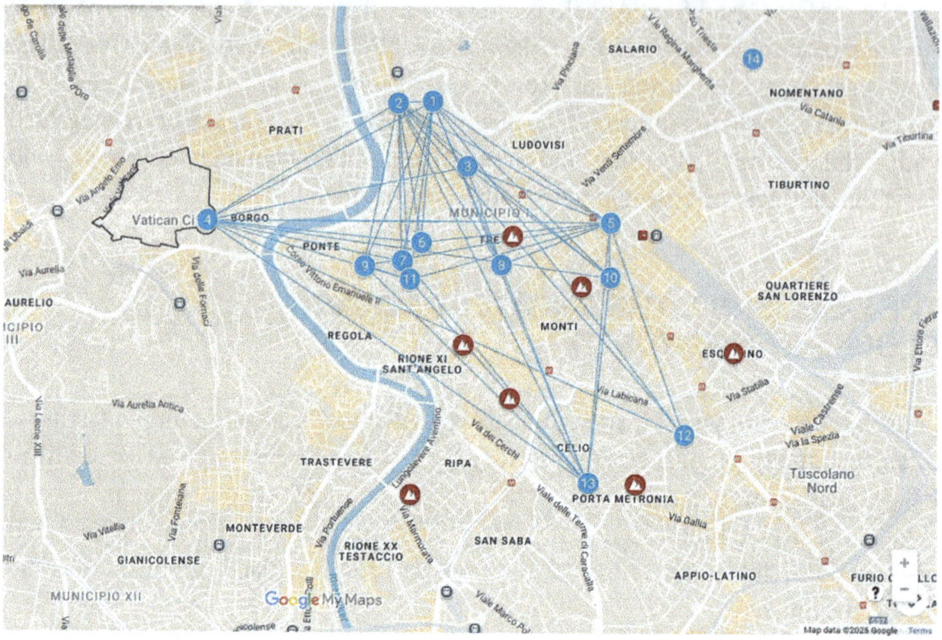

An elevation drawing by Giovanni Battista Cipriani (1823) (Figure 16), illustrating twelve Egyptian obelisks in Rome at the same scale, demonstrates their significant variation in height while maintaining a consistent typological proportion. The comparison reveals the vertical hierarchy embedded within Rome's skyline, where obelisks function not only as monuments but as calibrated urban instruments. The comparative elevation underscores the Roman practice of relocating and recontextualizing Egyptian monoliths within a new imperial and later papal cosmological framework.

Figure 16. Elevation of twelve Egyptian obelisks in Rome drawn to scale. From Cipriani (1823), Su i dodici obelischi egizj che adornano la città di Roma. Public domain.

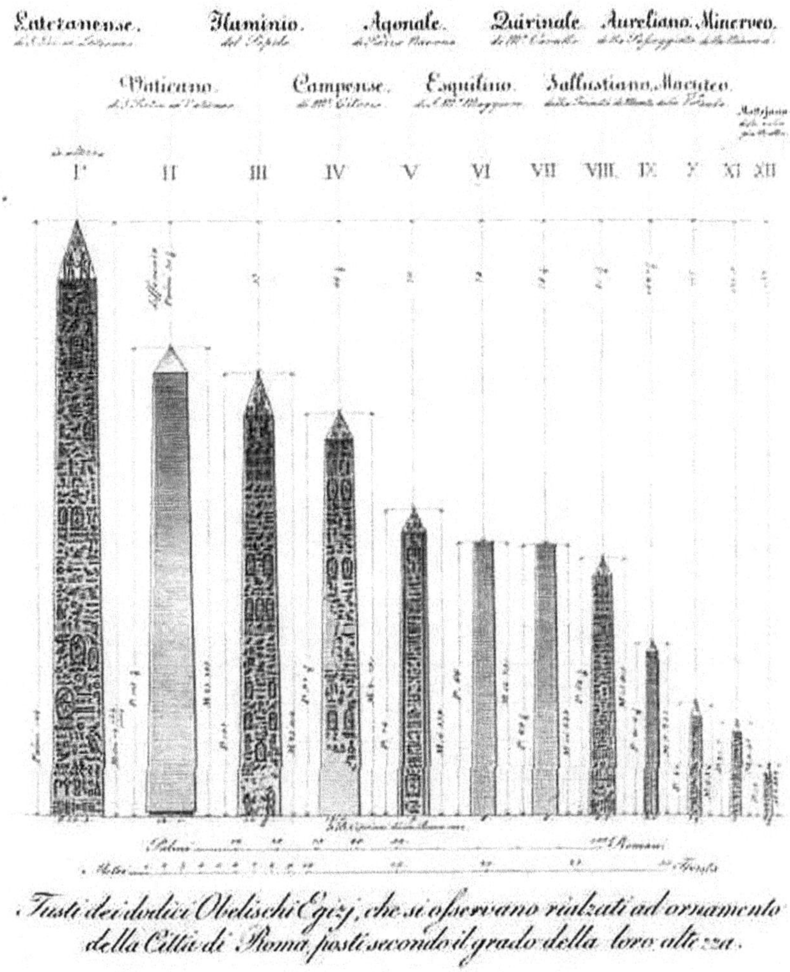

Obelisks as Urban Markers and Solar Instruments

Obelisks operate simultaneously as monuments and spatial markers. They often define nodal points within piazzas, anchor urban axes, and articulate figure–ground relationships in the city. Their square plan inherently suggests potential cardinal alignment, projecting axial geometry outward into surrounding space. However, cardinal orientation is not universal among Rome's obelisks.

The table below records the thirteen obelisks, listing latitude, longitude, azimuth (measured clockwise from true north), and approximate shaft height and base length.

Table 1. The thirteen ancient obelisks in Rome.

#	Name	Latitude	Longitude	Azimuth	Height (meters)	Base (meters)	Notes / Figures
1	Pinciano	41.9108°	12.47975°	344.8°	56.6	9	Figure 17
2	Flaminio	41.91072°	12.47634°	346.2°	150	12	Figure 18
3	Sallustiano	41.90617°	12.48322°	336.0°	100	8	Figure 19
4	Vaticano	41.90223°	12.45726°	0°	134.5	16	Section "Obeliscus Vaticanus and the Meridian Line"
5	Dogali	41.90214°	12.49746°	297.8°	20.8	8	Figure 20
6	Montecitorio	41.90068°	12.47867°	0°	111.5	8	Section "Obelisco di Montecitorio & Horologium Augusti"
7	Macuteo	41.89932°	12.47674°	357°	47.6	8	Figure 21
8	Quirinale	41.89907°	12.48666°	120°	95	8	Figure 22
9	Agonalis	41.8989°	12.4730°	359°	98.4	8	Figure 23
10	Esquilino	41.89816°	12.4975°	264°	83.8	8	Figure 24
11	Minerveo	41.89797°	12.47756°	359°	41.6	6	Figure 25
12	Lateranense	41.88688°	12.50481°	348.0°	150	12	Figure 26
13	Matteiano	41.88339°	12.49532°	346°	40	6	Figure 27

Figure 17. Obelisco Pinciano, Villa Borghese, looking northwest, 2025.05.0.

Figure 18. Obelisco Flaminio with cross at summit, looking northeast, 2023.05.04.

Figure 19. Obelisco Sallustiano and the Church of Trinità dei Monti. Vertical sundial visible on the right side of the façade, 2023.03.07.

Figure 20. Obelisco Dogali, Piazza della Repubblica, looking west, 2025.06.02.

Figure 21. Obelisco Macuteo at apparent solar noon, with the Pantheon in the background. The obelisk rises at the center of the Fontana del Pantheon, designed by Giacomo della Porta (1575), sculpted by Leonardo Sormani, and modified by Filippo Barigioni (1711). 2025.06.11.

Figure 22. Obelisco del Quirinale looking southeast, with the Fontana dei Dioscuri in the foreground. 2023.01.12.

Figure 23. Obelisco Agonale surmounted by the Fontana dei Quattro Fiumi by Gian Lorenzo Bernini (1651). 2025.06.13.

Figure 24 Obelisco Esquilino, with the Basilica di Santa Maria Maggiore in the background. 2025.06.17.

Figure 25. Obelisco Minerveo at solar noon; in the foreground the statue of "Elefantino della Minerva" by Gian Lorenzo Bernini (1667). 2025.06.13.

Figure 26. Obelisco Lateranense, with the rear façade of the Basilica di San Giovanni in Laterano in the background. 2024.06.07.

Figure 27. Obelisco Matteiano in Villa Celimontana. 2024.05.26.

Methodological Note

Coordinates were obtained through geospatial mapping tools. Azimuth values represent the orientation of the faces of the obelisk shaft relative to true north. Measurements are approximate and may vary slightly due to restoration phases or subsequent repositioning.

Observed Geometric Relations

Latitudinal Alignments

Comparison of latitude values reveals notable near-horizontal alignments:

- *Pinciano and Flaminio* lie at nearly identical latitudes, separated by approximately 926 feet. Their east–west alignment establishes a horizontal axis across the northern sector of the city.
- *Dogali and Vaticano* also share nearly equal latitude, separated by approximately 2.07 miles. However, the Dogali obelisk was installed in 1924, making intentional ancient alignment unlikely; the correspondence is probably coincidental.

These latitudinal relationships are measurable but must be interpreted cautiously in terms of intentionality.

Cardinally Oriented Obelisks

The azimuth values of the obelisks Vaticano, Montecitorio, Macuteo, Agonalis, and Minerveo approximate 0°, indicating that their faces align closely with the four cardinal directions. In such cases, the obelisk functions as a potential gnomon, capable of casting shadows that register solar movement.

Only the Vaticano and Montecitorio obelisks incorporate formal meridian lines, installed centuries after their original erection. In these cases, the obelisk became explicitly integrated into solar timekeeping systems, transforming monumental symbol into calibrated astronomical instrument.

Within the Earth–sky geometries framework, this represents a literal translation of celestial motion into terrestrial inscription.

Axial Urban Alignments

A significant visual axis links the obelisks Sallustiano and Esquilino along Via Sistina and Via delle Quattro Fontane (Figure 28). The distance between the two monuments is approximately 1.48 km (0.92 miles). The azimuth of this axis is 231.6° (clockwise from true north).

Figure 28. Alignments between obelisks and urban axes, screenshot from Google Earth.

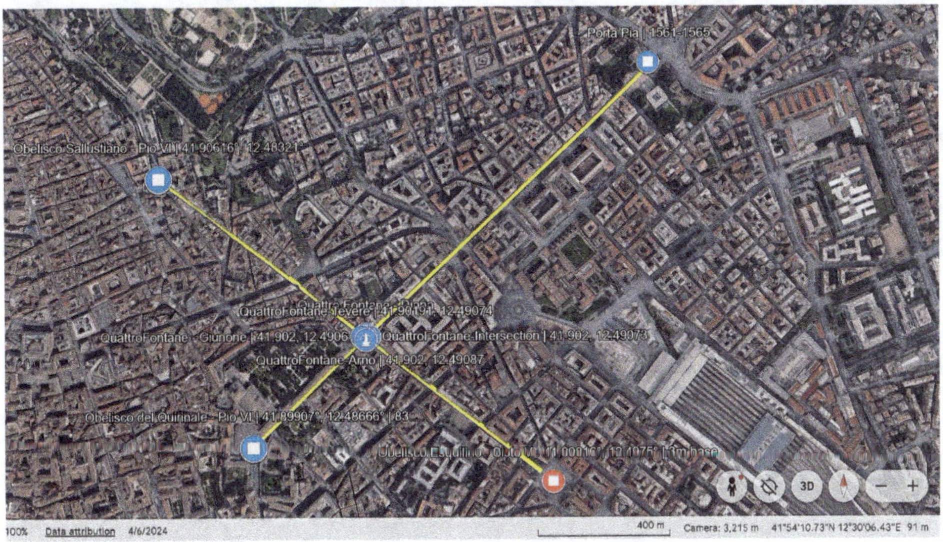

Via delle Quattro Fontane intersects Via del Quirinale to the southwest and Via XX Settembre to the northeast. The four corner fountains at this intersection, erected in the late sixteenth century, are themselves oriented toward the cardinal directions, reinforcing axial geometry at urban scale. Via XX Settembre has an azimuth of approximately 46.2°, forming a near-orthogonal relationship to the Sallustiano–Esquilino axis.

Here, obelisks operate not as isolated artifacts but as terminal markers of perspectival and geometric alignments embedded within the urban fabric.

Obelisco di Montecitorio and Horologium Augusti

The Obelisco di Montecitorio, also known as the Obelisco Solare, is currently located at latitude 41.90068° N and longitude 12.47867° E in Piazza Montecitorio, directly in front of the 17th-century Palazzo Montecitorio, seat of the Italian

Chamber of Deputies (Figures 29–30). The red granite monolith was commissioned under Pharaoh Psammetichus II in the 6th century BCE for the temple precinct of Heliopolis. The shaft measures approximately 21.8 meters; its present total height, including pedestal and bronze globe, is approximately 30 meters.

Figure 29. Obelisco di Montecitorio casting shadow, satellite imagery (Google Earth).

Figure 30. Obelisco di Montecitorio looking north, with Palazzo di Montecitorio in the background. 2025.02.26.

Transported to Rome in 10 BCE under Augustus, the obelisk was erected in the Campus Martius. A long-standing scholarly debate concerns its function within the so-called Horologium (or Solarium) Augusti. In the early 1980s, Edmund Buchner proposed that the obelisk served as the gnomon of a vast monumental sundial, casting its shadow across an extensive marble pavement. Subsequent reassessments, particularly by Heslin (2007) and Haselberger (2011), argue that the archaeological evidence more plausibly indicates a meridian line rather than a full radial dial. While consensus remains incomplete, current scholarship favors interpretation of the complex as a meridian instrument marking solar noon and possibly key calendrical dates.

Rediscovered in the Renaissance and fully excavated in 1748, the obelisk was found broken into five fragments. It was re-erected in its current location under Pope Pius VI, directed by the architect Giovanni Antinori (Cipriani, 1823; Haselberger, 2011).

A modern bronze meridian line, installed during the 1998 redesign of Piazza Montecitorio (Zagari, 1998), reactivates the obelisk's solar function; at local noon, the shadow intersects the line, symbolically reconnecting the monument to its Augustan astronomical program. Figures 31–32 document the obelisk casting its noon shadow along this modern line.

Figure 31. Obelisco di Montecitorio at apparent solar noon casting shadow on the meridian line, two days after the June solstice. 2025.06.23.

Figure 32. Palazzo di Montecitorio and the meridian line, looking north. 2025.06.23.

SAN PIETRO'S SUBLIME: HISTORY AND MEMORIES

I was born and raised less than one mile from the Basilica and Piazza San Pietro (St. Peter's Basilica and square). My childhood memories now intertwine with scholarly understanding of these sublime cosmological architectures. On Sundays, I attended the Angelus prayer in the piazza with my family, watching the Pope appear at the Apostolic Palace window. Passing through Bernini's colonnade, I felt both awe and intimidation at the monumental scale. Standing on one of the marble discs marking the focal points of the elliptical geometry generated early fascination with spatial order I could not yet articulate.

Once in a while we climbed the narrow spiral staircase between the two shells of the structure of the dome of the Basilica: the reward for the climb in the claustrophobic space was the sense of wonder at the top, looking down at the inside of the church from over one hundred meters. Another vivid memory is attending the midnight mass of Christmas Eve: the grand space of the Basilica was filled with live organ music and choir chanting and smell of incense. It was an immersive experience involving all the senses, in which the body, the architecture, and the sky were in relation.

These times are long gone; both the Basilica and the piazza in the past decades have been increasingly filled with tourists and access to the Basilica currently requires standing on long lines and going through several security checks. But when I am in the eternal city, I am still going to San Pietro as often as I can and still experience the sense of awe I was feeling as a child, but perhaps even more now that I can understand the uncanny geometries behind the unprecedented beauty and grandeur of the architecture and art.

Vatican City and Piazza San Pietro

The San Pietro complex lies within Città del Vaticano (Figures 33-34), an independent city-state entirely enclosed within Rome. With an area of approximately 44 hectares (110 acres) and a perimeter of roughly 3.2 kilometers, Vatican City is the smallest sovereign state in the world. It achieved independence through the Lateran Treaty of 1929 between the Kingdom of Italy and the Holy See.

Figure 33. Satellite imagery of Piazza San Pietro (in yellow) and Città del Vaticano, highlighted in grey (from Google My Maps).

Figure 34. Piazza San Pietro, viewed from the dome of the Basilica, looking east. 2016.10.29.

The fortifications known as the Mura Leonine (Figure 35) were constructed in the 9th century under Pope Leo IV to defend the Vatican hill from Saracen incursions.

These walls were later reinforced and integrated with the 16th-century bastions commonly associated with Michelangelo. The layered fortifications materially express the intertwining of spiritual authority and political sovereignty.

Figure 35. Mura Leonine at the entrance of the Vatican Museums, an access to the Vatican City. 2026.05.17.

Piazza San Pietro provides access to the homonymous Basilica. It is a major destination in Rome for mass tourism, visiting not only for religious purposes. Also here, as in many archaeoastronomical examples, the architecture of cosmology overlaps with religion and rituals. The square functions as a vast urban enclosure in which the sky operates not merely as a background but as architectural component, becoming figure in the composition: a dynamic roof with ever-changing light and patterns defined by the passage of clouds The elliptical colonnade frames the celestial vault, transforming atmosphere, light, and solar motion into active elements of the spatial composition. and marked by the apparent path of the Sun. The square celebrates God not only with its access to the Basilica but also with the framing of the sky, whole sacred nature is mirrored by the humanmade environment.

The Basilica

For the majority of visitors to Piazza San Pietro, the Basilica di San Pietro (St. Peter's Basilica) is the primary destination. The Basilica is located west of Piazza San Pietro, with an entrance at latitude 41.90219° N and longitude 12.45504° E. The basilica is named after Saint Peter (c. 1 CE – c. 64/67 CE) who, in the Christian tradition, was one of the Twelve Apostles of Jesus Christ and leader of the early Christian Church. The Basilica di San Pietro is not only a religious destination for Roman Catholics and Christians worldwide, but is considered a symbol of Roma, a masterpiece of art and architecture. I will present a summary of its 1,800 years history only in reference to the aim of this chapter—the study of the architecture of cosmology in an urban context.

The present Basilica stands on the same site as the *Antica Basilica di San Pietro* (*Old St. Peter's Basilica*), which is called Antica (Old) to differentiate it from the current one (Figure 36). This Paleochristian church was built under the emperor Constantine I between the years 326 and 333 CE; it presented a five-aisle layout plan (Encyclopædia Britannica, n.d.), following the architectural typology of an ancient Rome Basilica. A courtyard, called *Garden of Paradise* was added approximately two hundred years later. The Old St. Peter's Basilica site was the *Circo di Nerone*, place of martyrdom of St. Peter who, according to tradition, was buried in a necropolis next to it.

Figure 36. Drawing of Old St Peter's Basilica, Rome, c. 1450, drawn by H. W. Brewer, 1891. Source: RIBA Collections, RIBApix.

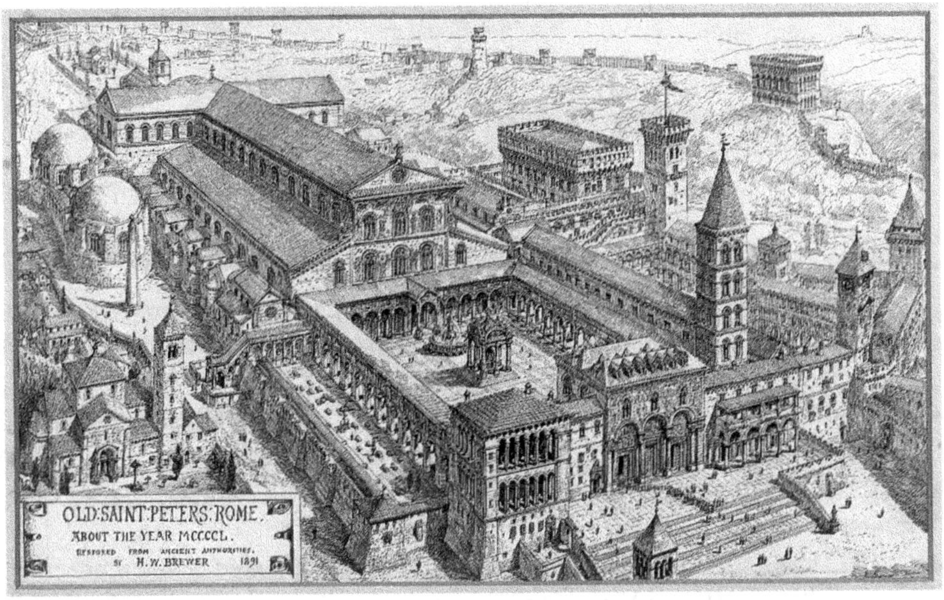

The Old St. Peter's Basilica was demolished in 1505 under the papacy of *Giulio II* (Julius II) to accommodate the construction of a new basilica responding to the need for a larger and more prominent building for expanding crowds of religious followers—and as a built expression of an ideological program establishing the power and glory of the Roman Catholic Church. The history of the construction of the new basilica is extremely complex and long; it took 120 years and several papacies, from the demolition of the Old one to the consecration of the new one. Many of the most prominent artists and architects of the Italian Renaissance and Baroque were involved. Noticeably, in that time geometry, astronomy, art, architecture, engineering and urban design were strictly linked in the creation of buildings and city spaces; the designers, whose expertise encompassed all these disciplines and practices, were often also ordered priests (Gerbino, 2014). The following timeline outlines the main dates of the long and laborious process, called *Fabbrica di San Pietro* (St. Peter's construction), which generated perhaps the greatest work of architecture, art, urban design and engineering ever built.

- **1506-1514 |** *Donato Bramante* is commissioned to design of the new basilica. His design was based on a *Greek Cross*, based on a square layout and surmounting a dome. The construction of four central piers and arches planned to support the dome began before his death in 1514.

- **1514-1520 |** Pope Leo X assigned the design to *Raffaello Sanzio* (Raphael), *Fra Giovanni Giocondo*, and *Giuliano da Sangallo*; Bramante's Greek cross is changed to a *Latin Cross* exhibiting three aisles (Editors of Encyclopædia Britannica, 2025)
- **1520 |** Following Raphael's death three architects are appointed by pope Leo X: *Antonio da Sangallo, Baldassarre Peruzzi,* and *Andrea Sansovino.*
- **1546 |** After the death of Antonio di Sangallo, *Michelangelo Buonarroti* was appointed by pope *Paolo III* (Paul III) as chief architect. His design followed the initial, Bramante's Greek cross concept.
- **1564 |** Michelangelo dies and only the dome drum is completed. The commission is transferred to the architects *Pirro Ligorio* and *Giacomo da Vignola.*
- **1572 |** Pope *Gregorio XIII* (Gregory XIII) assigns the direction of work to the architect and sculptor *Giacomo della Porta.*
- **1586 |** The Egyptian obelisk is moved to its current location under the direction of the architect *Domenico Fontana* during the papacy of *Sisto V.*
- **1590 |** The dome is completed under the direction of *Giacomo della Porta* and *Domenico Fontana.*
- **1603 |** The Baroque architect *Carlo Maderno* is appointed chief architect by pope *Paolo V* (Paolo V); he modifies again the plan layout from Greek cross to Latin cross and designs the façade.
- **1626 |** The basilica is consecrated on November 18 by pope Urbano VIII (Urban VIII).
- **1633 |** The bronze Baldacchino (canopy) by Gian Lorenzo Bernini is completed. The Baldacchino is a major feature of the basilica placed underneath the dome, above the main altar and the tomb of St. Peter, beneath it.

As outlined in the timeline, after several designs alternating between Greek and Latin cross (Figure 37) the basilica finally built presents a plan layout geometrically shaped as a Latin cross, with three aisles along the east-west longitudinal axis. The dome is supported by four piers and is located at the intersection between the arms of the cross, that is the main aisle and the transept. There are additional ten smaller domes: the four around the main dome are hemispherical, while the other six, located in symmetrical pairs along the two lateral aisles, are elliptical. From a plan view, the geometry is characterized by symmetry and hierarchical elements, resembling recursive forms. The dimensions of the basilica are outstanding: it measures approximately 218 meters (715 feet) along the central east-west axis, with a width of about 150 meters (492 feet); the central nave inside reaches a height of 46 meters (151 feet).

Figure 37. Basilica first designs, from Fletcher (Fletcher 1929, p. 586).

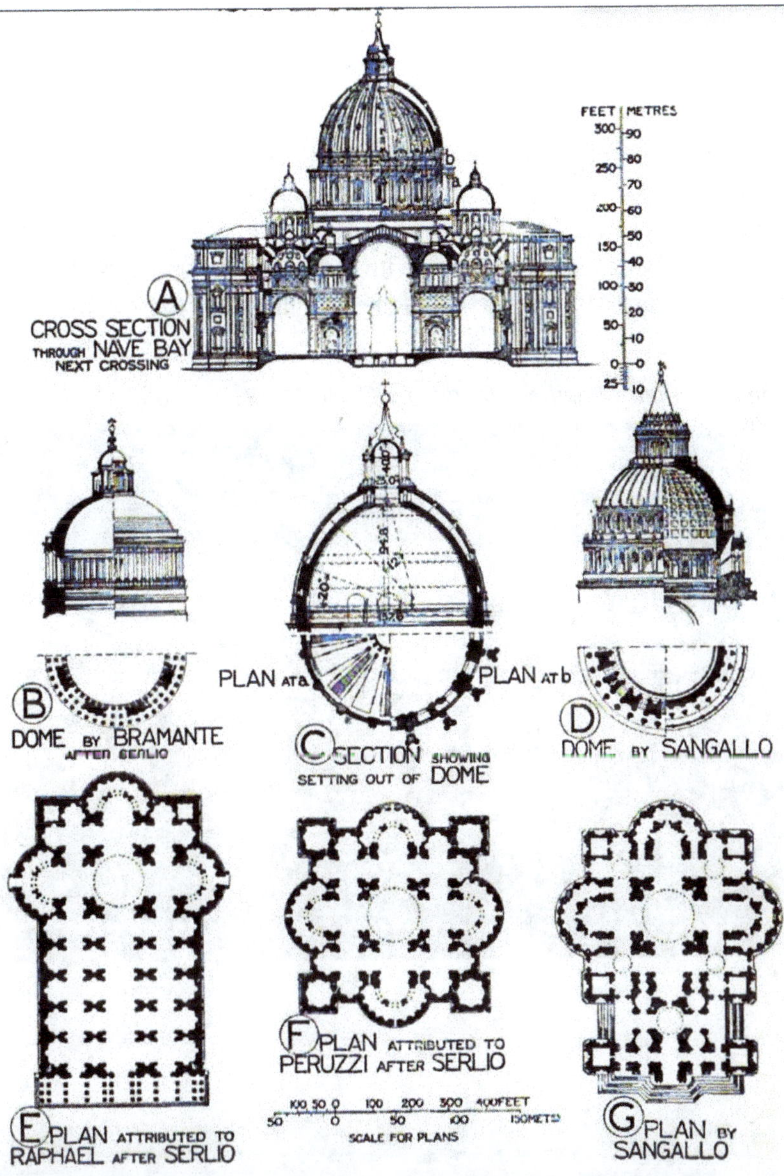

The dome was designed by Michelangelo and built up to the tambur (drum) during his lifetime; its construction was completed under the direction of Giacomo della Porta and Domenico Fontana in 1590; internally it has a diameter of about 42 meters and reaches a height of 117 meters (Fabbrica di San Pietro, n.d.). Its structure is based on a double-shell with an inner cavity, which can be climbed by visitors

(6.4.2-12) to reach the walkway on the tambur (drum), which is decorated with a gold mosaic, with contrasting blue letters stating in Latin aquote from the Bible, specifically the Gospel of Matthew 16:18-19: “tv es petrvs et svper hanc petram aedificabo ecclesiam meam et tibi dabo claves regni caelorvm”, which translates in “You are Peter and on this rock I will build my church and give you the keys to the Kingdom of Heaven” (Fabbrica di San Pietro, n.d.).

The photos in this section were taken on different dates and illustrate the approach to the Basilica, from outside the Piazza to the inside of the Basilica following a westward path (Figures 38-40) aligned with the main east-west axis of the building. The photos in Figure 41 are views from the catwalk inside the cupola.

Figure 38. San Pietro basilica: façade by Carlo Maderno (1603). 2016.10.30.

Figure 39. The main aisle, walking westward with the Baldacchino by Bernini as focal point. 2021.05.13.

Figure 40. Walking westward toward the Baldacchino, looking up at the dome. 2025.06.02.

Figure 41. Views from inside the Cupola. 2016.10.29.

Noticeable from the satellite imagery, the axis of the Basilica is slightly deviated from the east-west orientation (Figure 42). One theory is that the axis of the Basilica is aligned with Monte Guadagnolo, a mountain at a distance of about 24.5 miles from Piazza San Pietro (map 6.4.2-2). According to the astrophysicist Costantino Sigismondi, the axis of the Basilica of San Pietro kept the azimuth of the Old Basilica, which coincides with the sunrise azimuth on the 25th of March; Monte Guadagnolo is where St. Eustachius had a vision of Jesus Christ and the Cross at the beginning of the 2nd century CE, and where a sanctuary was built under the emperor Constantine (Sigismondi, 2021).

Figure 42. The Basilica and Piazza San Pietro; satellite imagery from Google Earth.

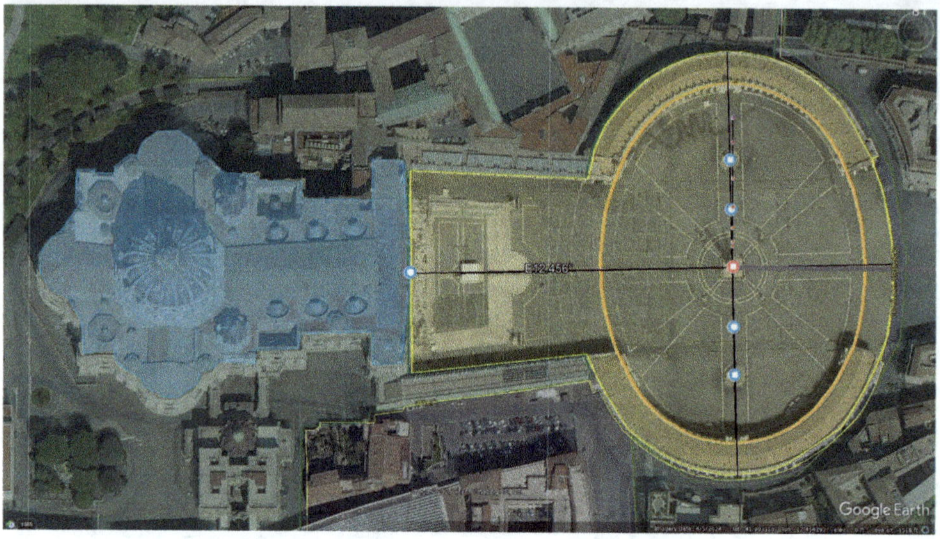

Figure 43. Alignment between Basilica di San Pietro and Monte Guadagnolo (Google My Maps).

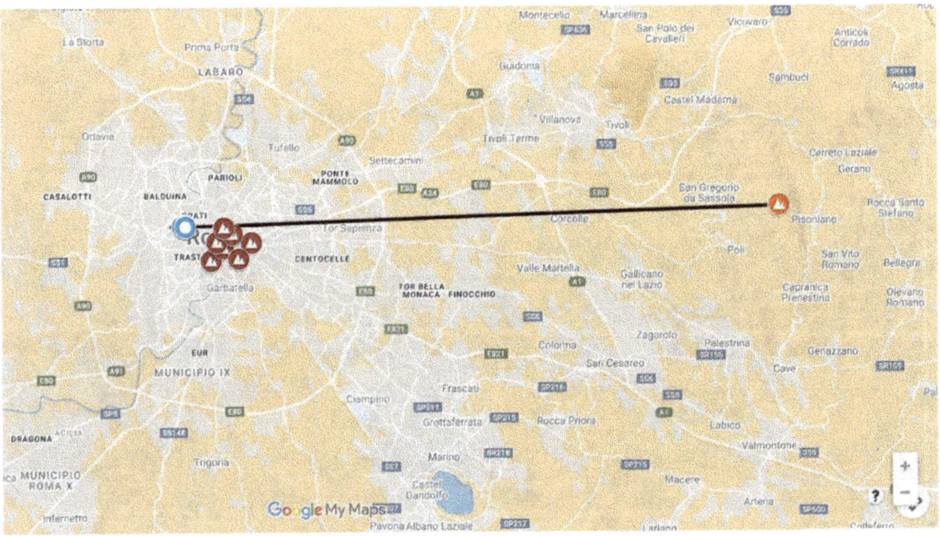

Obeliscus Vaticanus and the Meridian Line

An Egyptian obelisk, *Obeliscus Vaticanus* stands at the center of Piazza San Pietro square at latitude 41.9022° N and longitude 12.4573° E. The obelisk functions as a gnomon, casting a shadow on a meridian line measuring 233 feet (71 meters).

The Obeliscus Vaticanus is a red granite monolith without hieroglyphics, measuring 25.36 meters in height (Platner & Ashby, 1929: 371). It was built in the ancient Egyptian city of Heliopolis ("city of the sun" in Greek) around 1900 BCE. After its transportation to Rome, it was erected by Caligula in the circus Gai et Neronis; in the Middle Ages it was called Tomb of Iulius Caesar (Platner & Ashby, 1929, p. 370).

The obelisk was initially placed in the Circus of Nero, at latitude 41.901576° N and longitude 12.454122°; this former location is now marked by a plaque installed in the pavement. According to tradition, the obelisk was a witness to St. Peter's crucifixion (St Peters Basilica.info, n.d.). The current location is approximately 227 feet north and 861 feet east from the original one (Figure 44). It was moved to its present location in 1586, under the papacy of Sisto V and the direction of the architect Domenico Fontana (St Peters Basilica.info, n.d.).

Figure 44. Old and current location of the Obelisco Vaticano, satellite imagery from Google Earth.

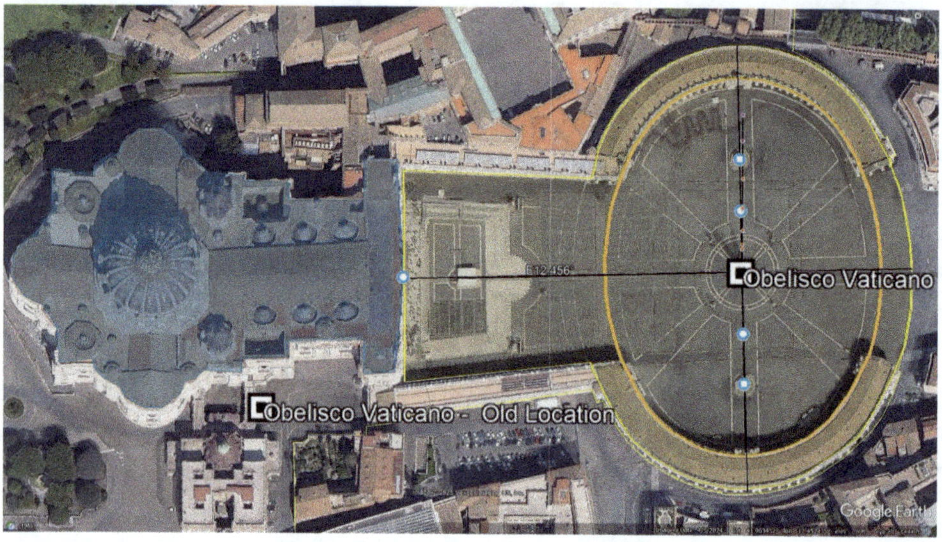

Of great interest for the architecture of cosmology is the paving of the piazza around the obelisk. The overall paving consists of *sanpietrini* (basalt cobblestones). The sanpietrini surface is interrupted by lines and concentric circles of travertine which follow a radial and circular geometry with the obelisk at the center. The innermost travertine annulus (or ring, region between concentric circles), has a width of about 3 feet, with a radius of 57 feet and a thickness of 3 feet; it is followed by a second ring with a radius of 68 feet. Eight lines of travertine radiate from the outer annulus, dividing the circle into four quadrants, whose chords are aligned to the four cardinal points.

A meridian line, measuring approximately 233 feet (71 meters), is located at the north side of the obelisk; it was realized in 1817, commissioned by pope Pio VII on the design by the official astronomer of the Vatican, monsignor Filippo Luigi Gigli, and financed by Pietro Maccarani. The meridian line is made of gray granite inserted in the paving and measures, at solar noon, the shadow of the Obeliscus Vaticanus (Vatican Radio, n.d.), which functions as a gnomon. Such measurements are provided by seven marble discs, measuring approximately 26 inches (66 cm) in diameter, and interrupting the meridian line. Each disc identifies the date of the year according to the shadow cast at solar noon by the tip of the cross—whose arms are aligned to the north and south—placed on top of the obelisk. The northernmost disc, at the north end of the meridian line, marks the length of the shadow when the Sun is at the lowest altitude, at the winter solstice on December 22; the southernmost disc instead marks the shadow of the sun at its highest altitude at the summer solstice on June 22. Between the two disc endpoints of the meridian, there are additional five marble discs, which identify—by marking the shadow of the cross at solar noon—the passage of Sun on the twelve constellations of the Zodiac; Each discs has incised the names of the pair of constellations associated with the shadows. After 1817 the d,iscs were replaced in 1852, in 1878 and finally in 1924 (Vatican Radio, n.d.). Table 2 lists sequentially, starting from north, the seven marble pavers, transcribing the text of the inscriptions, the latitude measured toward north from the equator, and the longitude, measured toward east from the Greenwich meridian.

Table 2. The seven discs along the meridian line.

	Text	Latitude	Longitude
1	SOLSTIZIO IN CAPRICORNO 22 DICEMBRE	41.90298°	12.45725°
2	SAGITTARIO 23 NOV. - ACQUARIO 21 GENNAIO	41.90286°	12.45725°
3	SCORPIONE 23 OTTOBRE - PESCI 19 FEBBR.	41.90267°	12.45725°
4	BILANCIA 23 SETTEMBRE - ARIETE 21 MARZO	41.90253°	12.45725°
5	VERGINE 23 AGOSTO - TORO 21 APRILE	41.90243°	12.45725°
6	LEONE 23 LUGLIO - GEMELLI 22 MAGGIO	41.90236°	12.45725°
7	SOLSTIZIO IN CANCRO 22 GIUGNO	41.90234°	12.45725°

A *Rosa dei Venti* (windrose) was also added in 1817, also under the direction of the astronomer Filippo Luigi Gigli. There are sixteen marble pavers (Figure 45), located in an annulus at about a 63-foot radius from the obelisk, with engraved the compass direction and the wind associated with it (The Holy See, n.d.). Table 3 lists clockwise, starting from north, the sixteen marble pavers, transcribing the text of the inscriptions, the latitude measured north from the equator, and the longitude, measured east from the Greenwich meridian.

Figure 45. Rosa dei Venti, satellite image from Google Earth.

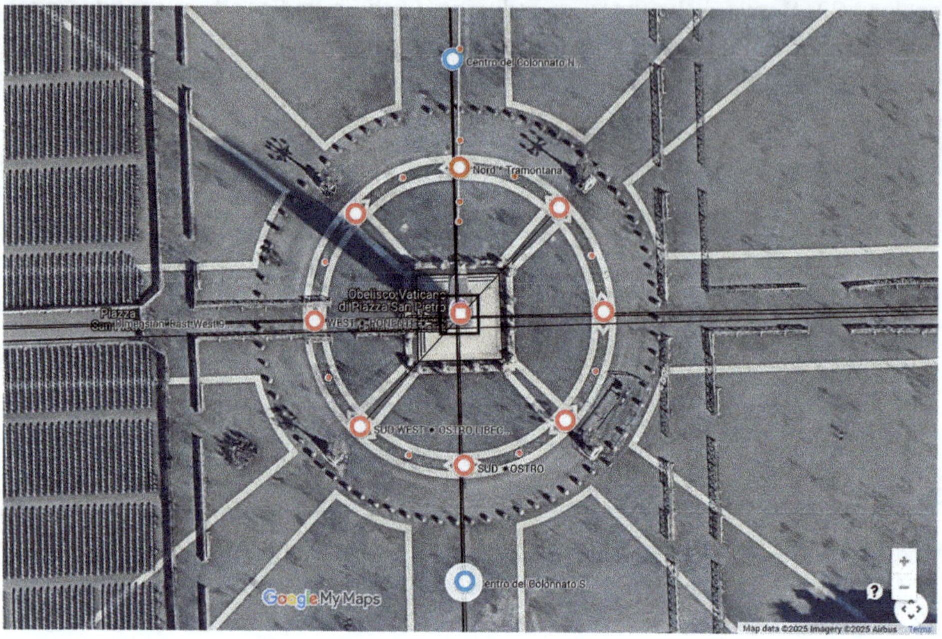

Table 3. The inscription and geolocation in each paver of Rosa dei Venti.

	Text	Latitude	Longitude
1	NORD ✷TRAMONTANA✷	41.9024°	12.45725°
2	NORD NORD EST ✷ TRAMONT GRECO✷	41.90239°	12.45734°
3	NORD EST ✷GRECO ✷	41.90236°	12.45741°
4	EST NORD EST ✷GRECO LEVANTE✷	41.9023°	12.45746°
5	EST ✷LEVANTE✷	41.90224°	12.45748°
6	EST SUD EST ✷LEVANTE SCIROCCO ✷	41.90217 °	12.45746°
7	SUD EST ✷ SCIROCCO ✷	41.90212°	12.45742°
8	SUD SUD EST ✷ OSTRO SCIROCCO ✷	41.90208°	12.45735°
9	SUD ✷OSTRO✷	41.90207°	12.45726°
10	SUD SUD WEST ✷OSTRO LIBECCIO ✷	41.90208°	12.45718°
11	SUD WEST ✷OSTRO LIBECCIO ✷	41.90211°	12.4571°
12	WEST[3] SUD WEST ✷ PONENTE LIBECCIO ✷	41.90216°	12.45705°
13	WEST ✷ PONENTE ✷	41.90223°	12.45703°
14	WEST WEST NORD ✷PONENTE ✷	41.90229°	12.45704°

continued on following page

Table 3. Continued

	Text	Latitude	Longitude
15	NORD WEST ✷ MAESTRO ✷	41.90235°	12.45709°
16	NORD NORD WEST✷TRAMONTANA MAESTRO ✷	41.90239°	12.45716°

Piazza San Pietro features also two fountains relating to the North and South Colonnade. The north fountain (Figure 46) is a few feet off the meridian line, located at latitude 41.90277° N and longitude 12.45723° E. It was realized by Carlo Maderno, under the papacy of Paolo V (Paul V) in the years between 1612 and 1614. The fountain on the south of the obelisk is located at latitude 41.9017° N and longitude 12.45727° E; it was designed by Gian Lorenzo Bernini and realized in the years 1667-1677, commissioned by pope Clemente X (Clement X). The two fountains are located on an axis passing through the entrance to the north and south colonnade and the obelisk in the middle. Such axis is not on the north-south axis—therefore the meridian line—but at an angle of approximately 3°, measured counterclockwise from the north; it is orthogonal instead to the longitudinal axis of the Basilica Such alignment has already been discussed in the section titled “The Basilica”.

Figure 46. North fountain by Carlo Maderno (1612-14), looking south at solar noon. 2024.02.01.

Geometry of the Piazza: As Above So Below

Piazza San Pietro was commissioned by pope Alessandro VII (Alexander VII) to the Baroque architect Gian Lorenzo Bernini; the construction took eleven years, from 1656 to 1667. Bernini is said to have designed the Piazza with a shape which would represent "the motherly arms of the church reaching out to embrace the faithful" (Harris & Zucker, 2015).

Geometrically, the Piazza is configured as an oval shape, called "ovato tondo" opened to its west to a trapezoidal area, named *Piazza Retta*, leading to the Basilica. Piazza Retta measures about 355 feet in the east-west direction, its east side is 315 feet and the west side, at the façade of the Basilica, is 370 feet. The oval measures 776 feet (237 meters) in the north-south direction, from the entrance to the north colonnade to the entrance of the south. The east-west direction measured from the easternmost point on the oval to the façade of the basilica is 901 feet (274.6 meters).

The oval shape, open to the east and west, is defined by two colonnades shaped as arcs of circles, each comprising four rows of Doric columns, measuring 13 meters in height (St Peters Basilica.info, n.d.) and defined by three centers. The Obeliscus Vaticanus (41.90224° N, 12.45725° E) is at the center of the overall shape. The center of the north colonnade is marked by a marble disc approximately 87 cm (2.91 ft); it is located at latitude 41.90252° N, and longitude 12.45724° E. The center of the south colonnade, also marked by a circular marble paver, is located at latitude 41.90194° and longitude 12.45726° E. As can be observed from the coordinates of the longitude, the three centers are slightly off the north-south axis; this discrepancy is visible by observing the meridian and the center of the north colonnade (Figure 47). This incongruity is however consistent with the alignment of the major axis of the Basilica, perpendicular to the major axis of the oval; a theory about this slight deviation, between the Basilica orientation and the cardinal directions, has already been introduced in the section "The Basilica."

Figure 47. The meridian line and the center of the north colonnade. 2021.05.08.

An interesting relationship between geometry and astronomical alignments can be observed in the endpoints of the colonnades (Figure 48); the orientations of the connecting diagonals are almost coincident with the azimuths of the sunrise and sunset at summer and winter solstices (Bauval et al., 2014).

Figure 48. Satellite imagery of piazza San Pietro from Google Earth, with superimposed geometrical relationships and alignments.

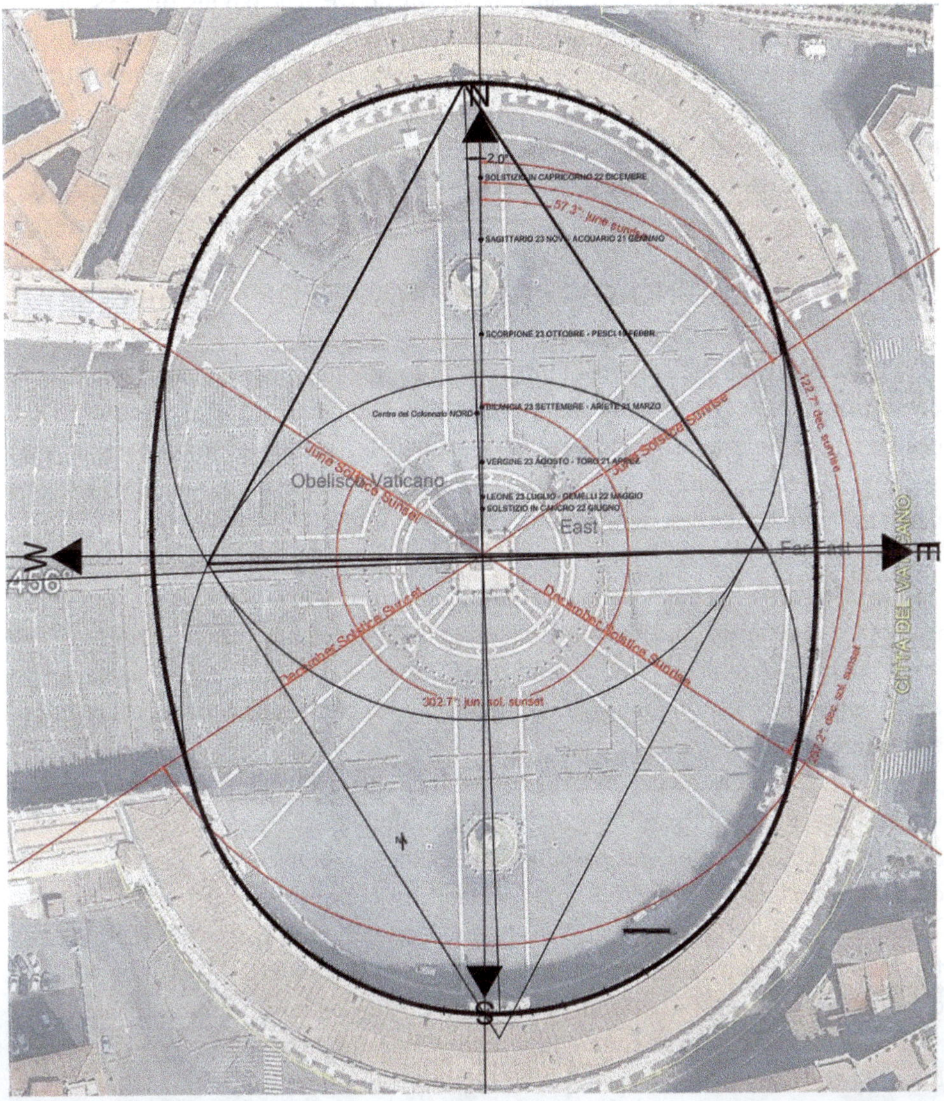

In the past decade, during my stay in Rome, Piazza San Pietro has become the destination of my daily "pilgrimage". Weather and tourist crowds permitting, I visit the Piazza during solar noon and take photographs of the shadow of the obelisk along the meridian line, to document the shadow of the cross on the top of the Obeliscus cast on the pavers on the meridian, which is inscribed with the name of the twelve constellations of the Zodiac (Table 2). And every time I am astonished by the precision of how this Piazza has a dialogue with the Sun. In a walkthrough of Piazza San Pietro, observations are not only about the constructions creating the enclosure of the square but also concern the sky above. The paving of the square is a mirror of the sky above, mainly of the Sun apparent path. Piazza San Pietro is the perfect example of architecture of cosmology and the Earth –sky geometries are embodied in the Piazza. The geometry is articulated in the flow between architectural and urban configuration, which express an ideological program as interface between the terrestrial and the celestial world, the temporal and spiritual sphere.

SOME CONCLUDING THOUGHTS: POWER OF THE SKY

This survey of Rome's architecture of cosmology—though necessarily selective—demonstrates how constructed space becomes an embodiment of metaphors of the cosmos. In the interface between sacred and profane, typologies transform: temples become churches; mausoleums become baptisteries; imperial monuments become papal instruments. City squares flow into basilicas and palaces, becoming courtyards and cloisters in undefined passages from urban to architectural spaces—and from public to private—in the ambiguous transitions between inside and outside. Public and private, exterior and interior, historical and legendary realms overlap. The boundary between history and legend is also blurred. The whole city is a narrative for the imaginary turning into reality. In the eternal city mythology meets religion, saints, angels, and popes

In Rome, the architecture of cosmology follows histories of power and ideology—imperial, ecclesiastical, civic—yet always in dialogue with greater forces: God and Sun. Egyptian obelisks offer paradigmatic examples. Raised by pharaohs in solar cities, transported by emperors, repositioned by popes, they persist as geometric anchors within shifting political theologies. Some align to cardinal directions and function as gnomons; others mark axes, centers, or thresholds. All operate within a continuous conversation between sacred geometry, celestial motion, and urban form.

Rome may be the most compelling case of how time and authority—divine and human—are inscribed into spatial construction. During my daily walks through the city, my steps traverse stones thousands of years old; the walls surrounding me bear witness to layered ideologies and transformations.

One of the most representative built examples of these millennial transitions is the Tempio di Antonino e Faustina (Temple of Antoninus and Faustina), located in the Roman Forum at latitude 41.89226° N and longitude 12.48674° E, erected in 141 CE by Emperor Antoninus Pius. In the 7th or 8th century it was converted into the church of San Lorenzo in Miranda (Figure 49). The stratification of pagan temple and Christian church materializes historical continuity and transformation within a single structure.

Figure 49. Tempio di Antonino e Faustina incorporated into the Church of San Lorenzo in Miranda. 2025.03.09.

Urban layers measure time no less precisely than the obelisks measure solar altitude. Above them all, the sky remains constant yet ever-changing—a cosmic backdrop for the theater of history. The Sun, with its cyclical movements, becomes the principal actor, defining the rhythm of architecture and the choreography of shadow.

In Rome, the power of the sky is not abstract. It is inscribed in stone, aligned in axes, measured in shadow, and lived in daily experience.

TRANSITION TO PART III: FROM EMBODIED COSMOGRAMS TO INSTRUMENTAL SKY

If Rome represents the culmination of monumental naked-eye cosmology in the Western world, it also marks the threshold of its transformation.

With the Scientific Revolution, the relationship between architecture and the sky began to shift. The obelisk as gnomon and the meridian line as public instrument gradually ceded their role to telescopes, observatories, and enclosed scientific devices. The sky moved from being spatially inscribed in the urban fabric to being accessed through increasingly complex instruments.

Where ancient and Renaissance constructions made celestial phenomena visible within civic space, modern astronomy internalized observation within laboratories and domes. The architectural intent of naked-eye observatories diminished as instruments became self-contained "black boxes" of optics and electronics. The sky lost neither scale nor power, but it lost public embodiment.

Part III examines this transition. It investigates how remote space is now mediated through scientific technologies and digital infrastructures rather than through monumental alignments. It also considers how artists—more often than architects—have re-engaged with cosmic scale through land art and phenomenological practices.

The section culminates in the introduction of S.U.N.F.A.R.M. in the Hudson Valley (New York, USA), an ongoing exploration of embodied awareness of space, time, and solar orientation. In this project, the dialogue between local place and remote space re-emerges—not through technological abstraction, but through renewed corporeal engagement.

If Part II traced the historical architectures that materialized cosmology, Part III investigates whether such embodiment can be reactivated in the contemporary world.

Figure 50. Tempio di Antonino e Faustina incorporated in the Church of San Lorenzo in Miranda. 2025.03.09.

REFERENCES

Barrett, C. E. (2017). Egypt in Roman visual and material culture. In Williams, G. (Ed.), *Oxford handbooks online in classical studies*. Oxford University Press., DOI: 10.1093/oxfordhb/9780199935390.013.18

Burn, R. (1880). *Old Rome: A handbook to the ruins of the city and the Campagna*. George Bell and Sons.

Cassirer, E. (1955). *The philosophy of symbolic forms* (Manheim, R., Trans.). Yale University Press.

Cipriani, G. B. (1823). *Su i dodici obelischi egizj che adornano la città di Roma*. [s.n.]

De Franceschini, M., & Veneziano, G. (2018). The symbolic use of light in Hadrianic architecture and the "Kiss of the Sun.". *Archaeoastronomy and Ancient Technologies, 6*(1), 111–137.

Dyson, S. L. (2010). *Rome: A living portrait of an ancient city*. Johns Hopkins University Press.

Fields, N., & Dennis, P. (2008). *Rome's ancient walls*. Osprey Publishing.

Fletcher, H. M. (1929). Chyluria.

Gerbino, A. (Ed.). (2014). *Geometrical objects: Architecture and the mathematical sciences 1400–1800*. Springer International Publishing.

Harris, B., & Zucker, S. (2015, November 18). Gian Lorenzo Bernini, Saint Peter's Square. *Smarthistory*. Retrieved October 30, 2025, from https://smarthistory.org/bernini-st-peters-square-2/

Haselberger, L. (2011). A debate on the Horologium of Augustus. *Journal of Roman Archaeology, 24*(1), 47–73.

Heslin, P. J. (2007). Augustus, Domitian, and the so-called Horologium Augusti. *Journal of Roman Studies, 97*, 1–20.

Hibbard, H. (1986). *The architecture of the new St. Peter's*. Pennsylvania State University Press.

Krautheimer, R. (1980). *Rome: Profile of a city, 312–1308*. Princeton University Press.

Magli, G. (2008). On the orientation of Roman towns in Italy. *Oxford Journal of Archaeology, 27*(1), 63–71.

Mercati, M. (1589). *De gli obelischi di Roma*. Domenico Basa.

Norberg-Schulz, C. (1971). *Existence, space & architecture*. Praeger.

Petrucci, A. (1993). *Roma nel Rinascimento: La città e il papato da Sisto IV a Clemente VIII*. Laterza.

Platner, S. B., & Ashby, T. (1929). *A topographical dictionary of ancient Rome*. Oxford University Press.

Quintela, M. V. G. (2022). Meton of Athens: An Astronomer in the Democratic Landscape of the Polis. *Gerión, 40*(2), 553.

Sigismondi, C. (2021). Calibrazione solstiziale della meridiana di Piazza S. Pietro. *Gerbertvs, 14*, 207–222.

Southern, P. (2001). *The Roman Empire from Severus to Constantine*. Routledge.

Turchin, P., Adams, J. M., & Hall, T. D. (2006). East–West orientation of historical empires. *Journal of World-systems Research, 12*(2), 219–229.

Ward-Perkins, J. B. (1994). *Roman imperial architecture*. Yale University Press.

Wittkower, R. (1999). *Art and architecture in Italy, 1600–1750* (6th ed.). Yale University Press.

Zagari, F. (1998). *Piazza Montecitorio: Progetto di riqualificazione ambientale 1996–1998*. Rome, Italy: Camera dei Deputati; Comune di Roma.

ADDITIONAL READING

Alberti, L. B. (1988). *On the art of building in ten books (De re aedificatoria)* (Rykwert, J., Leach, N., & Tavernor, R., Trans.). MIT Press. (Original work published 1485)

Baumgarten, P. M. (1913). Basilica of St. Peter. In Herbermann, C. (Ed.), *The Catholic Encyclopedia* (Vol. 13, pp. 483–494). Robert Appleton Company.

Bauval, R., Hohenzollern, C., & Zicari, S. (2014). *The Vatican heresy: Bernini and the building of the hermetic temple of the sun*. Bear & Company.

Brown, D. (2000). *Angels & demons*. Pocket Books.

Cibin, L. (2003). *Selciato romano: Il sampietrino: Materiale, lavorazione, evoluzione storica, tipologie, apparecchiature, posa in opera*. Gangemi Editore.

Fagiolo, M. (2000). Il segno del giovane Borromini nella "Città del Sole" di Urbano VIII (1624–1631). In Frommel, C. L., & Sladek, E. (Eds.), *Francesco Borromini* (pp. 215–232). Electa.

González-García, A. C., Quintela, M. V. G., Rodríguez Antón, A., & Espinosa Espinosa, D. (2022). The winter solstice as a Roman cultural fingerprint from the mythical origins of Rome to Augustus. *Environmental Archaeology*, *27*(3), 322–346. DOI: 10.1080/14614103.2022.2053825

Gregorovius, F. (1903). *History of the city of Rome in the Middle Ages* (Vol. 3; Hamilton, A., Trans.). G. Bell & Sons.

Howard, R. (Director). (2009). *Angels & demons* [Film]. Culver City, CA: Columbia Pictures; Imagine Entertainment.

MacDonald, W. L. (1982). *The architecture of the Roman Empire* (Rev. ed., Vol. 1). Yale University Press.

ENDNOTES

1. Throughout this chapter, Italian (or Latin) names of locations are used, with English translations provided in parentheses at first mention.

2. De Franceschini and Veneziano (2018) proposed that solar events carried symbolic meaning within imperial design. Their measurements indicate that while the monument is cardinally aligned and architecturally predisposed to seasonal light variation, direct equinoctial penetration into the *Sala delle Urne* is difficult to demonstrate conclusively in its current state. The interpretation of equinoctial illumination should therefore be approached as plausible within a cosmological framework rather than as empirically established calibration.

3. The inscriptions preserve the original historical spelling as engraved on site, including the use of "WEST."

Part III
Remote Space in the Contemporary World

Chapter 8
The Sky From Earth, Now: S.U.N.F.A.R.M.

ABSTRACT

This chapter investigates the architecture of cosmology in the contemporary world, more than four centuries after the invention of the telescope. Why construct built environments for naked-eye observation in the 21st century? What meaning can such structures assume in an era dominated by scientific instrumentation and digital mediation? Addressing these questions from both theoretical and personal perspectives, this chapter introduces S.U.N.F.A.R.M.—a phenomenological landscape and naked-eye observatory located in the Hudson Valley, New York (US). The project reactivates the relationship between local place and remote space through embodied awareness, solar orientation, and geometric inscription, proposing a renewed cosmological architecture for the present.

DOI: 10.4018/979-8-3373-3399-1.ch008

Figure 1. S.U.N.F.A.R.M. conceptual map, section, and iconological program; north oriented to the left. 2009.

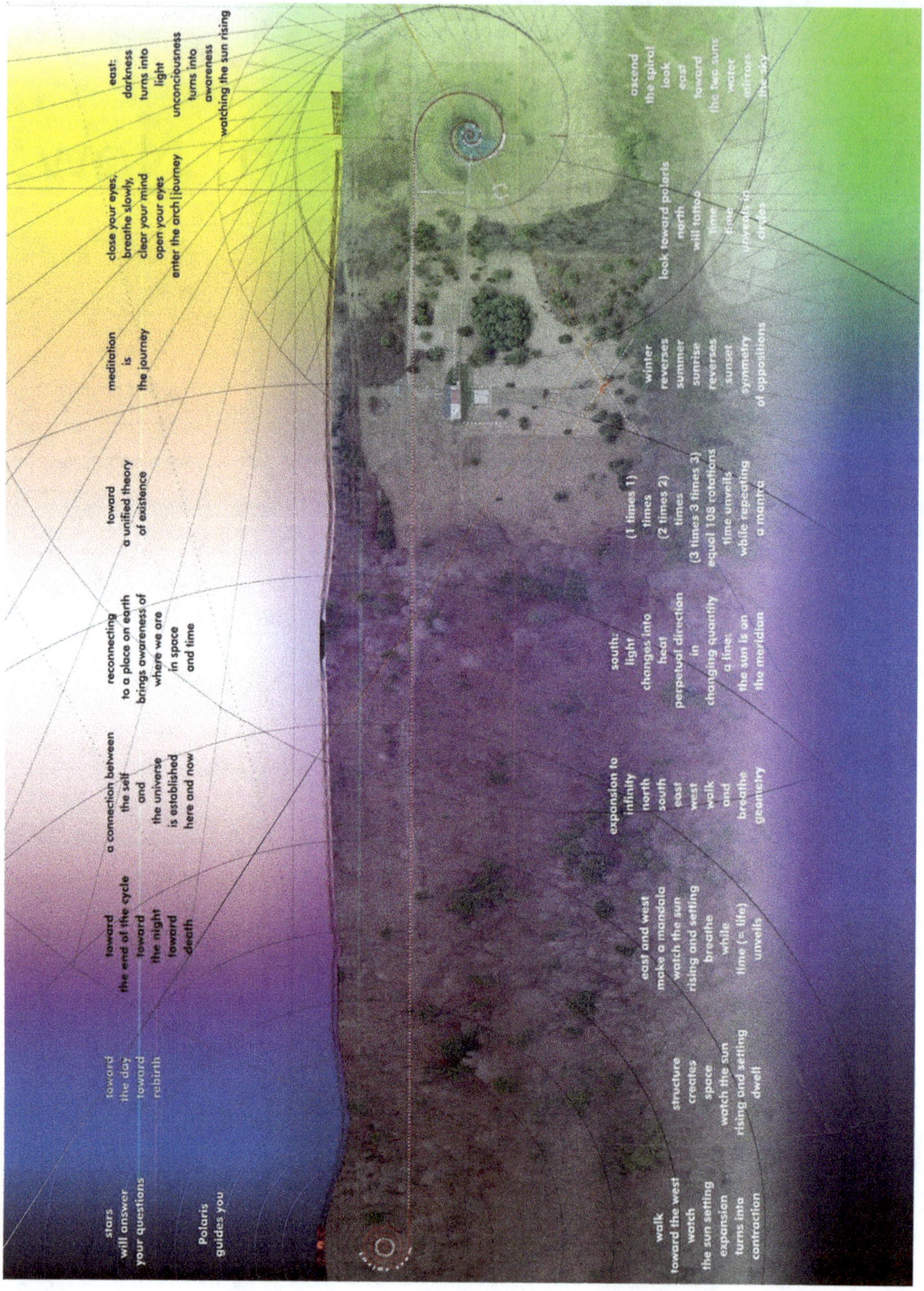

THE SKY AND THE TELESCOPE

The architecture of cosmology—once a defining presence in landscapes and cityscapes—appears to have largely disappeared in the 20th and 21st centuries. Monumental alignments, meridian infrastructures, and civic cosmograms no longer structure urban space as they did in antiquity or the Renaissance.

One explanation lies in the transformation of the contemporary skyscape itself. Light pollution has profoundly altered nocturnal perception; darkness, once the necessary ground for stellar observation, is increasingly rare in inhabited territories. The Milky Way, which shaped mythologies and architectural orientations for millennia, has become invisible to large portions of the global population.

Yet the more decisive shift may not be environmental, but epistemological: the invention of the telescope fundamentally transformed the relationship between human vision and the cosmos.

Before the telescope, astronomy relied exclusively on naked-eye observation. Architectural constructions—temples, pyramids, observatories, meridian lines, obelisks—served as instruments embedded in the built environment. Orientation, alignment, and geometric precision were spatial devices that anchored celestial phenomena within terrestrial coordinates.

With the telescope, observation became mediated. The sky was no longer measured against architectural axes; it was magnified through optical apparatus. The necessity for large-scale orientational structures diminished as precision migrated into the instrument itself.

The etymology of the word *telescope* clarifies its function: from the Greek τηλε (*tēle*), meaning "far," and σκοπός (*skopos*), meaning "watcher," "observer," or "one who looks." The instrument is designed to render distant objects visible by collecting and focusing light. In optical telescopes, an objective lens or mirror gathers light, forming an image that is then magnified through an eyepiece.

The telescope emerged in the Netherlands in the early 17th century. In 1608, the eyeglass maker Hans Lippershey applied unsuccessfully for a patent for a device capable of magnifying distant objects (King, 1955). Optical magnification instruments had already existed, but this configuration allowed for systematic astronomical use.

Initially, telescopes were employed primarily for terrestrial purposes such as surveying and military observation (Library of Congress, 2014). In 1609, Galileo Galilei (1564–1642) was among the first to direct the telescope toward the heavens (Library of Congress, 2014). His observations were published in *Sidereus Nuncius* (1610), written in Neo-Latin and later translated as *The Sidereal Messenger* or *The Starry Messenger* (Drake, 1957). In this treatise, Galileo documented the Moon's surface, the phases of Venus, and the satellites of Jupiter through text, drawings, and diagrams (Figure 2). The Moon was no longer a perfect metaphysical sphere; it became a topographical body.

Figure 2. *Pages from Galileo's Sidereus nuncius: on the left, a diagram of the telescope (Galilei, 1610, p. 11); on the right, a drawing of the Moon observed through Galileo's telescope (Galilei, 1610, p. 16).*

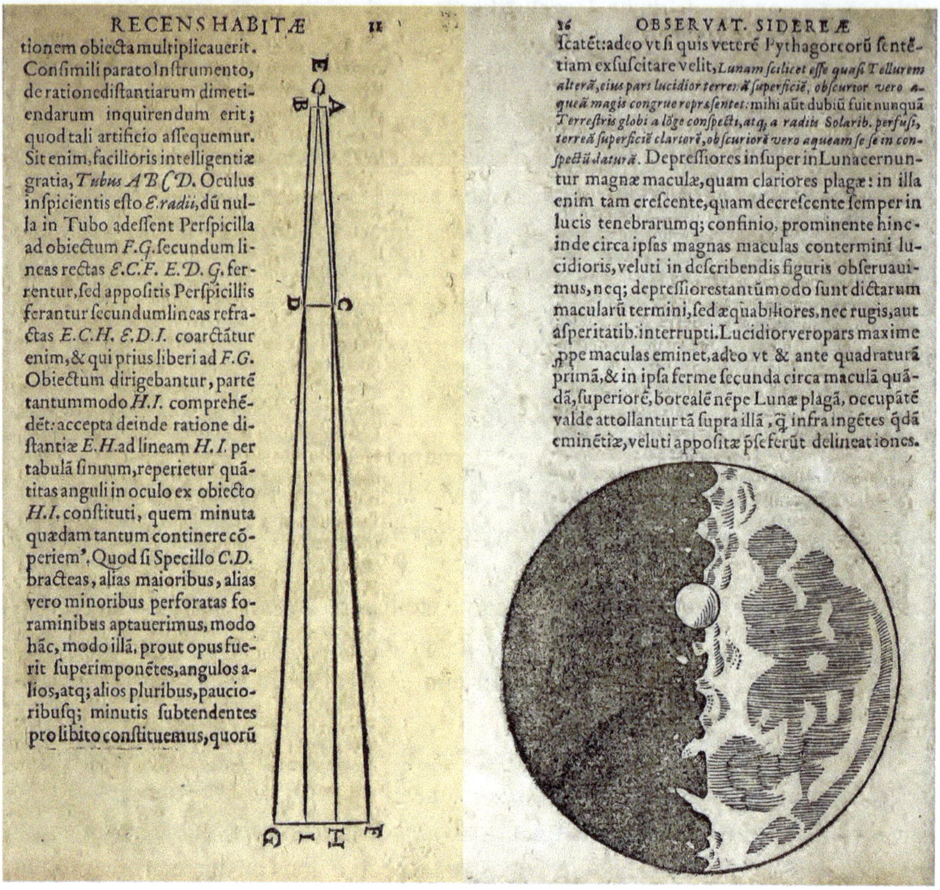

RECENS HABITÆ 11

tionem obiecta multiplicauerit. Conſimili parato Inſtrumento, de ratione diſtantiarum dimetiendarum inquirendum erit; quod tali artificio aſſequemur. Sit enim, facilioris intelligentiæ gratia, *Tubus A B C D.* Oculus inſpicientis eſto *E. radii*, dũ nulla in Tubo adeſſent Perſpicilla ad obiectum *F. G.* ſecundum lineas rectas *E. C. F. E. D. G.* ferrentur, ſed appoſitis Perſpicillis ferantur ſecundum lineas refractas *E. C. H. E. D. I.* coarctãtur enim, & qui prius liberi ad *F. G.* Obiectum dirigebantur, partẽ tantummodo *H. I.* comprehẽdẽt: accepta deinde ratione diſtantiæ *E. H.* ad lineam *H. I.* per tabulã ſinuum, reperietur quãtitas anguli in oculo ex obiecto *H. I.* conſtituti, quem minuta quædam tantum continere cõperiem'. Quod ſi Specillo *C. D.* bracteas, alias maioribus, alias vero minoribus perforatas foraminibus aptauerimus, modo hãc, modo illã, prout opus fuerit ſuperimponẽtes, angulos alios, atq; alios pluribus, paucioribuſq; minutis ſubtendentes pro libito conſtituemus, quorũ

16 OBSERVAT. SIDEREÆ

ſcatẽt: adeo vt ſi quis veterẽ Pythagoreorũ ſententiam exſuſcitare velit, *Lunam ſcilicet eſſe quaſi Tellurem alterã, eius pars lucidior terrenã ſuperficiẽ, obſcurior vero aqueã magis congrue repræſentet:* mihi aũt dubiũ fuit nunquã *Terreſtris globi a lõge conſpecti, atq; a radiis Solarib. perfuſi, terreã ſuperficiẽ clariorẽ, obſcuriorẽ vero aqueam ſe ſe in conſpectũ daturã.* Depreſſiores inſuper in Luna cernuntur magnæ maculæ, quam clariores plagæ: in illa enim tam creſcente, quam decreſcente ſemper in lucis tenebrarumq; confinio, prominente hincinde circa ipſas magnas maculas contermini lucidioris, veluti in deſcribendis figuris obſeruauimus, neq; depreſſiores tantũmodo ſunt dictarum macularũ termini, ſed æquabiliores, nec rugis, aut aſperitatib. interrupti. Lucidior vero pars maxime ppe maculas eminet, adeo vt & ante quadraturã primã, & in ipſa ferme ſecunda circa maculã quãdã, ſuperiorẽ, borealẽ nẽpe Lunæ plagã, occupãtẽ valde attollantur tã ſupra illã, q̃ infra ingẽtes q̃dã eminẽtiæ, veluti appoſitæ pſeferũt delineationes.

Over the subsequent four centuries, telescopic technology evolved dramatically. Contemporary telescopes are classified according to the wavelength of electromagnetic radiation they detect, ranging from gamma rays and X-rays to ultraviolet, visible light, infrared, microwaves, and radio waves (Las Cumbres Observatory, n.d.). Gamma rays possess the shortest wavelengths and highest energy, while radio waves have the longest wavelengths and lowest energy. The visible spectrum—accessible to the human eye—constitutes only a narrow band within this continuum.

Telescopes are also distinguished by location. Space-based observatories orbit Earth to avoid atmospheric interference and to detect wavelengths absorbed by the atmosphere. Observation has thus expanded beyond optical magnification into electronic detection and computational interpretation.

The contemporary use of telescopes ranges from amateur "backyard" astronomy to highly specialized research observatories. Modern astronomical observatories are typically designed around instrumental requirements—stability, enclosure, climate control—rather than cardinal orientation or solstitial alignment. Aside from rare exceptions, their architecture does not embody cosmological geometry in the manner of ancient or Renaissance constructions.

The architecture of cosmology, born from naked-eye observation and spatial alignment, appears largely displaced within modern scientific astronomy.

MODERN & CONTEMPORARY SCIENTIFIC SKY WATCHING

After the invention of the telescope, astronomical observatories started to rely on the use of scientific instruments. The buildings, which were facilitating the observation and interpretation of the celestial phenomena during naked-eye skywatching, lost their architectural features in terms of geometric shape, orientation and alignment to celestial objects. The different types of telescopes are the major astronomical instruments for scientific observations of the sky and define the relationship between observer and the objects of observation. Such relationship is often detached from the architectural space and place where it happens

Observatories & Optical Telescopes

Optical telescopes were the first type of instrument to study the sky beyond naked-eye observations. The Earth-based observatories housing optical telescopes, at least until the beginning of the 20th century, represent the first transition from the architecture of cosmology based on direct experience of the sky to a scientific approach disconnected from the places of observation. These observatories usually feature an operable hemispherical dome on the roof. The dome has the function to protect the telescope from the elements; it can be opened and rotated for the telescope to focus on the areas of observation. A hemispherical dome by geometric definition does not have a specific orientation.

Often existing buildings have been retrofitted to house domes for optical telescopes. An example is *Villa Mellini* (Figures 3-4) located at latitude 41.9225° N and longitude 12.45242° E, in Monte Mario, Rome at an altitude of 128 meters above sea level (International Working Group on Astronomy and World Heritage, n.d). Villa Mellini was the main site of OAR, *Osservatorio Astronomico di Roma* (Rome astronomical observatory) until 1988 and also hosts, since 1838, *the Museo Astronomico e Copernicano* (Astronomical and Copernican Museum) (Istituto Nazionale di Astrofisica, n.d.). It was built in the 15th century as residence for the cardinal Mario Mellini and presents three hemispherical domes which are hosting telescopes. The *Torre Solare* (Solar Tower), shown in Figures 8.4 and 8.6, is located north-east of Villa Mellini, at latitude 41.92306° N and longitude 12.45232° E. It was built in 1951, made of reinforced concrete, and includes several instruments for collecting Sun related data (International Working Group on Astronomy and World Heritage, n.d). The site of the Osservatorio Astronomico Romano in Monte Mario also features the *Primo Meridiano di Roma* (Rome First Meridian), located about 700 feet north of Villa Mellini, at longitude 12.4523333° E. The Primo Meridiano is marked by a tower located at longitude 41.9244° N, 12.45213° E and was defined by the Jesuit astronomer Angelo Secchi in 1870 (Altamore & Maffeo, 2012); it was used as reference point of longitude 0° for Italian cartography, until 1884, when the Greenwich meridian was adopted as prime meridian.

Figure 3. Osservatorio Astronomico di Roma, Monte Mario, 2023.02.19.

Figure 4. Sites of astronomical interest at Monte Mario, Rome; satellite imagery from Google Earth.

Figure 5. Osservatorio Astronomico di Roma, Villa Mellini, Monte Mario, satellite imagery from Google Earth.

Figure 6. Torre Solare, Monte Mario, Roma, 2023.02.19.

Radio Telescopes

Radio telescopes, which are currently some of the most powerful sources for collection of astronomical data, are self-standing structures of gigantic dimensions. These telescopes comprise a parabolic metal reflector dish for collection of radio waves, antennas receiving the radio waves and converting them into electric currents, amplifiers strengthening the current, and, of course, computer networks to analyze and process the data gathered by the various instruments (National Radio Astronomy Observatory, n.d.). Radio telescopes can be made of a single-dish or comprise an Interferometer array.

The size of radio telescopes is far beyond human scale: the largest single-dish radio telescope is the Five-hundred-meter Aperture Spherical Telescope (FAST) located in (Go-Astronomy.com, n.d.), at latitude 25.65292° N, and longitude 106.85669° E (Figure 7).

Figure 7. Fast radio telescope, Guizhou, China, Google My Maps.

Arrays of radio telescopes can be arranged in a single location or spread in a networked around the world, as in the *Event Horizon Telescope (EHT)*. The EHT started as an international collaboration in 2009 with for the study of *black holes* and related inquiries in the framework of the general relativity theory; it "has formed to continue the steady long-term progress on improving the capability of Very Long Baseline Interferometry" (Event Horizon Telescope, 2024). The EHT is constantly evolving in terms of participating countries and number of telescopes: "Observations in April 2017 were carried out with eight observatories in six geographical locations around the globe. For the observing campaign in 2018, one new telescope was added to the array, totaling nine observatories at seven sites" (Event Horizon Telescope, 2024).

Currently, the largest array of radio telescope in a single location is the *Very Large Array (VLA)*, in the Plains of San Agustin, Socorro, New Mexico (Go-Astronomy.com, n.d.). It comprises 27 radio antennas arranged in a Y-shape, with each antenna diameter measuring 25 meters and providing data which is "combined electronically to give the resolution of an antenna 36 km (22 miles) across, with the equivalent sensitivity of a dish 130 meters (422 feet) in diameter" (US National Radio Astronomy Observatory, n.d.). The center of Y is at latitude 34.07908° N, and longitude 107.61838° W. Geometrically, the arms of the Y-shaped configuration, resemble the angle bisectors of an equilateral triangle, connecting its center with the vertices: in fact, the axes along which the radio dishes are arranged, measure an angle of about 120° (Figure 8). One axis of the VLA is approximately aligned with a north-south axis.

Figure 8. VLA, Socorro, New Mexico diagram superimposed to satellite imagery of Google My Maps.

With the advent of these extremely large instruments, the observation of celestial objects is completely out of human sight for the unaided eye; therefore, a disconnect is created in the continuity between local and remote space. The vault of the celestial sphere, which, in naked-eye astronomy, was out of touch but in sight, becomes completely detached from the human experience of the sky in the contemporary world: such experience, in the past, had provided meaning and created belief systems for many people and cultures.

THE SKY FROM (LAND) ART

After the various developments of the telescope, astronomical practices do not require any longer the alignments of buildings to celestial objects and phenomena to guide the observations of remote space. The architecture of cosmology, which for thousands of years has represented the liaison between Earth and sky, may have started disappearing from landscapes and cityscapes. This disappearance goes hand-in-hand with the compartmentalization of knowledge and lack of interaction between disciplines. The organic, wholistic, Weltanschauung which have accompanied the flourish of the architecture of cosmology seems lost in the contemporary overspecialized, mechanistic era. But in the past decades art seems to have come to the rescue of the lost connection between Earth and sky: the architecture of cosmology has shifted from astronomical to art practices, more specifically to some land art projects.

The *land art* movement started in the 1960s, and it is also called art-in-nature, earthworks, landscape, environmental or Earth art. Despite some variations, these different names share the common definition of art using the land or landscape as a medium: gravel, stones, sand, soil, grass, mud, water are the main materials used in the artworks by this movement. Land art is considered a development from *conceptualism* and *minimalism,* recalling the art materials used by the Italian movement *Arte Povera* (The Art Story, 2025). Land art presents several variations, due to the different types of relationships the artwork has with the natural environment.

Earthworks are generated by an invasive relationship with the site, using excavators to alter dramatically the existing landscape. Perhaps the most famous example is *Spiral Jetty* (1970) by the late American artist Robert Smithson (1938-1973), located by Rozel Point, in the Great Salt Late, Utah, US. The center of the spiral is situated at latitude41.4377° N and longitude 112.66886° W (Figure 8-9). *Spiral Jetty* has cycles of disappearance, submerged by the water of the lake, and then re-emerging; it was submerged when I visited the site in April 1996. The disappearance responds to the notion of art conversant with nature and subject to entropy. *Pratt Farm* (1970-1982) is another early earthwork, now destroyed. Pratt Farm was on a 17-acre property along the Kennebec River in Maine, US at the approximate geolocation 44.66779° N, 69.6189° W; it was owned by the late American art historian and photographer James Pierce (1930-2010), who created a series of mounds he called "garden of history", including *Earth Woman, Turf Maze*, a triangular labyrinthine forms (Center for Land Use Interpretation, n.d.), and *Motte,* a raised mound inspired by the historic Norman motte-and-bailey castles (University of North Dakota, n.d.).

Figure 9. Spiral Jetty (1970), Great Salt Lake, Utah, from Google My Maps.

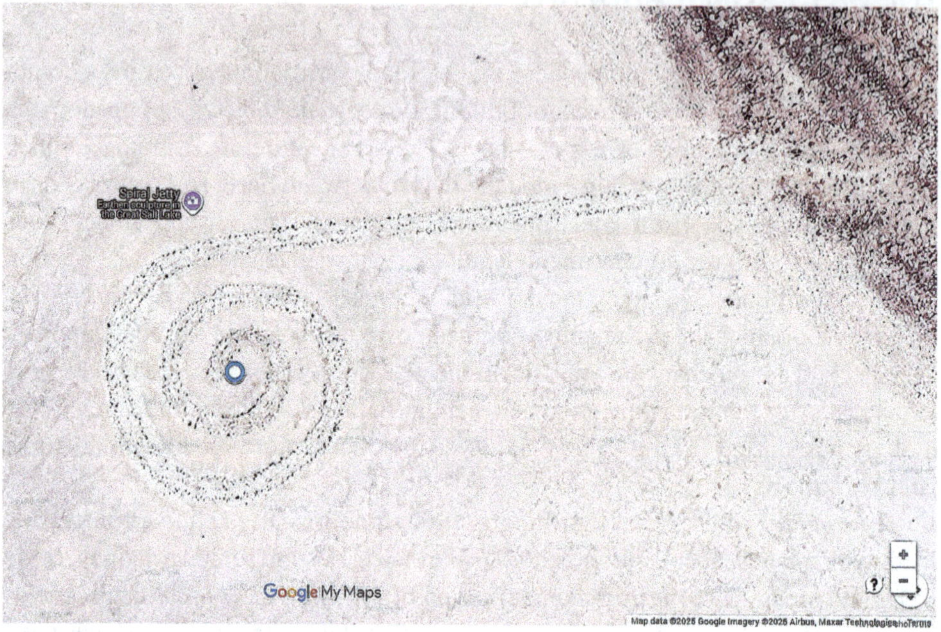

Other land art practices use an ephemeral approach, like walking, or creating lines by rearranging stones and other materials found on the site, like in the works by the British artists *Richard Long*[1]and *Hamish Fulton*, who defines himself as a "walking artist" (Tate Britain, 2002). Both artists, whose ephemeral work on the landscape can be seen through the lens of conceptualism, also use text to define their experience with the site they walked on. Text, photographs, and other documentation of the encounter with the landscape comprise the tangible artwork generated by the walks. *Andy Goldsworthy* is another British artist making very complex and beautiful works from materials found on the site, like stones, pebbles, branches, leaves.

The late European artists *Christo & Jeanne-Claude* are also considered major exponents of the environmental art movement. Their temporary and large-scale site-specific installations are located in natural landscapes, streetscapes and cityscapes, based on wrapping existing buildings, bridges, park and natural areas with sheets of fabric. Their environments are temporary, lasting usually for no longer than two weeks. I interviewed them in 2008 and when I asked them for a simple definition of their art they replied that it was "a work of joy and beauty" (Bertol, 2008).

The brief overview presented in the above paragraphs has outlined the major works of land art, which do not present specific reference to astronomical concepts. There are some works from this art movement which instead relate to celestial alignments or the cardinal orientation. Below is an outline listing the best known artworks in sequential chronological order.

- **1973–76 | *Sun Tunnels***, by **Nancy Holt**(1938-2014) is located in the Great Basin Desert of Utah at latitude 41.30352° N and longitude 113.86381° W (Figure 8-10). This work is made of based on four custom prefabricated concrete pipes measuring 9 feet in diameter and 18 feet in length. Their composition measures 86 feet along the diagonal. The tunnels are arranged according to an X shape, aligned to sunrise and sunset of summer and winter solstices. There are also representations of star constellations through holes drilled in the concrete pipes: "in the upper half of each tunnel are holes of four different sizes—seven, eight, nine, and ten inches in diameter. Each tunnel has a different configuration of holes corresponding to stars in four different constellations—Draco, Perseus, Columba, and Capricorn" (Holt/Smithson Foundation. n.d).
    - I visited Sun Tunnels in April of 1996, unfortunately it was a cloudy day and I could not see the sunlight entering the holes and projecting shadows on the inner surface of the pipes.

Figure 10. Nancy Holt Sun Tunnels (1973-76), Great Basin Desert, Utah, from Google My Maps.

- **1976–ongoing | Charles Ross *Star Axis*** (35.26423° N, 105.08721° W) is located on a mesa in the New Mexico. It is defined by the artist as "architectonic earth/star sculpture constructed with the geometry of the stars" with "shapes and angles are determined by earth-to-star alignments" (Star Axis, n.d.). This structure is "11 stories high and 1/10th of a mile across" (Star Axis, n.d.) and made of granite and sandstone. Its axis follows a north-south alignment (Figure 11); in elevation it is aligned to the latitude, therefore to the Earth's axis, "which now points toward our north star Polaris" (Star Axis, n.d.). *Star Axis* was conceived in 1971, construction began in 1976 completion is anticipated for 2027-2028.
    - o The site is closed to the public and opened only to donors via special arrangements to the Land Light Foundation

Figure 11. Charles Ross Sun Axis (1976 - ongoing), Las Vegas, New Mexico, from Google My Maps.

- **1977 | *Robert Morris: Observatorium*** (52.54963° N, 5.56592° E) by the late American Artists Robert Morris (1931-2018) is located in Lelystad, Flevoland, in the Netherlands. It consists of an earthwork aligned to the four cardinal points: a circular mound with three openings, framing the sunrise during the equinoxes and the solstices (Figure 12). Morris was inspired by Stonehenge in creating work dealing with cyclical time (Land Art Flevoland, n.d.).

Figure 12. Robert Morris: Observatorium (1976), from Google My Maps.

- **1977–ongoing | *Roden Crater*** by **James Turrell** (35.42555° N, 111.25861° W) is located in the Painted Desert, Arizona (Figure 13). This project, one of longest in terms of construction, and probably the most expensive and best funded in the history of land art, is based on an extinct volcanic cinder cone. It creates a naked-eye observatory of several celestial phenomena, using a complex system of tunnels, which frame the view of the Sun and moon.

Figure 13. James Turrell Roden Crater (1970), Arizona, from Google My Maps.

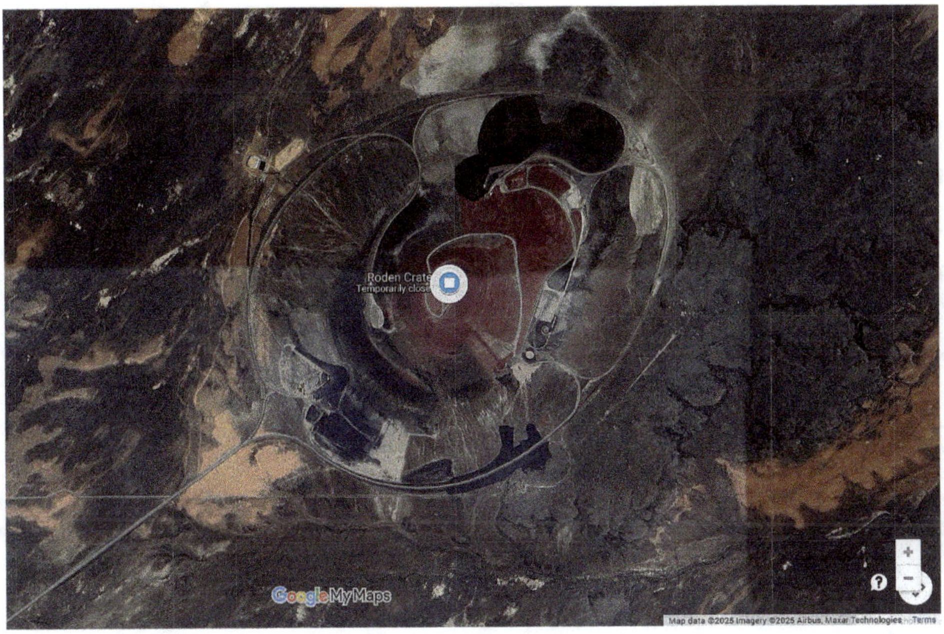

- **1977 | *The Lightning Field*** by the late American artist Walter de Maria (1935-2013). This work, located in Western New Mexico (US), was commissioned and maintained by the Dia Art Foundation; it consists of "400 polished stainless-steel poles installed in a grid array measuring one mile by one kilometer" and the poles measure "two inches in diameter and averaging 20 feet, 7½ inches in height—are spaced 220 feet apart and have solid, pointed tips that define a horizontal plane" (Dia Art Foundation, n.d.). Unfortunately, the location of the site is not disclosed and this work can be visited only by appointment with the Dia Art Foundation. The poles are aligned to four cardinal points (Elderfield, 2021).
- **1984-2015 | *Cretto di Burri***, translated in English as "crack of Burri" (37.78782° N, 12.97163° E) is located in the old town of Gibellina, Sicily. This work takes its name from the late Italian artist Alberto Burri (1915-1995) who conceive as a memorial to the Italian town of Gibellina, completely destroyed by the 1968 Belice earthquake. Although it was not meant to be oriented to astronomical alignments, it features an uncanny orientation to the four cardinal directions (Figure 14). The Cretto is made of concrete, poured on the rubble of the destroyed town, and measures approximately 1016 feet in the east-west direction and 906 feet along the north-south axis.

Figure 14. Cretto di Burri (1984-2015), Gibellina, Sicily, from Google My Maps.

S.U.N.F.A.R.M

S.U.N.F.A.R.M., located in the Hudson Valley, New York, at latitude 42.1729° N and longitude 73.67266° W (Figures 15-16), has motivated the inquiries of this book. The search for historical precedents—Roman meridians, obelisks, archaeoastronomical landscapes—originated from the desire to situate a contemporary project within a lineage of architectures of cosmology.

Figure 15. S.U.N.F.A.R.M. (2000–ongoing), Claverack, New York; satellite imagery (2025), from Google My Maps.

Figure 16. S.U.N.F.A.R.M., aerial view looking northeast, showing East Spiral (top right), Solstice Axis, Labyrinth, and Zen Square. 2007.06.06.

The project was initially titled *Sunset Farm*, a name tied to its agricultural past and western orientation. As the work evolved from site occupation to cosmological inquiry, the name shifted to *Sun Farm*, emphasizing solar alignment rather than agrarian inheritance. The later articulation as S.U.N.F.A.R.M.—both acronym and constructed compound—formalized its conceptual scope: a place structured by solar orientation, ecological attention, and embodied practice.

Conceived as both a contemporary cosmogram and "A Place for Art, Nature & Well-Being" (S-U-N-F-A-R-M., n.d.), the project has absorbed nearly all my time, energy, and financial resources over the past twenty-five years.

This chapter addresses the first phase of S.U.N.F.A.R.M., from the beginning of construction in 2000 to 2011. Comprehensive documentation of this phase is archived at the Center for Art & the Environment at the Nevada Art Museum (Bertol & Foell, n.d.). The subsequent phase (2012–present) will be the subject of the second volume of *Earth Sky Geometry*.

Overview

S.U.N.F.A.R.M. operates as an acronym and as a semantic condensation of Sun–Farm, fusing cosmology and agriculture into a single term:

Spacetime (Spiral System Survey) of Universe Nature in a Future Art Revolution (Mathematical) Movement

The landscape embodies a mathematical theory of the universe in which art becomes a vehicle for interpreting celestial events within a designed environment intended to cultivate awareness of where we are in space and time. Such awareness, I propose, is inseparable from well-being-in-the-world.

The project operates simultaneously at multiple scales:

- Ground experience through walking.
- Aerial comprehension through satellite imagery.
- Diagrammatic abstraction through digital models.
- Astronomical calibration through solar charts and time-lapse documentation.

The geometry is not fully legible from a single point of view. Walking reveals partial alignments; aerial imagery reveals the whole. Satellite cartography does not replace embodied experience but complements it.

Several narrative layers construct the vision of S.U.N.F.A.R.M.: diagrams, astronomical charts, geographic overlays, digital simulations, texts, and performances. Documentation is not secondary to the work; it is integral to its unfolding.

The body functions both as perceiver and as agent of transformation.

Movement is fundamental. From Earth, celestial bodies appear in constant motion—the Sun by day; the Moon and stars by night. The built structures at S.U.N.F.A.R.M. create calibrated frames through which these motions become perceptible.

Movement is also existential. As Heraclitus expressed, πάντα ῥεῖ (*panta rhei*), "everything flows." The landscape shifts seasonally; paths erode; vegetation transforms; entropy becomes visible. Built interventions are not conceived as static monuments but as reference systems within a continuously changing environment.

S.U.N.F.A.R.M. differs from most land art practices not only in its theoretical foundations but in its phenomenological intentionality. The project is not solely a sculptural intervention in terrain; it is an embodied practice.

Site, Latitude, and Orientation

At 42.1729° N latitude, the solar arc expands and contracts with marked seasonal amplitude. This geographic specificity is not incidental. The project is inseparable from the valley's geomorphology, light conditions, and seasonal solar arc.

At this latitude, the solar altitude varies dramatically between solstices. The summer sun rises and sets far north of east and west; the winter sun contracts toward the southern horizon, tracing a compressed arc. The long oblique light of late autumn extends shadows across the terrain in a manner that renders geometry visible not through drawing, but through duration.

The Hudson River itself forms a north–south axis of orientation. Glacial history shaped the valley's topography, producing subtle undulations that interrupt pure cardinal alignment. Thus, the site resists ideal geometry. It demands negotiation between astronomical abstraction and terrestrial irregularity.

Unlike monumental land works constructed in deserts or remote plateaus, S.U.N.F.A.R.M. exists within an inhabited agricultural landscape. Weather, frost heave, plant growth, erosion, and maintenance continuously alter the field. The cosmogram is therefore not imposed upon a neutral ground; it is recalibrated within a living environment.

Geographic coordinates, in this context, are not cartographic data alone. They define a relationship to the solar cycle, to agricultural timing, and to embodied orientation. The latitude situates the project within a specific cosmological envelope — one that cannot be transferred without transformation.

As described in "Sun Farm: Body, Self, Universe" (Bertol, 2015), visual astronomical observation gradually evolved into movement practices. Static observation became walking meditation (Figure 17) and yoga-based performance. Astronomical time—the measurable cycles generated by Earth's rotation and revolution—began to intersect with existential time—the lived duration of breath and consciousness.

Figure 17. Walking meditation during the spring equinox of 2006.

The intersecting spirals that structure the site eventually became a tattoo on my lower back: a permanent mark corresponding to ephemeral drawings in the landscape. The body became both archive and instrument.

Geometry and Alignments

Multiple geometric systems structure the site, each aligned to the four cardinal directions.

The primary configuration consists of two interlocking spirals connected by an east–west axis (Figure 18). The spirals function symbolically and mathematically:

- The East Spiral (logarithmic; approx. 42.172535° N, 73.673393° W) embodies expansion—metaphorically associated with cosmological expansion.
- The West Spiral (Archimedean; approx. 42.17315° N, 73.68306° W) embodies contraction.

Figure 18. Sun Farm spiral diagram.

Figure 19. (a) S.U.N.F.A.R.M., aerial view with conceptual overlay; (b) site plan with spiral system (2005).

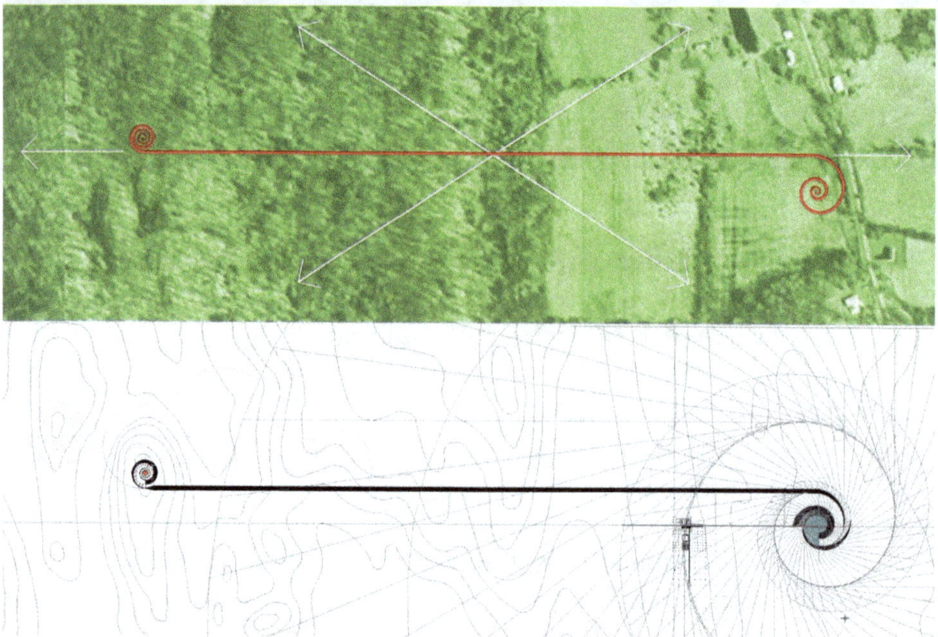

The spiral form was selected for its presence in mathematics and astronomy (Heath, 1921), in cosmology (Shinde, 2024), and in archaeoastronomical contexts across cultures (Bertol, 2006). It operates simultaneously as scientific form, cultural archetype, and metaphor bridging physics with history (Serres, 2018).

Intersecting the spirals is a Cartesian grid aligned north–south and east–west. Built elements—frames, paths, columns—are positioned either on spiral nodes or along grid axes. These constructions establish built azimuths calibrated to solar events and selected stellar alignments.

Geometry here is not decorative. It is operational.

Art + Place = (Space + Time) of Existence

In a 2007 bilingual interview with art critic Enrico Cocuccioni (Cocuccioni, 2007), S.U.N.F.A.R.M. was described as the convergence between digital abstraction and embodied landscape practice.

My early work centered on virtual architectural space and digital geometric modeling. The spirals of *Sky Spirals* originated as algorithmic expansions projected onto geographic and celestial maps. For years they existed only as digital diagrams.

In 1999, the purchase of a property in Claverack, New York transformed virtual spirals into physical interventions. Digital models became operational tools: topographic sections, sun-path simulations, and algorithmic form generation guided the placement of structures such as *Sunrise Trellis* and *Noon Columns*.

Algorithmic processes used solar position as input parameter, generating alternative three-dimensional configurations. The resulting models were exported in KML format for geospatial visualization in Google Earth™, enabling multi-scalar perception. This layered mediation recalls the shifting scales in *Powers of Ten* by Charles and Ray Eames. The movement between microscopic precision and planetary overview became not merely representational, but methodological: scale itself functioned as a conceptual hinge between body and cosmos.

The project thus represents a synthesis: not rejection of the digital, but reintegration of virtual abstraction with material presence.

Two temporalities guide the work:

1. Astronomical time — generated by Earth's rotation and revolution.
2. Existential time—experienced through breath, perception, and bodily awareness.

Structures such as *Sunrise Trellis* frame equinoctial sunrises, yet the framed sky continuously transforms. Light becomes the fourth coordinate in a four-dimensional spatial system. The structure is stable; illumination is not. Duration, therefore, completes geometry.

Time-lapse photography, repeated from identical viewpoints on solstices and equinoxes, produces animated sequences—a "tattoo of time" inscribed through light.

The repetition of viewpoint converts observation into metric record, and record into temporal drawing.

Embodiment, Systems, and Ecological Consciousness

The theoretical foundation of S.U.N.F.A.R.M. integrates system theory, constructivism (von Foerster, 1984), and the wholistic cosmology of David Bohm, alongside Eastern philosophies articulated by Fritjof Capra in *The Tao of Physics* (1975).

Observer and observed are not separate. Subject and universe form a system.

Breath—*prāṇa* in Sanskrit—becomes both physiological rhythm and cosmological metaphor. Expansion and contraction of breath mirror spiral expansion and contraction in the landscape. Performative works such as *Breath of the Earth* extend this inquiry into choreographic collaboration. The body does not symbolize cosmology; it enacts it.

Yantras and mandalas offer geometric precedents: square diagrams aligned to cardinal directions. *Square Field*, a white gravel intervention oriented to the four cardinal points, operates as a yantra at architectural scale. During the 2006 spring equinox, a Buddhist monk led a walking meditation along its perimeter.

Ecology, in its fullest meaning, denotes a system to which we belong. Archaeoastronomy provides historical precedent for integrated ecological consciousness: celestial awareness embedded in dwelling, ritual, and agriculture.

If contemporary specialization has estranged us from visceral cosmological awareness, art may assume the task of re-establishing ecological consciousness—not through scientific necessity, but through contemplative practice.

Initiation? Perhaps.

But not as doctrine—rather as immersion. Awareness of being in space and time requires sensory, physical, and emotional participation. Participation precedes interpretation. Experience precedes theory.

A Narrative Timeline (1997-2011)

Documentation is an essential component of land art practice, given the transitory nature of many projects within this movement—an ephemerality shared with performance art. Even before the digital era, early works of land art in the 1960s and 1970s relied on photography, video recordings, artist statements, journals, publications, and press coverage.

In the previous section, the ephemerality of land art was discussed as a result of entropy in nature—referencing *Spiral Jetty* and *Pratt Farm*—and also as an artistic choice, as in the walking practices of Richard Long and Hamish Fulton. For some works, storytelling and memory are integral components of the artwork itself.

Digital cartography and archival systems have rendered documentation more precise, more layered, and more accessible. The archive becomes not only repository but extension of the work's spatial field.

The Emergence of a Cosmogram (1997–2011)

The first phase of S.U.N.F.A.R.M. (1997–2011) did not unfold as a master-planned monument, but as an incremental cosmographic calibration. What began as speculative digital modeling gradually translated into geodetic inscription upon land. The project evolved from virtual spirals to cardinal alignments, from conceptual diagrams to embodied meridians. Each intervention functioned as both correction and continuation.

1997–1999: Digital Cosmography

The early "Sky Spirals" were conceived digitally before land was acquired. These spiral geometries—modeled in three-dimensional software—were not yet site-specific but already cosmological in ambition. They explored logarithmic curvature as metaphor for expansion, temporal unfolding, and cosmic emergence.

The 1999 acquisition of the farm transformed abstraction into responsibility. Topographic surveys and GPS measurements translated virtual geometry into terrestrial coordinates. The cosmogram would no longer float in digital space; it required latitude and longitude.

2000–2002: Establishing True North

The insertion of survey monuments defining true north marks the first decisive cosmographic act. The *North-South Monument* (2000), established via GPS survey, anchored the site to the meridian.

This moment is critical: before sculpture, before spiral, before aesthetic gesture, there was orientation.

The land was calibrated. Orientation preceded expression.

The *House* (2000), simultaneously under construction, functioned not merely as dwelling but as observational apparatus. Windows framed solar arcs; daily routines became solar measurement.

The *Sunrise Trellis* (2002) aligned explicitly with the equinox sunrise. With this intervention, celestial alignment entered built form.

2003–2004: Meridian and Spiral

The *Square Field* (2003) aligned to cardinal points, and Noon Columns (aligned to the meridian) formalized cardinal geometry in the landscape. These were not decorative markers but temporal instruments: shadows marked solar passage.

The *East Spiral* (2004) introduced curvature. If the meridian embodied linear cosmology, the spiral introduced cyclical cosmology.

Water and earth were shaped to articulate solar movement. The spiral was not symbolic; it was observational. Its orientation and proportions were derived from solar vectors and bodily scale.

Here the cosmogram began cohering.

2005–2006: Paths, Ritual, System

With the *Meditation Path* and *Labyrinth* (2005), movement entered the cosmological system. The body became the measuring instrument.

The *Spring Equinox Celebration* (2006) marked the transformation from installation to ritual space. S.U.N.F.A.R.M. became performative.

The *North Arrow* (2006), pointing toward Polaris, expanded orientation from solar to stellar. The site now contained:

- Meridian
- Cardinal grid
- Equinox alignment
- Stellar axis

The system was no longer fragmentary; it was cosmographic.

2007–2008: Axis Mundi and International Reflection

The *Solstice Monument* (2007) framed winter solstice sunrise and summer solstice sunset. This dual alignment integrated seasonal oscillation.

Axis Mundi (2007), located at the center of the East Spiral, symbolically and spatially established vertical cosmology — earth, horizon, sky.

The planned Rome exhibition (*Il Tempo Attraverso lo Spazio*) functioned as reflective mirror: the work returned to the city of meridians, creating dialogue between Roman cosmology and Hudson Valley cosmography.

2009–2010: Functionality and Compression

The *Orto Solare* (2009) integrated agriculture into cosmology. Solar orientation informed cultivation.

The *Working Sun Pyramid* (2010), later called *Mandala Compass* (2020), introduced functional orientation — sculpture became instrument. It was not only contemplative; it was directional.

The cosmogram now contained:

- Agriculture
- Architecture
- Ritual
- Sculpture
- Alignment
- Documentation
- Publication

It had become an ecosystem. Not metaphorically, but operationally—interdependent elements regulating one another through solar, seasonal, and human rhythms.

2011: Digitization and Augmented Mapping

The introduction of Augmented Reality POI development marked a new phase: cosmography entered digital overlay. The physical landscape acquired informational layers.

This is a crucial theoretical shift:

The cosmogram extended beyond physical matter into digital space.

The first phase concludes not with monumentality, but with informational expansion. The site acquired a second topology—material and virtual—coexisting in parallel.

Archival Timeline of the First Phase

Table 1 presents the timeline of selected works in *Sun Farm*—as the project was titled at the end of this phase in 2011. As shown in the table, unlike traditional object-based two- and three-dimensional art, land art includes performative actions, writings, photography, and documentation processes; the recording of the work's making may itself constitute artwork.

Table 1. Timeline of selected works at S.U.N.F.A.R.M., 1997–2011.

Year(s)	Spatial Intervention	Cosmological Function
1997–1999	Sky Spirals (digital models)	Virtual exploration of spiral cosmology
2000	North-South Monument	Establishment of true north / meridian
2002	Sunrise Trellis	Equinox sunrise alignment
2003	Square Field	Cardinal orientation
2003	Noon Columns	Solar meridian inscription
2004	East Spiral	Cyclical cosmology; solar vector integration
2005	Meditation Path / Labyrinth	Embodied cosmology
2006	North Arrow	Stellar orientation toward Polaris
2007	Solstice Monument	Seasonal solar extremes framing
2009	Orto Solare	Solar agriculture integration
2010	Working Sun Pyramid	Functional solar compass
2011	AR Mapping Projects	Digital cosmogram expansion

From Instrument to Lived Field: Beyond Object-Based Land Art

Twentieth-century land art frequently constructed instruments for perceiving celestial phenomena. In works such as *Sun Tunnels* by Nancy Holt or the ongoing *Roden Crater* by James Turrell, the sky is framed, isolated, intensified. These works function as optical devices: they choreograph vision.

Similarly, the earth/star alignments of Charles Ross articulate astronomical precision through sculptural form. The visitor encounters a constructed apparatus designed to reveal celestial cycles.

S.U.N.F.A.R.M. diverges from this paradigm in a fundamental way.

It does not construct a singular viewing instrument. It establishes a distributed field of orientation integrating:

- Solar alignment
- Agricultural cultivation
- Architectural framing
- Movement practice
- Duration across decades

The project is not entered as a spectator; it is inhabited as a condition.

The difference is ontological. In instrument-based land art, the artwork mediates between viewer and cosmos. At S.U.N.F.A.R.M., the mediation becomes continuous and lived. Solar geometry informs planting cycles. Construction schedules respond to seasonal light. The body's vulnerability and strength influence spatial alignment.

The cosmogram is therefore ecological rather than monumental. It is composed of relationships rather than isolated forms.

Geometry at S.U.N.F.A.R.M. is never insulated from weather. Frost heave displaces stakes; heavy rains redraw drainage lines; drought alters soil compaction. Alignment is therefore not a single act of measurement but a seasonal recalibration. The cosmogram exists within climatic time. Precision here is dynamic rather than fixed.

IN FIERI

This chapter has traced a movement from past cosmological architectures to a contemporary enactment.

Unlike earlier chapters, where my voice was that of historian and observer, here it becomes participant and builder. Writing about S.U.N.F.A.R.M. is not solely documentation of built geometry. It is also a record of daily labor—financial constraints, manual work, persistence, and doubt—intertwined with astronomical calculation and hope.

The project remains in fieri: unfinished, evolving, subject to entropy and transformation.

If the history of the architecture of cosmology culminated in scientific abstraction, S.U.N.F.A.R.M. proposes neither regression nor opposition. It proposes a parallel path: a contemporary cosmogram grounded in embodiment.

Perhaps this is an apparent oxymoron: a conclusion that remains open.

Yet the architecture of cosmology has never been static. It has always been a negotiation between Earth and sky.

S.U.N.F.A.R.M. continues that negotiation. Not as monument, but as practice.

REFERENCES

Altamore, A., & Maffeo, S. (Eds.). (2012). *Angelo Secchi: L'avventura scientifica del Collegio romano*. Quater.

Bertol, D. (2006). Framing the land and sky: Art meets cosmology in a sustainable environment. *Leonardo*, *39*(2), 125–130, 133.

Bertol, D. (2008, June 17). A work of joy and beauty: Christo and Jeanne-Claude's art interview. *Axesmundi*. https://axesmundi.blogspot.com/2008/06/work-of-joy-and-beauty-christo-and.html

Bertol, D. (2015). Sun farm: Body, self, universe. *PAJ a Journal of Performance and Art*, *37*(1), 36–47.

Bertol, D., & Foell, D. (n.d.). *Sun Farm* [Environmental installation]. Center for the Arts and the Environment Archive, Nevada Museum of Art, Reno, NV, United States. Retrieved October 20, 2025 from https://www.nevadaart.org/art/collections/the-archive/CAE1115/finding-aids

Center for Land Use Interpretation. (n.d.). *Pratt Farm*. https://clui.org/ludb/site/pratt-farm

Cocuccioni, E. (2007). « *Art + Place = (Space + Time) of Existence* » *Intervista con Daniela Bertol*. Retrieved October 19, 2025, from https://www.lacritica.net/bertol.htm

Dia Art Foundation. (n.d.). *Walter De Maria The Lightning Field* [Brochure]. https://diaart.org/media/_file/brochures/24_0909_lightning%20field%20brochure_web.pdf

Drake, S. (1957). *Discoveries and opinions of Galileo* (1st ed.). Anchor Books.

Elderfield, J. (2021, March 26). *A day in the life of the Lightning Field*. Retrieved October 25, 2025, from Gagosian Quarterly. https://gagosian.com/quarterly/2021/03/26/essay-day-life-lightning-field/

Event Horizon Telescope. (2024). *Event Horizon Telescope*. Retrieved on October 29 from https://eventhorizontelescope.org/

Galilei, G. (1610). *Sidereus nuncius*. Apud Thomam Baglionum.

Go-Astronomy.com. (n.d.). *10 largest radio telescopes in the world*. Retrieved October 29, 2025, from https://www.go-astronomy.com/10-largest-radio-telescopes.htm

Heath, T. L. (1921). *A history of Greek mathematics*. The Clarendon Press.

Holt/Smithson Foundation. (n.d.). *Sun tunnels*. Retrieved October 31 2025 from https://holtsmithsonfoundation.org/sun-tunnels-0

International Working Group on Astronomy and World Heritage. (n.d). *Osservatorio Astronomico di Roma Sede di Monte Mario, Rome, Italy*.Retrieved October 6, 2025 from https://web.astronomicalheritage.net/show-entity?identity=268&idsubentity=1

Istituto Nazionale di Astrofisica. (n.d.). Museo di Roma. Beni culturali INAF. Retrieved October 6 2025 from https://www.beniculturali.inaf.it/musei/roma/#presentazione

King, H. C. (1955). *The history of the telescope*. With a foreword by Sir Harold Spencer Jones. Dover Publications.

Land Art Flevoland. (n.d.). *Robert Morris: Observatorium*. Retrieved October 21 2025 from https://www.landartflevoland.nl/en/land-art/robert-morris-observatorium/

Las Cumbres Observatory. (n.d.). *Telescopes*. SpaceBook. Retrieved October 30 2025 from https://lco.global/spacebook/telescopes

Library of Congress. (2014). Galileo and the telescope. In *Finding our place in the cosmos: From Galileo to Sagan and beyond*.Retrieved October 6 2025 from https://www.loc.gov/collections/finding-our-place-in-the-cosmos-with-carl-sagan/articles-and-essays/modeling-the-cosmos/galileo-and-the-telescope

National Radio Astronomy Observatory. (n.d.). *Radio telescopes*. Retrieved October 6, 2025 from https://public.nrao.edu/telescopes/radio-telescopes/

National Radio Astronomy Observatory. (n.d.). *Very Large Array*.Retrieved October 6, 2025 from https://www.vla.nrao.edu/

Richard Long. (n.d.).*Richard Long Official*. Retrieved October 30, 2025, from http://www.richardlong.org/

S-U-N-F-A-R-M. (n.d.). *A Place for Art, Nature & Well-Being*. Retrieved October 31, 2025, from https://s-u-n-f-a-r-m.org/

Serres, M. (2018). *The birth of physics*. Rowman & Littlefield Publishers.

Shinde, T. (2024). The spiral aspects of multiverse formation: Continuous big bang and everyverse connection. *International Journal for Research in Applied Science and Engineering Technology*, *12*(III), 48–58. DOI: 10.22214/ijraset.2024.58697

Star Axis. (n.d.). *Star Axis: An architectonic earth/star sculpture*. https://www.staraxis.org/

Tate Britain. (2002, March 23–June 4). Hamish Fulton: Walking journey [Exhibition]. Retrieved September 21 2025 from https://www.tate.org.uk/whats-on/tate-britain/hamish-fulton-walking-journey

The Art Story. (2025.). *Earth Art Movement Overview and Analysis*. Retrieved October 29, 2025, from https://www.theartstory.org/movement/earth-art/

University of North Dakota. (n.d.). *Pf mound*. UND Commons. Retrieved October 25, 2025, from https://commons.und.edu/pf-mound/

Von Foerster, H. (1984). Principles of self-organization—in a socio-managerial context. In *Self-organization and management of social systems: Insights, promises, doubts, and questions* (pp. 2-24). Berlin, Heidelberg: Springer Berlin Heidelberg.

Chapter 9
The Awareness of the Sky

ABSTRACT

This book is the first in a projected series. It concludes by reflecting on how the architecture of cosmology fosters awareness of the interconnectedness between the remote space of the sky and local places on Earth. The conclusion remains intentionally open-ended. It proposes that the sky can be experienced by anyone, across cultures, times, and geographies, and that such experience can awaken awareness of human presence within the universe. Ancient wisdom reminds us that we are part of an integrated system. Even in contemporary conditions, this awareness can encourage shifts toward more sustainable individual practices and public policies.

DOI: 10.4018/979-8-3373-3399-1.ch009

Figure 1. A representation of the Emerald Tablet in Heinrich Khunrath's Amphitheatrum Sapientiae Aeternae. Source: Wellcome Collection. Licensed under CC BY 4.0.

EVERY EXIT IS AN ENTRANCE

Writing a book is an act of contribution to shared knowledge. It is also an act of positioning—situating oneself within a lineage of inquiry while extending it forward. In my experience, writing emerges as a hybrid process shaped by creative practice, scholarly research, drawing, travel, teaching, and sustained questioning. It does not proceed linearly. It accumulates.

The development of this work spanned several years. Its present form condenses hundreds of pages of research notes and travel journals, alongside drawings and thousands of photographs. Each journey—whether to observatories, archaeological sites, or contemporary installations—added fragments to a larger constellation of ideas. Other projects and personal events inevitably intervened, altering rhythm and direction. Nevertheless, the attempt to establish a consistent thread—the geometry of the architecture of cosmology—has been sustained. The inquiry remains incomplete, yet coherence has gradually emerged through recurrence: alignment, orientation, measure, horizon, light.

For me, a conclusion has never been a terminus but an opening. It is not closure; it is reorientation. A conclusion gathers what has been explored and turns it outward. In this case, it offers a "takeaway": the possibility of experiencing cosmological architecture directly, without mediation by monumentality or travel.

There is no necessity to journey to Stonehenge, the Egyptian pyramids, or Rome. Awareness of cosmic alignment can begin anywhere—outdoors, along a city street, in a suburban backyard, or even at a window. The first step is simple: noticing the most visible star from Earth, the Sun.

Observation begins with figure–ground relationships. Objects in our field of vision become figures against the comparatively stable background of the sky—the perceptual surface of the celestial sphere. Even the Sun is perceived as a moving figure, shifting throughout the day and across the year. Its apparent trajectory is not random but patterned. Tracking the Sun's position relative to a fixed object—a building edge, a tree trunk, a window frame—reveals Earth–sky geometries otherwise unnoticed.

What initially appears ordinary becomes structured. The shadow cast at noon shortens and lengthens seasonally. The point of sunrise migrates along the horizon. The setting Sun briefly aligns with urban corridors or rural clearings. These recurring phenomena transform familiar spaces into instruments of orientation. Geometry, here, is not imposed; it is discovered.

This practice is possible in any environment—natural or built. Urban conditions, often perceived as obstacles to celestial awareness, provide their own reference systems. The orthogonal grid of Manhattan, for example, becomes a contemporary cosmographic device during Manhattanhenge—the phenomenon in which the setting

Sun aligns with the east–west street grid (American Museum of Natural History, n.d.). On specific days each year, the solar disk frames itself between architectural facades, momentarily transforming infrastructure into horizon marker (Figures 2–3). The city, typically associated with abstraction and artificial illumination, briefly reveals its embedded astronomical orientation.

Figure 2. Manhattanhenge from West 76 Street between Tenth Avenue and Riverside Drive, 2020.07.09.

Figure 3. Manhattanhenge from West 25 Street between Tenth and Eleventh Avenue, 2025.06.25.

A parallel lesson emerges from Michel de Certeau's *The Practice of Everyday Life* (1984). In "Walking in the City," de Certeau contrasts the detached, totalizing view from skyscrapers with the lived experience of pedestrians navigating streets. Ordinary spatial practices reclaim agency from imposed systems. Similarly, cosmological awareness does not arise from satellite imagery or astronomical data alone; it emerges through embodied looking. To look upward attentively is to shift from passive reception to active participation. Seeing becomes intentional. Orientation becomes chosen rather than assumed.

Tom Stoppard's phrase, "Look on every exit as being an entrance somewhere else" (1968), captures the generative nature of conclusions. Questions do not terminate inquiry; they extend it. Each answer reframes the horizon of the next question. Doors close and open simultaneously.

Yet something has shifted historically. Cosmological architecture, once visibly inscribed into landscapes and cityscapes, appears attenuated in the twentieth and twenty-first centuries. This attenuation is not solely architectural; it is perceptual. One contributing factor may be the transformation of the skyscape itself. Light pollution drastically reduces night-sky visibility, especially in urban regions. Darkness—the essential ground against which celestial bodies become legible—has become scarce.

When the ground disappears, the figures fade.

Where stars once structured myth, ritual, navigation, and settlement, artificial illumination now interrupts perceptual continuity. The horizon is no longer a threshold between Earth and sky but a luminous barrier. Orientation shifts from embodied recognition to technological mediation. We consult devices rather than directions; we measure without necessarily perceiving.

The consequence is not only astronomical but ontological. If cosmological awareness depends upon relational perception—Earth below, sky above—then its diminishment alters our sense of emplacement. Architecture loses one of its primordial dialogues: the negotiation between terrestrial stability and celestial movement.

Reclaiming even a modest practice of observation—tracking a sunset between buildings, noting the seasonal migration of sunrise, observing the measured contraction and expansion of shadows—reintroduces that dialogue. It restores alignment not as spectacle but as participation. The city street becomes meridian. The window frame becomes horizon. The body becomes instrument.

Every exit, then, is an entrance. The end of this book does not conclude cosmological architecture. It relocates it—from monument to perception, from distant site to immediate environment, from abstraction to embodied awareness.

The architecture of cosmology persists wherever Earth and sky are consciously held in relation.

MICROCOSM AND MACROCOSM

Observing the sky is not merely an optical act; it is a spatial positioning of the self within a larger order. The gesture of looking upward situates the body between ground and firmament, activating an awareness of relational scale. In that moment, the distinction between observer and observed begins to soften. The sky is no longer distant backdrop but dynamic counterpart.

What emerges is a phenomenological recognition of interconnectedness between microcosm and macrocosm—not as metaphor alone, but as lived orientation. This relationship operates simultaneously on contemplative and intellectual levels. Perception becomes reflection; reflection becomes structure. The body measures celestial motion through shadow, horizon, and duration, even before mathematics formalizes it.

In contemporary scholarship, such relational awareness resonates with systems thinking and transdisciplinary approaches to knowledge (Carayannis, 2020). A transdisciplinary mode does not fragment inquiry into isolated domains; rather, it recognizes that complex phenomena—cosmos, environment, culture—require integrated methodologies. What I describe as a "mindfulness turn" in knowledge production extends beyond academic frameworks. It concerns a mode of inhabiting the world in which observation, reflection, and action are understood as interdependent.

Although the term *transdisciplinary* is recent, systemic thinking has deep roots. Aristotle's proposition that the whole exceeds the sum of its parts anticipates later holistic models of understanding (Aristotle, 1908). A wholistic mode of knowing—applied individually and collectively—may be essential in rebalancing the relationship between humankind and the natural environment. Cosmological architecture historically functioned within such a framework: it united geometry, astronomy, ritual, proportion, and lived space into coherent systems of meaning.

Throughout recorded human knowledge, numerous treatises have articulated this wholistic vision. Text and illustration operate together in cosmographic atlases, architectural manuals, astronomical diagrams, and esoteric compendia. Many of these works explore celestial phenomena not in isolation, but as structurally linked to terrestrial life. They form an intellectual genealogy relevant to this book and to the broader series.

Vitruvius' *De Architectura* (ca. 25 BCE; Vitruvius, 1914) stands as an early and foundational example. Architecture, astronomy, and timekeeping are treated as interdependent disciplines. Orientation, solar exposure, and climatic knowledge are not decorative considerations but structural principles. In the 1556 Italian translation *I Dieci Libri dell'Architettura* (Vitruvius, 1556), Daniele Barbaro reinforced these concepts with additional illustrations, clarifying the geometric and mathematical foundations of architecture. With Vitruvius, the human body becomes an integral mediator between mathematics and construction. Human proportions are articulated geometrically and serve as modulatory systems for architectural design. Leonardo da Vinci's *Vitruvian Man* (ca. 1490) visually synthesizes this principle: the body inscribed within circle and square embodies the correspondence between microcosm and macrocosm.

In the sixteenth century, Peter Apian's *Cosmographicus Liber* (1524) extended this integrative approach. Apian's diagrams translate celestial motions into geometric projections, rendering the heavens measurable and representable within printed space. His volvelles—rotating paper instruments—demonstrate the dynamic relation between Earth-based observation and celestial mechanics. Geometry becomes both explanatory tool and epistemological bridge: it mediates between observation and abstraction.

The seventeenth century intensifies this synthesis in the work of Robert Fludd. His monumental treatise *Utriusque Cosmi Maioris Scilicet et Minoris Metaphysica, Physica atque Technica Historia* (1617–1621) explicitly articulates the dual structure of macrocosm and microcosm. The title itself announces the ambition: a history—metaphysical, physical, and technical—of the greater and the lesser worlds.

The first volume (*Tomus Primus*) addresses the macrocosm—the "cosmi maioris." Through elaborate symbolic engravings (Figures 4–10), Fludd explores cosmogenesis, celestial hierarchies, harmonic proportion, arithmetic, geometry, architecture, music, and surveying. Text and image operate inseparably. Concentric spheres organize planetary order; triangular diagrams encode theological structure; geometric forms articulate cosmic harmony. While esoteric symbolism permeates the imagery, mathematical reasoning and observational frameworks are equally present. Geometry functions simultaneously as metaphysical diagram and practical instrument.

Figure 4. Macrocosm and microcosm in the title page of Utriusque cosmi maioris scilicet et minoris metaphysica, physica atque technica historia (Fludd, 1617–1621).

Figure 5. Concentric spheres representing planets, nature, animals, and humans (Fludd, 1617, p. 5).

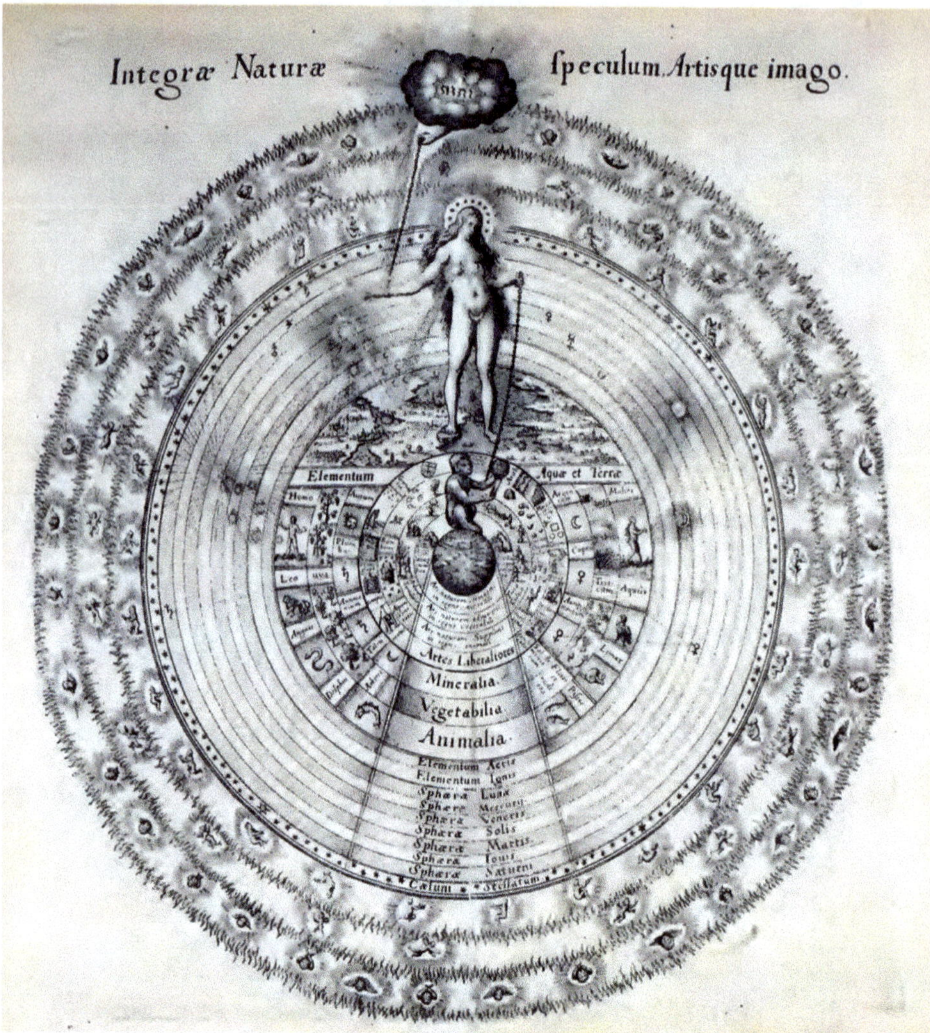

Figure 6. Origin and evolution (Fludd, 1617, p. 9).

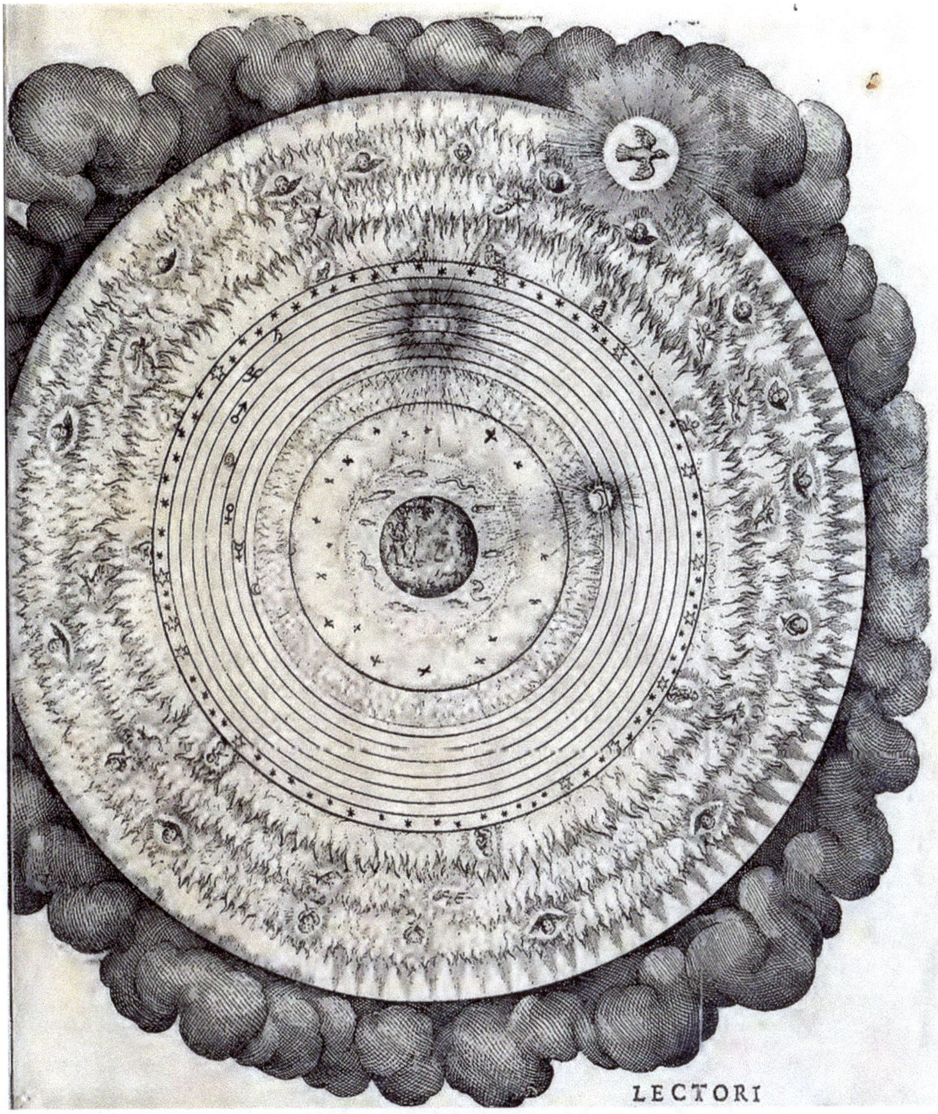

Figure 7. Triangle as diagram of religion and heavens (Fludd, 1617).

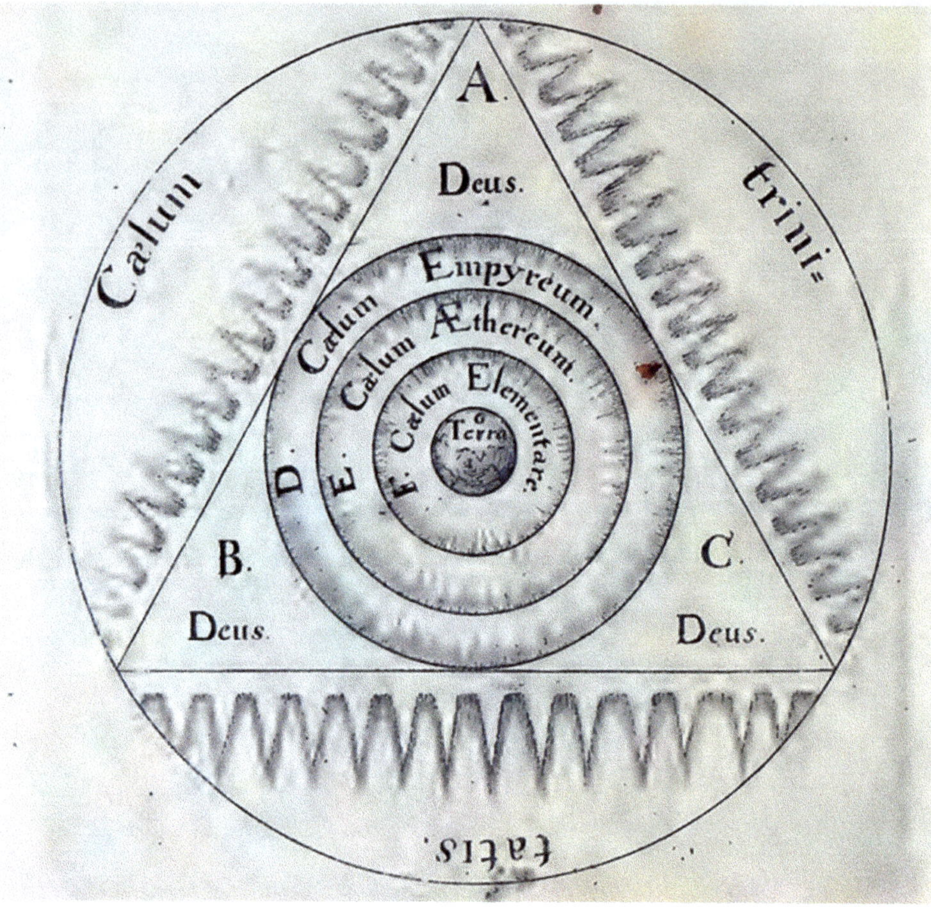

Figure 8. Elements of number theory (Fludd, 1617, p. 20).

20 TRACT. II. PART. I. LIB. I.

NVMERORVM COSSICORVM SPECVLVM,

in quo valor illorum cernitur, ab iis, qui radicum eorum naturas magis respiciunt, quàm formas.

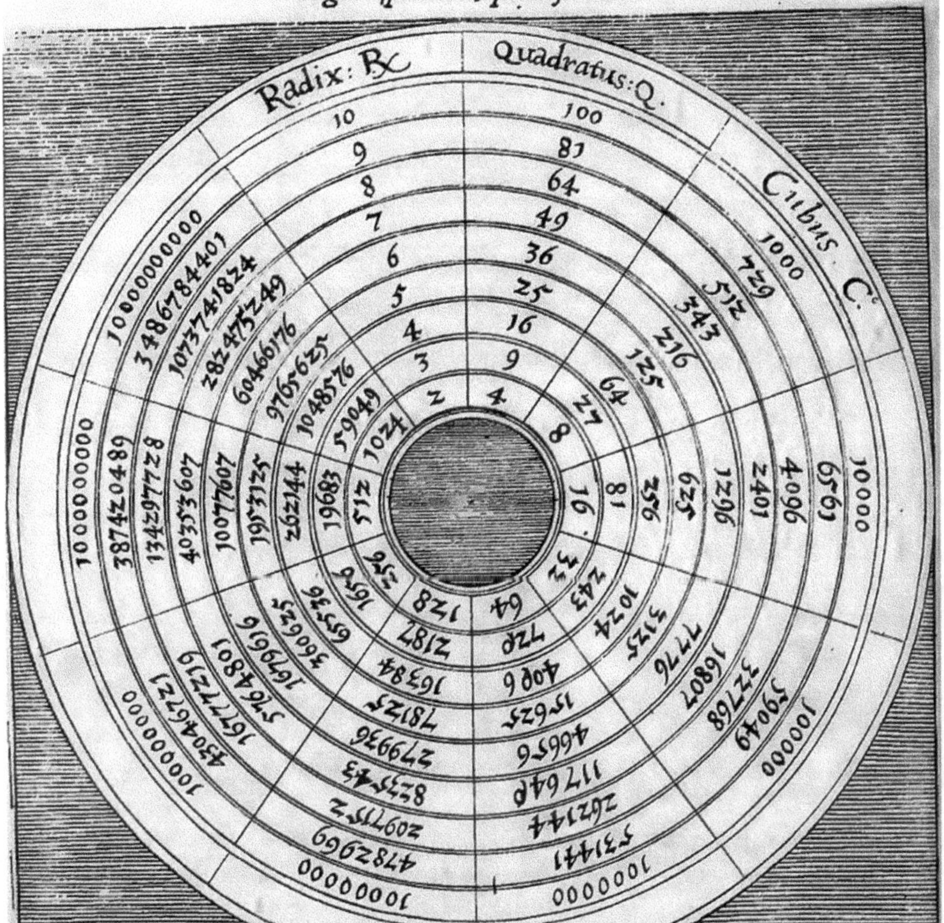

Regula I.

Eodem modo progrediendum erit cum radicibus mixtis, quo cum digitis sive radicibus simplicibus.

Figure 9. Architecture, geometry, and music in De Templo Musicae (Fludd, 1617).

Figure 10. Polygons, circles, and spirals in De Arithmetica Geometrica (Fludd, 1617, pp. 81, 93).

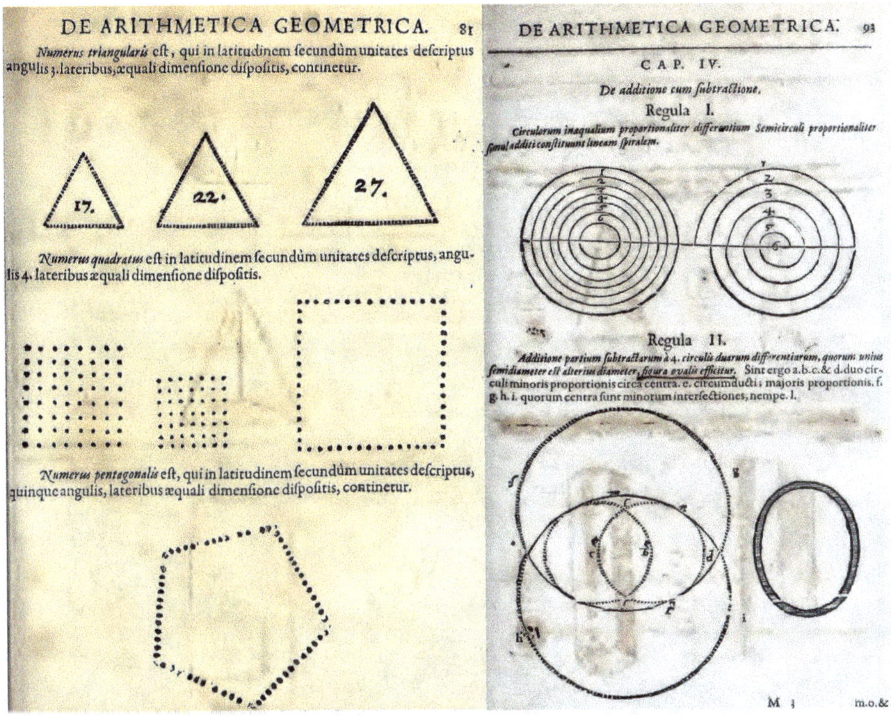

The second volume, *Tomus Secundus de Supernaturali, Naturali, Praeternaturali et Contranaturali Microcosmi Historia* (1619), shifts focus to the microcosm: the human being. The title page itself (Figure 11) visually mirrors the macrocosmic structure of the first volume, reinforcing the correspondence between universal and individual orders. Anatomical studies, geometric human proportions, alchemical processes, memory theaters, and divinatory systems coexist within a unified visual language (Figures 11–13). Scientific inquiry and symbolic imagination are not opposed; they are interwoven. What modern discourse might classify as pseudoscience nonetheless reflects a transdisciplinary ambition: to reconcile empirical observation with metaphysical interpretation.

Figure 11. Title page of Tomus Secundus de supernaturali, naturali, praeternaturali et contranaturali microcosmi historia (Fludd, 1619).

Figure 12. Memory theater (Fludd, 1619, p. 55).

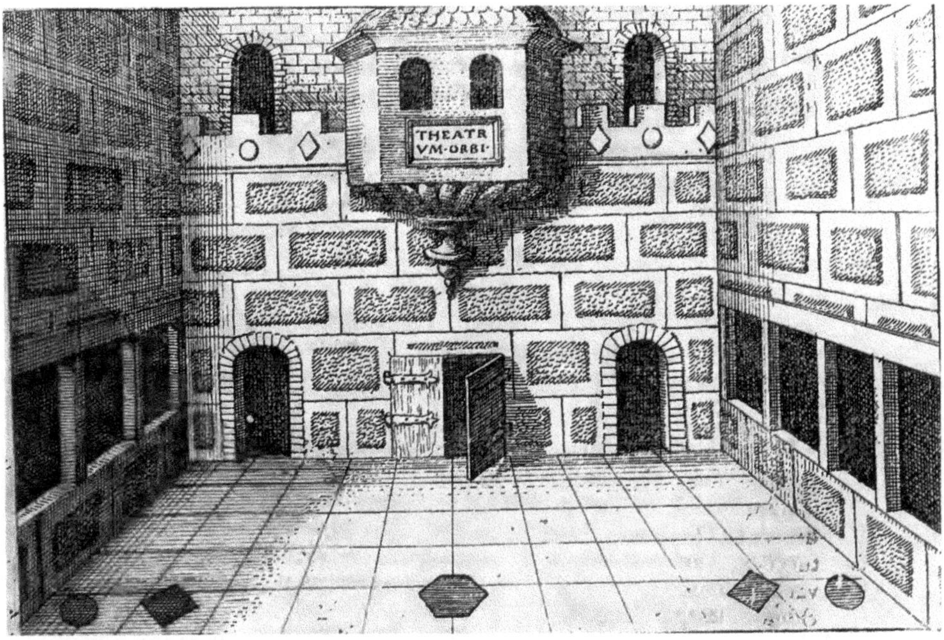

Figure 13. De technica Microcosmi historia (Fludd, 1619, p. 62).

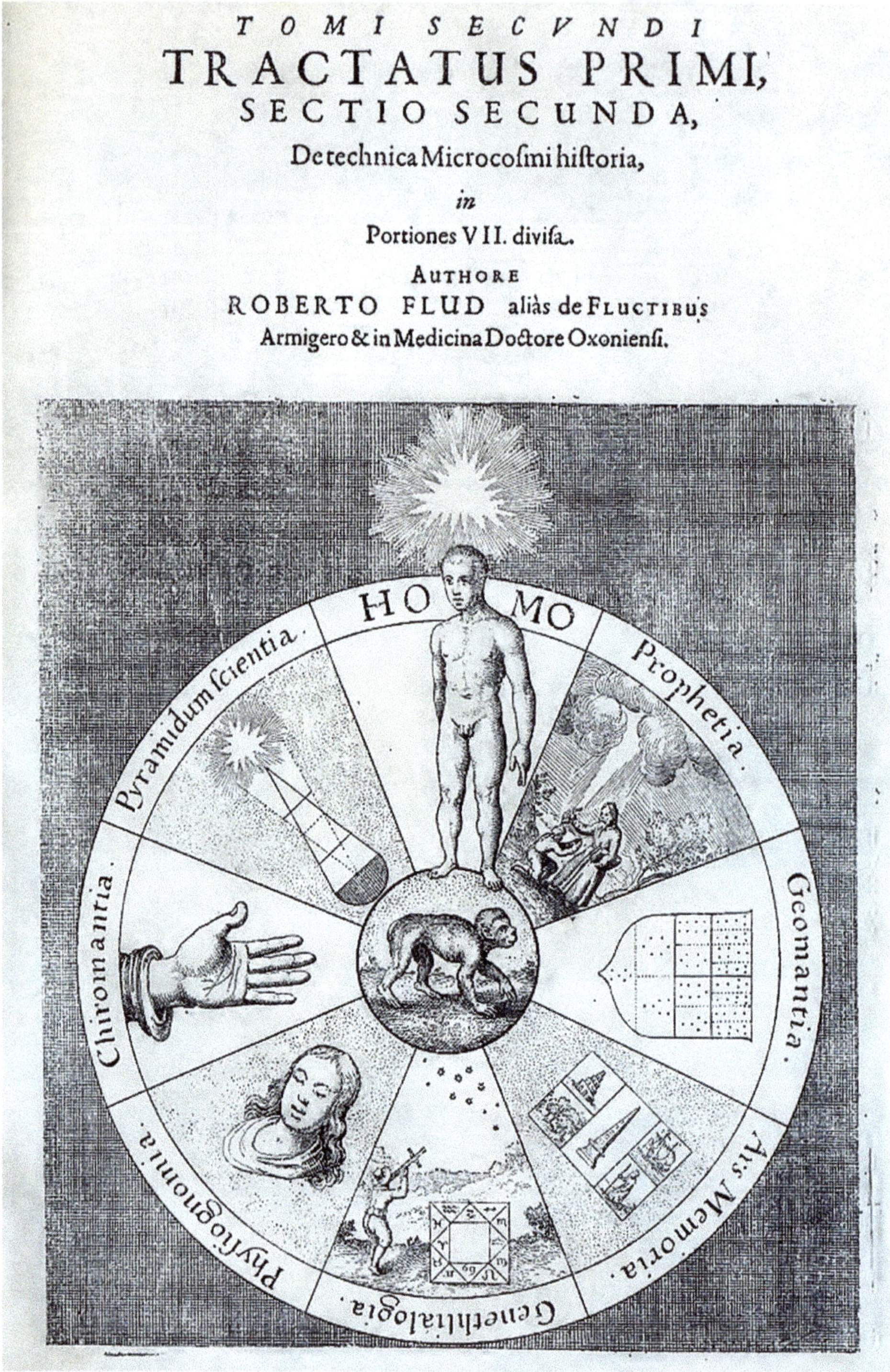

TOMI SECVNDI

TRACTATUS PRIMI,

SECTIO SECUNDA,

De technica Microcoſmi hiſtoria,

in

Portiones VII. diviſa.

AUTHORE

ROBERTO FLUD aliàs de FLUCTIBUS

Armigero & in Medicina Doctore Oxonienſi.

The figures presented in this section—excerpts from *Utriusque Cosmi*—demonstrate how a geometric understanding of the world operates across domains. Arithmetic, music, architecture, anatomy, and cosmology converge within shared proportional systems. Fludd's work exemplifies an epistemological condition in which disciplinary boundaries remain porous. Knowledge is structured relationally rather than hierarchically.

From this perspective, the architecture of cosmology reveals itself as inherently transdisciplinary—not as an intellectual fashion, but as a structural condition of knowing. Its mathematical foundations, embodied in constructed Earth–sky geometries, do not stand apart from symbolic systems or spiritual cosmologies; rather, they intersect with them. Measurement and meaning are not mutually exclusive. Geometry becomes the medium through which empirical observation and metaphysical speculation coexist.

Within this relational field, the observer is not external to the system described. The body stands between terrestrial gravity and celestial motion, mediating scale through perception. The microcosm does not merely reflect the macrocosm metaphorically; it participates in it spatially and temporally. Proportion, orientation, rhythm, and harmonic correspondence are experienced before they are theorized.

In this sense, geometry is not simply a technique of measurement. It is a mode of alignment—an ordering principle through which Earth and sky are consciously held in relation. The architecture of cosmology, whether expressed in ancient treatises, early modern engravings, or contemporary spatial practices, persists wherever this alignment is enacted. The dialogue between microcosm and macrocosm remains unfinished—not because it lacks coherence, but because it unfolds continuously within lived experience.

METHODOLOGICAL REFLECTIONS: BETWEEN EMBODIMENT AND SATELLITE VISION

The architecture of cosmology presents a dual character: material and visual. It is, first, a physical construction—embedded in material culture through stone, wood, metal, earth, and built form. Yet it is also a visual structure, dependent upon the act of observing the sky. The architectural artifact is terrestrial; its referent is celestial. It occupies a specific site on Earth, yet it derives meaning from its relationship to remote space.

Such location is never neutral. An architectural structure is shaped by its environment—topography, horizon line, climate, neighboring constructions—and in turn shapes the way that environment is perceived. The link between local space and the distant sky is therefore reciprocal. The sky is visually apprehended, but that apprehension is always mediated by constructed thresholds, axes, apertures, and alignments. The architecture of cosmology exists precisely in this tension between the grounded and the infinite.

Based on this duality, the present work has required what may appear as an unusual methodology: the integration of qualitative lived experience with digital and quantitative spatial tools. A phenomenological approach has guided the selection of case studies. Being there—walking, standing, orienting, waiting for the light to shift—has preceded verification. First-person experience has then been followed by consultation of maps, satellite imagery, astronomical data, and existing scholarship.

The geometry of the architecture of cosmology is not always immediately perceptible through walkthroughs. Certain alignments exceed the limits of embodied perception; others are obscured by urban accretions or partial ruin. In such cases, satellite imagery and geospatial coordinates provide essential insight. The analysis of geographic data—latitude, longitude, azimuth, altitude—allows the reconstruction of spatial relationships that cannot be apprehended at ground level.

Scholarship in the twenty-first century is profoundly enhanced by interactive maps and location-based media. Tools such as satellite imagery, historical overlays, and astronomical simulators make it possible to examine almost any site on Earth from multiple vantage points. These tools do not replace lived experience; rather, they extend it. They allow a dialogue between embodied perception and planetary-scale visualization.

The IRL (in real life) experience of ancient sites is frequently constrained by historical loss and contemporary regulation. Archaeoastronomical locations often survive only as fragments. Ruins do not always preserve the complete volumetric or visual conditions under which alignments were originally perceived. Furthermore, access is frequently restricted. Unless one is an archaeologist or granted special academic permissions, visiting hours rarely coincide with the astronomical moments for which many of these constructions were designed—for example, sunrise at the summer solstice or sunset at the winter solstice.

This limitation was evident in my visits to the Neolithic temples of Malta. Protective tensile structures—necessary for conservation—altered the perception of the horizon and partially obscured the original relationship between interior space and exterior landscape. While physically present, I was not able to fully experience the intended visual dialogue between temple axis and celestial event. In this case, examining historical satellite imagery through Google Earth, combined with astronomical calculations, allowed a clearer understanding of the site's geometric orientation.

Even in contemporary urban settings, access can be limited. Piazza San Pietro in Rome provides a current example. The year 2025 coincides with the Catholic Jubilee (Iubilaeum or Holy Year). Since the first Jubilee proclaimed by Pope Boniface VIII in 1300, the Catholic Church has celebrated Jubilees at intervals of 25 or

50 years as periods of remission and pilgrimage. For the 2025 “Jubilee of Hope,” an estimated 32 million pilgrims are expected (U.S. Embassy & Consulates in Italy, n.d.). During this period, daily liturgical events and extraordinary gatherings have rendered Piazza San Pietro densely populated and frequently inaccessible for sustained spatial observation. The IRL verification of the astronomical alignments described in Chapter 6 has therefore been significantly constrained. Digital cartography, satellite imagery, and solar calculation tools have been indispensable for confirming azimuths and solstitial relationships.

Figure 14 illustrates a simulation generated by the online tool SunCalc (SunCalc, n.d.), showing the Sun’s azimuth and altitude for December 21, 2025, calculated at the center of Piazza San Pietro, marked by the Obeliscus Vaticanus. The diagram indicates the solar path relative to the geometry of the square and highlights the alignment of solstitial sunrise toward the southeast and solstitial sunset toward the southwest, corresponding to the edges of the southern colonnade. Such simulation does not replace embodied experience; however, it allows a precise verification of geometrical relationships embedded within the urban fabric.

Figure 14. Simulation of Sun azimuth and altitude at the center of Piazza San Pietro (Obeliscus Vaticanus), 2025.12.21 generated with SunCalc (SunCalc, n.d.).

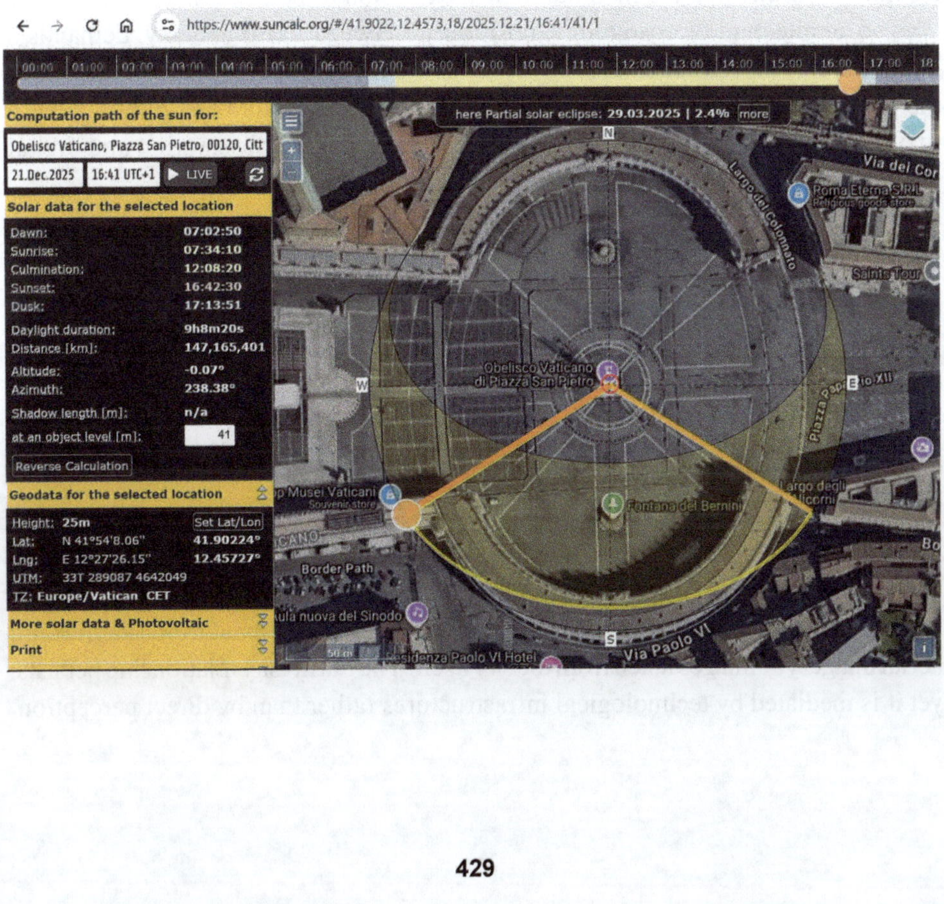

Digital cartography has also revealed alignments that cannot be detected through ground-level perception alone. In Chapter 6, satellite imagery of the Egyptian pyramids confirmed their precise orientation toward the cardinal directions and allowed additional consideration of diagonal alignments between pyramidal complexes. Similarly, the well-documented orientation of the Karnak complex toward solstitial sunrise and sunset—impossible to experience personally due to restricted access at those hours—was verified through the superimposition of calculated solar axes on satellite images.

In Rome, several alignments between monuments are not visually perceivable in situ yet become evident through cartographic analysis. The spatial relationships between Egyptian obelisks relocated within the city, and the axial geometry connecting Castel Sant'Angelo with the Basilica di San Pietro, reveal a network of orientations embedded within the urban plan. These examples demonstrate that the architecture of cosmology often operates at scales exceeding immediate sensory awareness.

Contemporary digital tools do not diminish the phenomenological dimension of this research; they recalibrate it. They allow the reconstruction of sightlines no longer accessible, the measurement of angles imperceptible at ground level, and the visualization of alignments extending beyond immediate sensory awareness. When integrated with lived experience, digital cartography becomes not merely a technical instrument but a methodological extension of perception.

The architecture of cosmology, positioned between Earth and sky, demands a methodology capable of inhabiting both the tactile and the orbital, the local and the planetary. The body stands on site; the satellite hovers above it. Between these two vantage points, geometry becomes legible. Embodied observation and digital verification do not compete; they form a reciprocal system of inquiry.

In this sense, methodology mirrors subject: just as architecture mediates between terrestrial matter and celestial motion, research must mediate between experience and abstraction. To study the architecture of cosmology is therefore to operate simultaneously at human scale and planetary scale—aligning perception with calculation, presence with projection, horizon with map.

WE ARE UNDER THE SAME SKY

In the digital age, we daily witness how culture defines our experience of space beyond its perceptual characteristics. Our location on Earth is tracked by GPS; web-mapping platforms guide our navigation through geodata derived from satellite and aerial imagery—images taken from the sky itself. The Earth–sky relationship persists, yet it is mediated by technological infrastructures rather than by direct perception.

The sky, once a primary orienting framework for human settlement and ritual, now functions invisibly as a platform for satellites, signals, and telecommunications.

The transformation is subtle yet profound. The connection between local and remote space no longer depends primarily on horizon lines, celestial cycles, or embodied observation. It is constructed through screens. In the contemporary world, where interaction increasingly revolves around digital technologies, locality and proximity are experienced differently from previous generations. A new spatial condition emerges. What was once "space" in its entirety—the visible field circumscribed by our cone of vision—now becomes merely local space. Remote space, once understood as distant and unreachable beyond the horizon, is instead encountered through live broadcasts, video calls, satellite feeds, and virtual classrooms.

The expansion of spatial boundaries extends beyond human perception. We not only see and communicate across thousands of miles; we interact with outer space through remote sensing, planetary missions, radio telescopes, and orbital observation systems. The sky is no longer only observed; it is instrumented.

Is this shift ontological, or does it represent a different mode of access to what has always been remote? The relationship between local and distant space has existed since the earliest civilizations. Ancient cultures interpreted the horizon as threshold. Celestial bodies—the Sun, Moon, stars, and planets—became deities, temporal markers, cosmological references, and political symbols. The remote was never merely distant; it was meaningful.

Architecture embodied this meaning. Priests stood before apertures framing planetary risings; temple axes were aligned with solstitial events; urban layouts encoded celestial hierarchies. The architecture of cosmology materialized the relationship between Earth and sky, transforming astronomical observation into cultural form.

In contrast, contemporary society often experiences an expanded horizon that is technologically amplified yet spatially dislocated. The artificial extension of the five senses—through cameras, sensors, digital networks—has not necessarily deepened the relationship with local environment. There is an increasing disjunction between the place where perception occurs and the object perceived. Space exploration is mediated by highly specialized scientific communities; telescopes and observatories operate beyond public participation. Even when astronomical data is accessible online, the act of viewing an image on a smartphone lacks the architectural and spatial framework that once connected observation to place.

The architecture of cosmology historically provided such frameworks. Naked-eye observatories, temple courtyards, megalithic alignments, and solar markers gave physical structure to celestial experience. They established continuity between the ground beneath one's feet and the sky overhead. When that continuity dissolves, the perception of remote space risks becoming abstract, detached from embodied awareness.

Can the architecture of cosmology regain presence in the human-made environment—not as replication of ancient forms, but as renewed consciousness of alignment? This book suggests that our existential space possesses a cognitive dimension beyond its immediate sensory experience. Culture interprets space; architecture materializes that interpretation. The relationship between local and remote space therefore becomes expressive of human presence in the world.

Archaeoastronomy offers lessons not as nostalgia, but as orientation. Ancient built environments demonstrate that geometry, celestial cycles, and human perception once formed an integrated system. Their architectural typologies extended beyond functional necessity to embody metaphysical understanding. Through solar and stellar alignments, geometry shaped awareness of humanity's position within the cosmos. Such awareness cultivated a harmonious relationship with landscape and environment.

A reflection on the architecture of cosmology ultimately leads to an understanding that life may hold a dimension of meaning exceeding material function. This higher purpose need not be confined to specific religions or belief systems; it may also arise through individual contemplation. The awareness of the sky evokes what Immanuel Kant described as the sublime (Kant, 1790/1914): a simultaneous sense of awe and humility before immensity.

The experience of the sky is universally shared. It is intangible, unreachable, yet visible from nearly every outdoor space. It transcends social class, nationality, religion, and political boundary. The Sun shines—or withdraws behind clouds—for all, as it once did for pharaohs, emperors, and kings. These reflections invite appreciation and responsibility. The sky is what all humans, and indeed all living beings on Earth, share.

Ancient cultures celebrated this shared condition by constructing temples, markers, and ritual spaces—material embodiments of cosmic awareness. There is no necessity to reproduce temples today. The architecture of cosmology can begin with attention. Respecting the sky implies environmental consciousness. The contemporary skyscape differs from that of ancient civilizations not only astronomically but environmentally. Light pollution and atmospheric degradation obscure celestial visibility. Reducing light pollution, improving air quality, and addressing climate change are not solely ecological imperatives; they are acts that restore our perceptual relationship with the cosmos. Honoring remote space through environmental stewardship simultaneously protects local Earth.

Environmental action can be accompanied by contemplative practices. Indigenous traditions continue to honor the sky through storytelling and ceremony. The Tibetan Buddhist maxim, "Body like a mountain, Breath like the wind, Mind like the sky," expresses an experiential analogy between human interiority and natural elements. Individual awareness becomes the seed of collective change. Solar and celestial alignments orient not only monuments but consciousness.

We are all under the same sky, yet we are often unaware of this shared condition. Perception is distracted; attention fragmented. To look upward intentionally is to re-situate oneself within space and time. Observing the movement of the Sun, Moon, planets, and stars—witnessing changing light and cloud formations—cultivates awareness of duration and orientation. Looking at the sky restores spatial humility.

The reference to the *Emerald Tablet* (Hermes Trismegistus, ca. 800/1923) in this context is not literal endorsement of esoteric doctrine. Rather, it functions as metaphor. The well-known phrase

As Above So Below

(and)

So Below So Above

encapsulates a relational principle central to this book. It expresses, in condensed symbolic language, the reciprocity between macrocosm and microcosm, between remote and local space. Whether interpreted through alchemical tradition, philosophical analogy, or poetic intuition, the statement articulates a structural correspondence.

From ancient esoteric texts to architectural treatises and cosmographic diagrams, humankind has repeatedly sought a wholistic understanding of reality. That search was often materialized through built forms—structures, monuments, tombs, temples, markers—that embodied the Earth–sky relationship. This embodiment is what has been defined throughout this work as the architecture of cosmology.

One need not be a scholar of alchemy or mysticism to grasp this connection. Mindfulness of action and awareness of experience are sufficient. Even the simplest architectural gesture—a window aligned to sunrise, a marker casting a shadow, a plaza oriented to cardinal directions—can become a reminder of where we stand in space and time.

We are under the same sky. To recognize this is to acknowledge both shared fragility and shared responsibility.

The architecture of cosmology begins not with construction, but with awareness.

Figure 15. Woodcut of Hermes Trismegistus holding an armillary sphere, from Daniel Stolcius, Viridarium Chymicum (1624). Source: Wikimedia Commons, public domain. The image represents a wholistic synthesis of human wisdom, scientific knowledge (armillary sphere), terrestrial landscape, celestial background, and symbolic representations of Sun and Moon.

REFERENCES

American Museum of Natural History. (n.d.). *Manhattanhenge*. Retrieved October 29, 2025, from https://www.amnh.org/research/hayden-planetarium/manhattanhenge

Aristotle, . (1908). *Metaphysics* (Ross, W. D., Trans.). Clarendon Press. (Original work published 350 BCE)

Carayannis, E. G. (Ed.). (2020). *Encyclopedia of creativity, invention, innovation and entrepreneurship*. Springer., DOI: 10.1007/978-3-319-15347-6

Cocuccioni, E. (2007). « Art + Place = (Space + Time) of Existence »: Intervista con Daniela Bertol. Retrieved October 19, 2025, from https://www.lacritica.net/bertol.htm

De Certeau, M. (1984). *The practice of everyday life* (Rendall, S., Trans.). University of California Press.

European Space Agency. (n.d.). *Safety tips for observing the Sun*. Retrieved October 29, 2025, from https://www.esa.int/Science_Exploration/Space_Science/Safety_tips_for_observing_the_Sun

Fludd, R. (1619). *Tomus secundus de supernaturali, naturali, praeternaturali et contranaturali microcosmi historia*. Hieronymus Gallerus for Johann Theodor de Bry.

Fludd, R. (1617–1621). *Utriusque cosmi maioris scilicet et minoris metaphysica, physica atque technica historia* [The metaphysical, physical, and technical history of the two worlds, namely the greater and the lesser]. Oppenheim, Germany: Hieronymus Gallerus for Johann Theodor de Bry.

Hermes Trismegistus. (1923). The emerald tablet (E. J. Holmyard, Trans.). *Nature, 112*(2814), 525–526. (Original work ca. 800)

Kant, I. (1914). *Kant's critique of judgement* (Bernard, J. H., Trans.). Macmillan. (Original work published 1790)

Silva, F., & Campion, N. (Eds.). (2015). *Skyscapes: The role and importance of the sky in archaeology*. Oxbow Books.

St. Fleur, N. (2025, July 10). Manhattanhenge is here again: How to see New York City's best sunset. *National Geographic*. Retrieved October 29, 2025, from https://www.nationalgeographic.com/science/article/manhattanhenge-new-york-city-sunset

Stolcius, D. (1624). *Viridarium chymicum figuris cupro incisis adornatum, et poeticis picturis illustratum*. L. Jennis.

Stoppard, T. (1968). *Rosencrantz and Guildenstern are dead*. Grove Press.

SunCalc. (n.d.). *Solar and lunar data for 41.9022° N, 12.4573° E on December 21, 2025*. Retrieved October 26, 2025, from https://www.suncalc.org

U.S. Embassy & Consulates in Italy. (n.d.). *2025 Catholic Jubilee in Rome—Pilgrims in hope*. Retrieved October 29, 2025, from https://it.usembassy.gov/jubilee2025/

Vitruvius Pollio. (1914). *The ten books on architecture* (Morgan, M. H., Trans.). Harvard University Press.

Vitruvius Pollio, M. (1556). *I dieci libri dell'architettura* (Barbaro, D., Ed. & Trans.). Francesco Marcolini. (Original work published 15 BCE)

Compilation of References

Albert, A. A. (1949). *Solid analytic geometry* (1st ed.). McGraw-Hill.

Altamore, A., & Maffeo, S. (Eds.). (2012). *Angelo Secchi: L'avventura scientifica del Collegio romano*. Quater.

American Museum of Natural History. (n.d.). *Manhattanhenge*. Retrieved October 29, 2025, from https://www.amnh.org/research/hayden-planetarium/manhattanhenge

Apian, P. (1524). *Cosmographicus liber Petri Apiani mathematici studiose collectus*.

Aquilina, J. (1987–1990). *Maltese-English dictionary* (Vols. 1–2). Midsea Books Ltd.

Aristotle, . (1908). *Metaphysics* (Ross, W. D., Trans.). Clarendon Press. (Original work published 350 BCE)

Arnheim, R. (1974). *Art and visual perception: A psychology of the creative eye*. University of California Press.

Arya, R. (2000). *Vaastu: The Indian art of placement: Design and decorate homes to reflect eternal spiritual principles*. Destiny Books.

Aveni, A. F. (1980). *Skywatchers of ancient Mexico*. University of Texas Press.

Ball, R. S. (1908). *A treatise on spherical astronomy*. Cambridge University Press.

Barratt, R. P. (2022). The crux of astronomical alignment in Neolithic Malta: Using 3D simulation to produce new data. *Digital Applications in Archaeology and Cultural Heritage, 26*, e00229.

Barrett, C. E. (2017). Egypt in Roman visual and material culture. In Williams, G. (Ed.), *Oxford handbooks online in classical studies*. Oxford University Press., DOI: 10.1093/oxfordhb/9780199935390.013.18

Belmonte, J. A. (2001). On the orientation of Old Kingdom Egyptian pyramids. *Journal for the History of Astronomy, 32*(1), 1–20.

Belmonte, J. A., & Shaltout, M. (2005). On the orientation of ancient Egyptian temples. *Journal for the History of Astronomy*, *36*(3), 273–298.

Bertol, D. (2008, June 17). A work of joy and beauty: Christo and Jeanne-Claude's art interview. *Axesmundi*. https://axesmundi.blogspot.com/2008/06/work-of-joy-and-beauty-christo-and.html

Bertol, D., & Foell, D. (n.d.). *Sun Farm* [Environmental installation]. Center for the Arts and the Environment Archive, Nevada Museum of Art, Reno, NV, United States. Retrieved October 20, 2025 from https://www.nevadaart.org/art/collections/the-archive/CAE1115/finding-aids

Bertol, D. (1994). *Visualizing with CAD: An AutoCAD exploration of geometric and architectural forms*. TELOS.

Bertol, D. (2006). Framing the land and sky: Art meets cosmology in a sustainable environment. *Leonardo*, *39*(2), 125–130, 133.

Bertol, D. (2015). Sun farm: Body, self, universe. *PAJ a Journal of Performance and Art*, *37*(1), 36–47.

Bertol, D., & Foell, D. (1997). *Designing digital space: An architect's guide to virtual reality*. John Wiley & Sons.

Blaeu, J. (1665). *Atlas Maior, sive cosmographia Blaviana* (Vol. 1). Joan Blaeu.

Blyth, E. (2006). *Karnak: Evolution of a temple*. Routledge.

Bohm, D. (1980). *Wholeness and the implicate order*. Routledge & Kegan Paul.

Brahe, T. (1602). *Astronomiæ instauratæ mechanica*. Noribergae [Nuremberg], Germany: Levinum Hulsium.

Brandt, Y., Dieterich, M., & Huppert, D. (2024). Human senses and sensors from Aristotle to the present. *Frontiers in Neurology*, *15*, 1404720. PMID: 39022724

British Sundial Society. (n.d.). *How sundials work*. Retrieved September 12, 2025, from https://web.archive.org/web/20130802033857/http://www.sundialsoc.org.uk/HDSW.php

Brough, J. B., Dowden, B., & Fieser, J. (n.d.). Phenomenology and time-consciousness. In J. Fieser & B. Dowden (Eds.), *Internet encyclopedia of philosophy*. Retrieved September 6, 2025, from https://iep.utm.edu/phe-time/

Brú na Bóinne World Heritage Site. (n.d.). *Newgrange*. Retrieved September 25, 2025, from https://www.newgrange.com/

Brünnow, F. (1865). *Spherical astronomy*. Ferd. Dümmler.

Burgess, E. (Ed.). (1935). *Translation of the Sūrya-Siddhānta: A text-book of Hindu astronomy* (P. Gangooly, Ed.). Calcutta, India: University of Calcutta. (Original work published ca. 4th century CE)

Burn, R. (1880). *Old Rome: A handbook to the ruins of the city and the Campagna*. George Bell and Sons.

Carayannis, E. G. (Ed.). (2020). *Encyclopedia of creativity, invention, innovation and entrepreneurship*. Springer., DOI: 10.1007/978-3-319-15347-6

Cassirer, E. (1955). *The philosophy of symbolic forms* (Manheim, R., Trans.). Yale University Press.

Center for Land Use Interpretation. (n.d.). *Pratt Farm*. https://clui.org/ludb/site/pratt-farm

Chauvenet, W. (1906). *A manual of spherical and practical astronomy*. J. B. Lippincott.

Cipriani, G. B. (1823). *Su i dodici obelischi egizj che adornano la città di Roma*. [s.n.]

City Planning of Jaipur. (n.d.). *City planning of Jaipur*. Retrieved September 17, 2025, from https://www.gopalancolleges.com/gsap/pdf/jaipur.pdf

Cocuccioni, E. (2007). *« Art + Place = (Space + Time) of Existence » Intervista con Daniela Bertol*. Retrieved October 19, 2025, from https://www.lacritica.net/bertol.htm

Cocuccioni, E. (2007). « Art + Place = (Space + Time) of Existence »: Intervista con Daniela Bertol. Retrieved October 19, 2025, from https://www.lacritica.net/bertol.htm

Copernicus, N. (1566). De revolutionibus orbium coelestium [). Basileae] [Basel] [, Switzerland: Ex officina Henricpetrina.]. *Libri*, VI.

Cosgrove, D. E. (2007). Images of Renaissance cosmography, 1450–1650. In Woodward, D. (Ed.), *The history of cartography* (Vol. 3, pp. 55–98). University of Chicago Press.

de Certeau, M. (1984). *The practice of everyday life*. University of California Press.

De Franceschini, M., & Veneziano, G. (2018). The symbolic use of light in Hadrianic architecture and the "Kiss of the Sun.". *Archaeoastronomy and Ancient Technologies*, *6*(1), 111–137.

De Soldanis, G. P. F. (1995). *Gozo, ancient and modern, religious and profane* (Aquilina, J., Trans.). Malta University Press. (Original work published 1746)

Derrida, J. (1978). *Edmund Husserl's* Origin of geometry*: An introduction* (J. P. Leavey, Trans.). University of Nebraska Press.

Dia Art Foundation. (n.d.). *Walter De Maria The Lightning Field* [Brochure]. https://diaart.org/media/_file/brochures/24_0909_lightning%20field%20brochure_web.pdf

Dirac, P. A. M. (1967). *The principles of quantum mechanics*. Oxford University Press.

Drake, S. (1957). *Discoveries and opinions of Galileo* (S. Drake, Trans.). Doubleday & Company, Inc. (Original work published 1610-1623).

Drake, S. (1957). *Discoveries and opinions of Galileo* (1st ed.). Anchor Books.

Dyson, S. L. (2010). *Rome: A living portrait of an ancient city*. Johns Hopkins University Press.

EBSCO. (n.d.). Time measurement. In *Research starters: Science*. Retrieved September 9, 2025, from https://www.ebsco.com/research-starters/science/time-measurement

Elderfield, J. (2021, March 26). *A day in the life of the Lightning Field*. Retrieved October 25, 2025, from Gagosian Quarterly. https://gagosian.com/quarterly/2021/03/26/essay-day-life-lightning-field/

Eliade, M. (1951). Images and symbols (P. Mairet, Trans.). Harvill Press.

Eliade, M. (1961). *The sacred and the profane*. Harper & Brothers.

Embree, L. (Ed.). (1997). *Encyclopedia of phenomenology*. Kluwer Academic Publishers.

Encyclopædia Britannica. (n.d.). Malta. In *Britannica academic*. Retrieved October 10, 2025, from https://www.britannica.com/place/Malta

Encyclopædia Britannica. (n.d.). Mathematics in the Islamic world (8th–15th century). In *Britannica.com*. Retrieved September 9, 2025, from https://www.britannica.com/science/mathematics/Mathematics-in-the-Islamic-world-8th-15th-century

European Space Agency. (n.d.). *Safety tips for observing the Sun*. Retrieved October 29, 2025, from https://www.esa.int/Science_Exploration/Space_Science/Safety_tips_for_observing_the_Sun

Evans, J. (1998). *The history and practice of ancient astronomy*. Oxford University Press.

Event Horizon Telescope. (2024). *Event Horizon Telescope*. Retrieved on October 29 from https://eventhorizontelescope.org/

Fields, N., & Dennis, P. (2008). *Rome's ancient walls*. Osprey Publishing.

Fletcher, H. M. (1929). Chyluria.

Fludd, R. (1617–1621). *Utriusque cosmi maioris scilicet et minoris metaphysica, physica atque technica historia* [The metaphysical, physical, and technical history of the two worlds, namely the greater and the lesser]. Oppenheim, Germany: Hieronymus Gallerus for Johann Theodor de Bry.

Fludd, R. (1619). *Tomus secundus de supernaturali, naturali, praeternaturali et contranaturali microcosmi historia*. Hieronymus Gallerus for Johann Theodor de Bry.

Foderà Serio, G., Hoskin, M., & Ventura, F. (1992). The orientations of the temples of Malta. *Journal for the History of Astronomy, 23*(2), 107–119.

Fraknoi, A., Morrison, D., & Wolff, S. C. (2022). *Astronomy* (2nd ed.). OpenStax.

Galilei, G. (1610). *Sidereus nuncius*. Apud Thomam Baglionum.

Galilei, G. (1623). *Il saggiatore*. Appresso Giacomo Mascardi. (Original work published 1623)

Gerbino, A. (Ed.). (2014). *Geometrical objects: Architecture and the mathematical sciences 1400–1800*. Springer International Publishing.

Gibson, W. (1984). *Neuromancer*. Ace Books.

Glossary. In (2023). *Astronomical Almanac Online*. United States Naval Observatory.

Go-Astronomy.com. (n.d.). *10 largest radio telescopes in the world*. Retrieved October 29, 2025, from https://www.go-astronomy.com/10-largest-radio-telescopes.htm

Goswami, D. Y. (2015). *Principles of solar engineering* (3rd ed.). CRC Press.

Government of India. (2009). *Nomination of the Jantar Mantar, Jaipur, for inclusion on the World Heritage List*. Paris, France: UNESCO World Heritage Centre. https://whc.unesco.org/uploads/nominations/1338.pdf

Griffero, T. (2014). Atmospheres and lived space. *Studia Phaenomenologica, 14*, 29–51.

Griffero, T. (2021). Sniffing atmospheres: Observations on olfactory being-in-the-world. In *Olfaction: An interdisciplinary perspective from philosophy to life sciences* (pp. 75–90). Springer.

Harris, B., & Zucker, S. (2015, November 18). Gian Lorenzo Bernini, Saint Peter's Square. *Smarthistory*. Retrieved October 30, 2025, from https://smarthistory.org/bernini-st-peters-square-2/

Haselberger, L. (2011). A debate on the Horologium of Augustus. *Journal of Roman Archaeology*, *24*(1), 47–73.

Heath, R. (2003). *Alexander Thom: Cracking the Stone Age Code*. Bluestone Press.

Heath, T. L. (1913). *Aristarchus of Samos, the ancient Copernicus: a history of Greek astronomy to Aristarchus, together with Aristarchus's Treatise on the sizes and distances of the sun and moon: a new Greek text with translation and notes*. Clarendon Press.

Heath, T. L. (1921). *A history of Greek mathematics*. Clarendon Press.

Heat-Moon, W. L. (1991). *PrairyErth (A Deep Map)*. Houghton Mifflin.

Heidegger, M. (1962). *Being and time* (J. Macquarrie & E. Robinson, Trans.). Harper & Row. (Original work published 1927)

Heidegger, M. (1971). Building, dwelling, thinking. In *Poetry, language, thought* (A. Hofstadter, Trans.). Harper & Row.

Heisenberg, W. (1958). *Physics and philosophy: The revolution in modern science*. Harper & Brothers.

Heritage, E. (n.d.). *Research on Stonehenge*. English Heritage. Retrieved on September 21, 2025 from https://www.english-heritage.org.uk/visit/places/stonehenge/history-and-stories/history/research/

Hermes Trismegistus. (1923). *The emerald tablet* (E. J. Holmyard, Trans.). *Nature, 112*(2814), 525–526. (Original work ca. 800)

Heslin, P. J. (2007). Augustus, Domitian, and the so-called Horologium Augusti. *Journal of Roman Studies*, *97*, 1–20.

Hibbard, H. (1986). *The architecture of the new St. Peter's*. Pennsylvania State University Press.

Holt/Smithson Foundation. (n.d.). *Sun tunnels*. Retrieved October 31 2025 from https://holtsmithsonfoundation.org/sun-tunnels-0

Husserl, E. (1970). *The crisis of European sciences and transcendental phenomenology* (D. Carr, Trans.). Northwestern University Press.

Husserl, E. (2001). *Logical investigations* (J. N. Findlay, Trans.). Routledge. (Original work published 1901–1902)

Husserl, E. G. (1991). *On the phenomenology of the consciousness of internal time (1893–1917)* (Brough, J. B., Trans.). Kluwer Academic Publishers.

International Working Group on Astronomy and World Heritage. (n.d). *Osservatorio Astronomico di Roma Sede di Monte Mario, Rome, Italy*.Retrieved October 6, 2025 from https://web.astronomicalheritage.net/show-entity?identity=268&idsubentity=1

Istituto Nazionale di Astrofisica. (n.d.). Museo di Roma. Beni culturali INAF. Retrieved October 6 2025 from https://www.beniculturali.inaf.it/musei/roma/#presentazione

Jaipur Municipal Corporation. (n.d.). Our heritage. Retrieved September 18, 2025, from https://jaipurmc.org/Presentation/AboutJaipur/OurHeritage.aspx

Jantar Mantar. (n.d.). Digamsa Yantra. Retrieved September 19, 2025, from https://jantarmantar.org/learn/observatories/instruments/digamsa/index.html

Johnson, W. (1912). *Byways of British Archaeology*. Cambridge University Press.

Kant, I. (1914). *Kant's critique of judgement* (Bernard, J. H., Trans.). Macmillan. (Original work published 1790)

Kepler, J. (1609). *Astronomia nova*. G. Voegelinus.

Kepler, J. (1619). *Harmonices mundi*. Johann Planck.

King, H. C. (1955). *The history of the telescope*. With a foreword by Sir Harold Spencer Jones. Dover Publications.

Krautheimer, R. (1980). *Rome: Profile of a city, 312–1308*. Princeton University Press.

Krupp, E. C. (1988). Light in the temples. In *Echoes of the ancient skies*. Harper & Row.

Krupp, E. C. (1988). Light in the Temples. In Ruggles, C. L. N. (Ed.), *Records in Stone: Papers in Memory of Alexander Thom* (pp. 473–499). CUP.

Land Art Flevoland. (n.d.). *Robert Morris: Observatorium*. Retrieved October 21 2025 from https://www.landartflevoland.nl/en/land-art/robert-morris-observatorium/

Lapaine, M., & Frančula, N. (2022). Map projections classification. *Geographies*, 2(2), 274–285. DOI: 10.3390/geographies2020019

Las Cumbres Observatory. (n.d.). *Telescopes*. SpaceBook. Retrieved October 30 2025 from https://lco.global/spacebook/telescopes

Lehner, M. (1997). *The complete pyramids*. Thames & Hudson.

Leybourne, W., & Morden, R. (1675). *An introduction to astronomy and geography: Being a plain and easie treatise of the globes*. Printed for R. Morden and W. Berry.

Library of Congress. (2014). Galileo and the telescope. In *Finding our place in the cosmos: From Galileo to Sagan and beyond*.Retrieved October 6 2025 from https://www.loc.gov/collections/finding-our-place-in-the-cosmos-with-carl-sagan/articles-and-essays/modeling-the-cosmos/galileo-and-the-telescope

Liddell, H. G., Scott, R., Jones, H. S., & McKenzie, R. (1940). *A Greek-English lexicon*. Clarendon Press.

Lockyer, N. (1894). *The dawn of astronomy*. Macmillan.

Long, A. A. (1986). *Hellenistic philosophy: Stoics, Epicureans, sceptics*. University of California Press.

Ludwig Wittgenstein Project. (n.d.). *Tractatus Logico-Philosophicus* (English). Retrieved October 28, 2025, from https://www.wittgensteinproject.org/w/index.php/Tractatus_Logico-Philosophicus_(English)

Magli, G. (2008). On the orientation of Roman towns in Italy. *Oxford Journal of Archaeology*, *27*(1), 63–71.

Magli, G. (2013). *Architecture, astronomy and sacred landscape in ancient Egypt*. Cambridge University Press.

Malta, H. (2007, May). *Megalithic Temples of Malta: Description*. https://web.archive.org/web/20071020170410/http://www.heritagemalta.org/Consultation/Documents/Temples%20-%20Description.pdf

Malta, H. (2023). *The Equinox & Solstice at Ħaġar Qim and Mnajdra* [Brochure]. https://heritagemalta.mt/app/uploads/2023/06/HQMnajdra-leaflet.pdf

Malta, H. (n.d.). *Ġgantija Archaeological Park*. Retrieved October 10 2025 from https://heritagemalta.mt/explore/ggantija-archaeological-park/

Malta, H. (n.d.). *Ħaġar Qim and Mnajdra Archaeological Park*. Retrieved October 9, 2025, from https://heritagemalta.mt/explore/hagar-qim-and-mnajdra-archaeological-park/

Malta, H. (n.d.). *Ħal Tarxien Prehistoric Complex*. Retrieved October 12, 2025, from https://heritagemalta.mt/explore/hal-tarxien-prehistoric-complex/

Malville, J. M., Wendorf, F., Mazar, A. A., & Schild, R. (1998). Megaliths and Neolithic astronomy in southern Egypt. *Nature*, *392*(6674), 488–491.

Mercati, M. (1589). *De gli obelischi di Roma*. Domenico Basa.

Merleau-Ponty, M. (1962). *Phenomenology of perception* (Smith, C., Trans.). Routledge & Kegan Paul. (Original work published 1945)

Merleau-Ponty, M. (1962). Un inédit de Maurice Merleau-Ponty. *Revue de Metaphysique et de Morale*, *67*(4), 401–409.

Merleau-Ponty, M., & Lingis, A. (1968). *The visible and the invisible*. Northwestern University Press.

Micallef, C. (2001). The Tal-Qadi Stone: A Moon Calendar or Star Map. *The Oracle. Journal of the Grupp Arkeologiku Malti*, *2*, 1–9.

NASA. (2021, July 28). *What is the North Star and how do you find it?* NASA Science. Retrieved August 25, 2025, from https://science.nasa.gov/solar-system/what-is-the-north-star-and-how-do-you-find-it/

National Geographic Society. (2023, October 19). *Equator.* Retrieved August 25, 2025, from https://education.nationalgeographic.org/resource/equator/

National Library of Scotland. (n.d.). *Atlas Maior, sive cosmographia Blaviana* (Vol. 1) [Digital reproduction of Blaeu, 1665]. Edinburgh, UK: Author. Retrieved from https://maps.nls.uk/atlas/blaeu-maior/vol/1

National Museums Scotland. (n.d.). *Casing stone from the Great Pyramid of Giza* [Stone]. Museum reference A.1955.176. Edinburgh, Scotland. https://www.nms.ac.uk/discover-catalogue/casing-stone-from-the-great-pyramid-of-giza

National Radio Astronomy Observatory. (2024, October 29). *What are constellations?* Retrieved August 24, 2025, from https://public.nrao.edu/ask/what-are-constellations/

National Radio Astronomy Observatory. (n.d.). *Radio telescopes*. Retrieved October 6, 2025 from https://public.nrao.edu/telescopes/radio-telescopes/

National Radio Astronomy Observatory. (n.d.). *Very Large Array*.Retrieved October 6, 2025 from https://www.vla.nrao.edu/

Needham, J., Beer, A., & Pulleyblank, E. G. (1961). An 8th-century meridian line: I-Hsing's chain of gnomons and the pre-history of the metric system. *Vistas in Astronomy*, *4*, 3–26.

Neolithic temples of Malta. (n.d.). In *Encyclopædia.com*. Retrieved October 3, 2025, from https://www.encyclopedia.com/humanities/encyclopedias-almanacs-transcripts-and-maps/neolithic-temples-malta

Newton, I. (1687). *Philosophiæ naturalis principia mathematica*.

Newton, R. R. (1980). The sources of Eratosthenes' measurement of the Earth. *Quarterly Journal of the Royal Astronomical Society, 21*(4), 379–387.

Norberg-Schulz, C. (1971). *Existence, space and architecture*. Praeger.

Norberg-Schulz, C. (1979). *Genius loci: Towards a phenomenology of architecture*. Rizzoli.

North American Sundial Society. (2020). *Sundial mathematics*. Retrieved September 11, 2025, from https://www.sundials.org/index.php/teachers-corner/sundial-mathematics

Norton, W. A. (1872). *A treatise on astronomy, spherical and physical*. John Wiley & Son.

Odyssey: Adventures in Archaeology. (n.d.). *Tarxien Temples*. Retrieved October 12, 2025, from https://www.odysseyadventures.ca/articles/malta_temples/maltaTemples03h_tarxien.htm

Orofino, V., & Bernardini, N. (2016). New evidence on the Orion Correlation Theory. *Mediterranean Archaeology & Archaeometry. International Journal, 16*(4), 1–12.

Oxford University Press. (n.d.). Archaeoastronomy. In *Oxford English dictionary*. Retrieved September 14, 2025, from https://www.oed.com/dictionary/archaeoastronomy_n

Oxford University Press. (n.d.). Time. In *Oxford English Dictionaries*. Retrieved from https://web.archive.org/web/20120704084938/http://oxforddictionaries.com/definition/time

Pallasmaa, J. (2005). *The eyes of the skin: Architecture and the senses*. John Wiley & Sons.

Patel, K. (2019, August 23). The first planned city in India. *NASA Earth Observatory*. https://earthobservatory.nasa.gov/images/145509/the-first-planned-city-in-india

Petrucci, A. (1993). *Roma nel Rinascimento: La città e il papato da Sisto IV a Clemente VIII*. Laterza.

Platner, S. B., & Ashby, T. (1929). *A topographical dictionary of ancient Rome*. Oxford University Press.

Ptolemy, . (1984). *Ptolemy's Almagest* (Toomer, G. J., Trans.). Princeton University Press.

Puranavedas. (2021, February 26). Vedic paintings and sculpture. Retrieved September 10, 2025, from https://www.puranavedas.com/paintings-and-sculpture/

Quintela, M. V. G. (2022). Meton of Athens: An Astronomer in the Democratic Landscape of the Polis. *Gerión*, *40*(2), 553.

Rajawat, D. S. (1980). *Astronomical observatory of Jaipur*. Delta Publishers.

Reel, B. B. C. (2022, February 23).*The mysteries of Portugal's 7,000-year-old 'Stonehenge'*. BBC. Retrieved on August 26 20025 from https://www.bbc.com/reel/video/p0bbrfm1/the-mysteries-of-portugal-s-7-000-year-old-stonehenge-

Reisch, G. (1525). *Margarita philosophica*. Johann Schott.

Richard Long. (n.d.).*Richard Long Official*. Retrieved October 30, 2025, from http://www.richardlong.org/

Roberts, L. (2016). Deep mapping and spatial anthropology. *Humanities (Washington)*, *5*(1), 5.

Robinson, A. H., Sale, R., Morrison, J., & Muehrcke, P. (1985). *Elements of cartography* (5th ed.). John Wiley & Sons.

Rountree, K. (2002). Re-inventing Malta's neolithic temples: Contemporary interpretations and agendas. *History and Anthropology*, *13*, 31–51.

Roy, A. E., & Clarke, D. (2003). *Astronomy: Principles and practice* (4th ed.). Institute of Physics Publishing.

Royal Museums Greenwich. (n.d.). *Astronomy with the naked eye*. Retrieved August 20, 2025, from https://www.rmg.co.uk/stories/space-astronomy/astronomy-naked-eye

Ruggles, C. L. N. (2005). *Ancient astronomy: An encyclopedia of cosmologies and myth*. ABC-CLIO.

Ruggles, C. L. N. (2015). *Handbook of archaeoastronomy and ethnoastronomy*. Springer.

Ruggles, C., Prendergast, F., & Ray, T. (Eds.). (2001). *Astronomy, cosmology and landscape*. Ocarina Books.

Sacrobosco, J. (1531). *Sphaera tractatus*. Lucantonio Giunta.

Science, N. A. S. A. (2024, October 22) *Our Solar System*. Retrieved August 25, 2025, from https://science.nasa.gov/resource/our-solar-system-2

Science, N. A. S. A. (2025, January 16). *Chapter 2: Reference systems. Basics of Spaceflight*. Retrieved August 25, 2025, from https://science.nasa.gov/learn/basics-of-space-flight/chapter2-1

Seidelmann, P. K. (Ed.). (1992). *Explanatory supplement to the Astronomical Almanac* (Rev. ed.). University Science Books.

Serres, M. (2018). *The birth of physics*. Rowman & Littlefield Publishers.

Sharma, V. N. (1991). The Kapāla Yantras of Sawai Jai Singh. *Indian Journal of History of Science, 26*(2), 173–190.

Shaw, I. (Ed.). (2000). *The Oxford history of ancient Egypt*. Oxford University Press.

Shinde, T. (2024). The spiral aspects of multiverse formation: Continuous big bang and everyverse connection. *International Journal for Research in Applied Science and Engineering Technology, 12*(III), 48–58. DOI: 10.22214/ijraset.2024.58697

Shukla, D. N. (1993). *Vastu-sastra: Hindu science of architecture*. Munshiram Manoharlal Publishers.

Sigismondi, C. (2021). Calibrazione solstiziale della meridiana di Piazza S. Pietro. *Gerbertvs, 14*, 207–222.

Silva, F., & Campion, N. (Eds.). (2015). *Skyscapes: The role and importance of the sky in archaeology*. Oxbow Books.

Silva, F., & Lomsdalen, T. (2025). 'No easy way from the earth to the stars': A new statistical approach to the orientation of the Maltese temples. *Archaeological and Anthropological Sciences, 17*(96). Advance online publication. DOI: 10.1007/s12520-025-02208-4

Silva, I. (2015). *Latin American Perspectives on Science and Religion*. Routledge.

Smart, W. A. M. (1936). *Text-book on spherical astronomy*. Cambridge University Press.

Smith, D. W. (2018). Phenomenology. In E. N. Zalta (Ed.), *The Stanford encyclopedia of philosophy* (Summer 2018 ed.). Retrieved June 26, 2025, from https://plato.stanford.edu/entries/phenomenology/

Smith, M. E. (2005). Did the Maya build architectural cosmograms? *Latin American Antiquity, 16*(2), 217–224.

Snyder, J. P. (1987). *Map projections: A working manual*. Washington, DC: United States Government Printing Office.

Southern, P. (2001). *The Roman Empire from Severus to Constantine*. Routledge.

St. Fleur, N. (2025, July 10). Manhattanhenge is here again: How to see New York City's best sunset. *National Geographic*. Retrieved October 29, 2025, from https://www.nationalgeographic.com/science/article/manhattanhenge-new-york-city-sunset

Stanford Solar Center. (n.d.). Viewing and understanding the analemma. Stanford University. https://solar-center.stanford.edu/art/analemma.html

Star Axis. (n.d.). *Star Axis: An architectonic earth/star sculpture*. https://www.staraxis.org/

Stetson University. (n.d.). *Trollasten Dös (Dolmen), Stora Köpinge, near Ystad, Skåne, Sweden*. Neolithic Studies. Retrieved September 25, 2025 fromhttps://www2.stetson.edu/neolithic-studies/simple-dolmens/trollasten-dos-dolmen-stora-kopinge-near-ystad-skane-sweden/

Stolcius, D. (1624). *Viridarium chymicum figuris cupro incisis adornatum, et poeticis picturis illustratum*. L. Jennis.

Stoppard, T. (1968). *Rosencrantz and Guildenstern are dead*. Grove Press.

SunCalc. (n.d.). *Solar and lunar data for 41.9022° N, 12.4573° E on December 21, 2025*. Retrieved October 26, 2025, from https://www.suncalc.org

S-U-N-F-A-R-M. (n.d.). *A Place for Art, Nature & Well-Being*. Retrieved October 31, 2025, from https://s-u-n-f-a-r-m.org/

Superintendence of Cultural Heritage. (2012, March 30). *NICPMI Inv. No. 00059* [Data set]. https://ps/neolithic-temples-malta

Superintendence of Cultural Heritage. Malta. (2012). *Tarxien Temple*. http://www.culturalheritage.gov.mt/filebank/inventory/00046.pdf

Taki, T. (2008, March 2). Sundial of Trinità dei Monti in Rome. http://ss977526.stars.ne.jp/sundial/sundial_trinita.htm

Tanton, J. S. (2005). *Encyclopedia of mathematics*. Infobase Publishing.

Tate Britain. (2002, March 23–June 4). Hamish Fulton: Walking journey [Exhibition]. Retrieved September 21 2025 from https://www.tate.org.uk/whats-on/tate-britain/hamish-fulton-walking-journey

Taylor, K. (2012). *Celestial geometry: Understanding the astronomical meanings of ancient sites*. Watkins Publishing.

The Art Story. (2025.). *Earth Art Movement Overview and Analysis*. Retrieved October 29, 2025, from https://www.theartstory.org/movement/earth-art/

The Astronomical Observatories of Jai Singh. (n.d.). The observatory sites. Retrieved September 10, 2025, from https://www.jantarmantar.org/learn/observatories/sites/index.html

The Editors of Encyclopædia Britannica. (2025, July 31). Pyramids of Giza. In *Encyclopædia Britannica*. Retrieved September 21, 2025, from https://www.britannica.com/topic/Pyramids-of-Giza

The Editors of Encyclopædia Britannica. (2025, March 7). *Karnak*. In Encyclopædia Britannica. Retrieved August 31, 2025 from https://www.britannica.com/place/Karnak

The Editors of Encyclopædia Britannica. (2025, September 3). *Maltese language*. In Encyclopædia Britannica. Retrieved October 10, 2025, from https://www.britannica.com/topic/Maltese-language

Thom, A., & Thom, A. S. (1971). The Astronomical Significance of the Large Carnac Menhirs. *Journal for the History of Astronomy, 2*(3), 147-160. https://doi.org/ (Original work published 1971).DOI: 10.1177/002182867100200301

Thompson, R. F., & Cornet, J. (1981). *The four moments of the sun*. National Gallery of Art.

Thoren, V. E. (1990). *The lord of Uraniborg: A biography of Tycho Brahe*. Cambridge University Press.

Tilley, C. (2004). *The materiality of stone: Explorations in landscape phenomenology*. Routledge.

Times of Malta. (2006, October 3). "17th century chapel restored".*Times of Malta*. Retrieved October 14 2025. https://timesofmalta.com/article/17th-century-chapel-restored.39531

Tresch, J. (2005). Cosmogram. In Ohanian, M., & Royoux, J. C. (Eds.), *Cosmograms* (pp. 67–76). Lukas & Sternberg.

Trump, D. H. (1999). The architecture of the Maltese temples. In Mifsud, A., & Savona Ventura, C. (Eds.), *Facets of Maltese prehistory* (pp. 91–99). Prehistoric Society of Malta.

Trump, D. H., & Cilia, D. (2002). *Malta, Prehistory and Temples*. Midsea Books.

Tuan, Y.-F. (1974). *Topophilia*. Prentice-Hall.

Tuan, Y.-F. (1977). *Space and place*. University of Minnesota Press.

Turchin, P., Adams, J. M., & Hall, T. D. (2006). East–West orientation of historical empires. *Journal of World-systems Research, 12*(2), 219–229.

U.S. Embassy & Consulates in Italy. (n.d.). *2025 Catholic Jubilee in Rome—Pilgrims in hope.* Retrieved October 29, 2025, from https://it.usembassy.gov/jubilee2025/

U.S. Naval Observatory, Astronomical Applications Department. (n.d.). *Computing approximate solar coordinates.* Retrieved September 15, 2025, from https://aa.usno.navy.mil/faq/sun_approx

UNESCO Portal to the Heritage of Astronomy. (n.d.). Retrieved from https://web.astronomicalheritage.net/index.php/show-entity

United Nations Educational, Scientific and Cultural Organization. (n.d.). *Megalithic Temples of Malta.* UNESCO World Heritage Centre. Retrieved October 3, 2025, from https://whc.unesco.org/en/list/132/

University of North Dakota. (n.d.). *Pf mound.* UND Commons. Retrieved October 25, 2025, from https://commons.und.edu/pf-mound/

Varela, F. J., Thompson, E., & Rosch, E. (1991). *The embodied mind: Cognitive science and human experience.* MIT Press.

Verner, M. (2001). *The pyramids: The mystery, culture, and science of Egypt's great monuments.* Grove Press.

Vitruvius Pollio, M. (1557). *I dieci libri dell'architettura* (D. Barbaro, Trans. & Ed.). Venice, Italy: Francesco de' Franceschi & Giovanni Chrieger. (Original work published ca. 15 BCE)

Vitruvius Pollio. (1914). *The ten books on architecture* (Morgan, M. H., Trans.). Harvard University Press.

Vollmer, M. (2025). Naked-eye celestial objects and phenomena: How far can we see at night? *European Journal of Physics, 46*(3), 035301.

Von Foerster, H. (1984). Principles of self-organization—in a socio-managerial context. In *Self-organization and management of social systems: Insights, promises, doubts, and questions* (pp. 2-24). Berlin, Heidelberg: Springer Berlin Heidelberg. Fuller, R. B. *R. Buckminster Fuller papers (M1090)* [Archival collection]. Special Collections and University Archives, Stanford University Libraries.

Ward-Perkins, J. B. (1994). *Roman imperial architecture.* Yale University Press.

Wells, O. (2021, February 17). *The megalithic temples of Malta.* World History Encyclopedia. Retrieved October 6 2025 from https://www.worldhistory.org/article/1678/the-megalithic-temples-of-malta/

Wever, E. G. (1927). Figure and ground in the visual perception of form. *The American Journal of Psychology*, *38*(2), 194–226.

Wilkinson, R. H. (2000). *The complete temples of ancient Egypt.* Thames & Hudson.

Wittgenstein, L. (1922). Tractatus logico-philosophicus (C. K. Ogden, Trans.). Kegan Paul, Trench, Trubner & Co.

Wittkower, R. (1999). *Art and architecture in Italy, 1600–1750* (6th ed.). Yale University Press.

Zagari, F. (1998). *Piazza Montecitorio: Progetto di riqualificazione ambientale 1996–1998.* Rome, Italy: Camera dei Deputati; Comune di Roma.

Zammit, T. (1927). *The Neolithic Temples of Hal-Tarxien, Malta: A short description of the monuments.* Empire Press.

Zodiac. (n.d.). In *Encyclopedia.pub.* Retrieved August 25, 2025, from https://encyclopedia.pub/entry/31827

About the Author

Daniela Bertol, Ph.D., is a research artist, scholar, author, and designer whose work investigates geometry as a perceptual, spatial, and cosmological structure linking the human body, the built environment, and the sky. In the early 1990s she founded SpaceInk, a meta-disciplinary art, architecture, and design practice developing cultural and artistic production across physical and digital environments. Her work spans computational design, video, land and performance art, artist books, and experiential landscapes, with a sustained focus on phenomenology, embodiment, and spatial cognition. She earned a Ph.D. by research in Art and Education from Deakin University (Australia). Her doctoral research, *Form Mind Body Space Time The Geometry of Human Movement*, examined how bodily movement generates spatial and temporal configurations, leading to the development of *Movement Infrastructure*, a sculptural and performative framework used in artistic research and practice. Bertol is the author of *Visualizing with CAD: An AutoCAD Exploration of Geometric and Architectural Forms* (Springer, 1994), *Designing Digital Space* (Wiley, 1996), among the earliest books on virtual reality in architecture, and *Form Geometry Structure: From Nature to Design* (Bentley Institute Press, 2011). In 2022 she founded the *Sky Spirals Institute*, dedicated to theoretical research and art-in-nature projects related to naked-eye astronomy. Its first site, S.U.N.F.A.R.M., is a phenomenological sky–earth observatory and landscape project in the Hudson Valley, New York. Her current research focuses on phenomenological and symbolic relationships between landscape, astronomy, and human perception through art-in-nature environments.

Index

A

B

C

D

E

F

G

H

J

L

M

N

O

P

R

www.ingramcontent.com/pod-product-compliance
Lightning Source LLC
LaVergne TN
LVHW062329300926
849161LV00007B/1630